CHARLES S. PEIRCE
and the Philosophy of Science

CHARLES S. PEIRCE

and the Philosophy of Science

Papers from the Harvard Sesquicentennial Congress

Edited by Edward C. Moore

The University of Alabama Press

Tuscaloosa and London

Manufactured in the United States of America

∞

The paper on which this book is printed meets the minimum requirements of American National Standard for Information Science-Permanence of Paper for Printed Library Materials, ANSI Z39.48-1984.

Library of Congress Cataloging-in-Publication Data

Charles S. Peirce and the philosophy of science : papers from the Harvard Sesquicentennial Congress / edited by Edward C. Moore.

p. cm.

Includes bibliographical references and index.

ISBN 0-8173-0665-X 978-0-8173-5416-9 (pbk: alk. paper)

1. Science—Philosophy—Congresses. 2. Peirce, Charles S. (Charles Sanders), 1839–1914—Congresses. I. Moore, Edward C. (Edward Carter), 1917– .

Q174.C46 1993

501—dc20 92-28877

British Library Cataloguing-in-Publication Data available

To Arthur W. Burks

with many thanks

I no longer think, as I once did, that there is a difference between science and metaphysics . . . as long as a metaphysical theory can be rationally criticized.

—Karl R. Popper

Contents

Part 2. The Physical Sciences

Part 3. The Life of the Mind

Preface

The Charles S. Peirce Sesquicentennial International Congress opened at Harvard University on September 5, 1989, and concluded on the tenth, Peirce's birthday. The congress had been convened by the Charles S. Peirce Society, following an idea by Michael Shapiro. The organizing institutions were Harvard University and Texas Tech University. Congress headquarters was Sever Hall in Harvard Yard, formerly the site of the residence of Benjamin Peirce, a mathematics professor at Harvard, where his son Charles (1839–1914) grew to adulthood.

During the five days of its existence, the congress was host to approximately 450 scholars from 26 different nations. The present collection is a portion of the papers that were presented at the congress. A group of papers, from which those in this volume have been selected, were assigned to the editor of the present volume by the organizing committee. Other papers have been given to other editors.

One of the sustained efforts of Peirce scholars over the years since 1931–36, when the bulk of the Peirce papers was first published, has been to stimulate a wider interest in Peirce's work. Peirce was known to only a few as recently as 1962, when the journal, *Transactions of the Charles S. Peirce Society*, was founded. That journal today has over 550 subscribing members. Papers concerning Peirce's philosophy of science were given at the congress by representatives from Italy, France, Sweden, Finland, Korea, India, Denmark, Greece, Brazil, Belgium, Spain, the Netherlands, Germany, and the United States. Thus a main point of this present volume is to show how worldwide the interest in Peirce is today. Ernest Nagel of Columbia University wrote in 1959 that "there is a fair consensus among historians of ideas that Charles Sanders Peirce remains the most original, versatile, and comprehensive philosophical mind this country has yet produced." The breadth of topics discussed in the present volume suggests that this is as true today as it was in 1959.

The congress was directed by an organizing committee consisting of the following persons: Hilary Putnam (president of the congress), Kenneth Laine Ketner (chairperson of the organizing committee), Hanna Buczynska-Garewicz (USA and Poland), Gerard Deledalle (France), Umberto Eco (Italy), Takashi Fujimoto (Japan), Susan Haack (United Kingdom), Jaako Hintikka (USA and Finland), Christian J. W. Kloesel (USA), Dan Nesher (Israel), Klaus Oehler (Federal Republic of Germany), Nicholas Rescher (USA), David Savan (Canada), Israel Scheffler (USA), and Michael Shapiro (USA).

Administrative assistants for the congress were, in Lubbock, Joyce Abbott (Division of Continuing Education), Texas Tech University; and, in Cambridge, Alexandra Collins (Programs in Professional Education), Harvard University. Ms. Collins was in charge of local arrangements.

The congress was supported by the following individuals and organizations: Claude Ventry Bridges Memorial Fund (Texas Tech University), National Endowment for the Humanities, Charles S. Peirce Foundation, Charles S. Peirce Society, College of Arts and Sciences (Texas Tech University), Division of Continuing Education (Texas Tech University), Institute for Studies in Pragmaticism (Texas Tech University), Library of Texas Tech University, Department of Philosophy (Harvard University), Programs in Professional Education (Harvard University), Mr. W. B. Rushing (Lubbock, Texas), Mary Baker Rumsey Foundation (Lubbock, Texas), and Society for the Advancement of American Philosophy.

The editor of the present volume wishes also to thank the Museum of Natural History at the University of Alabama for providing support for him during the preparation of this volume. Given the interest that Charles Peirce had in all the natural sciences, this support seems particularly appropriate.

Tuscaloosa, Alabama

Acknowledgments

The editor acknowledges, with gratitude, the Department of Philosophy at Harvard University for permission to use quotations from the various manuscripts of Charles Peirce in the Widener Library at Harvard and from the eight volumes of the *Collected Papers of Charles Sanders Peirce*, published by Harvard University Press; the MIT Press, Cambridge, Massachusetts, for permission to use the illustration on page 413 from Gary Lynch, ed., *Synapses, Circuits, and the Beginnings of Memory*; the Royal Society of London for permission to use the illustration on page 80 from David Marr, *Proceedings of the Royal Society of London*, B 200 (1982), page 269; and the McGraw-Hill Book Co., New York, N.Y., for permission to use the illustration on page 409 from Francis Jenkins, *Fundamentals of Optics* (1957), page 575.

Abbreviations

Standard editions of Peirce's works are referred to with the following notations:

CP *Collected Papers of Charles Sanders Peirce*, ed. C. Hartshorne, P. Weiss, and A. Burks, 8 vols. (Cambridge: Harvard University Press, 1935–58). References are cited by volume and paragraph number: *CP* 5.119, for example, refers to volume 5, paragraph 119.

HP *Historical Perspective on Peirce's Logic of Science*, ed. Carolyn Eisele, 2 vols. (The Hague: Mouton, 1985). References give volume and page.

MF *The Published Works of Charles Sanders Peirce* (Bowling Green, Ohio: Philosophy Documentation Center, 1986), a large microfiche edition of all of Peirce's lifetime publications, cataloged in *A Comprehensive Bibliography of the Published Works of Charles Sanders Peirce*, by K. L. Ketner (with the assistance of C. V. Bridges and A. Stewart), 2d rev. ed. (Bowling Green: Philosophy Documentation Center, 1986).

MS *The Charles S. Peirce Papers*, microfilm edition (Cambridge: Harvard University Library Photographic Service, 1966). References to this film set of Peirce's unpublished papers held in the Houghton Library at Harvard employ the numbering system for manuscripts developed by Richard S. Robin in his *Annotated*

Catalogue of the Papers of Charles S. Peirce (Amherst: University of Massachusetts Press, 1967), as emended by Robin in "The Peirce Papers: A Supplementary Catalogue," *Transactions of the Charles S. Peirce Society* 7 (1971):37–57. For example, *MS* 325:3 indicates Robin's catalog manuscript number 325, page 3.

N *Charles S. Peirce: Contributions to The Nation*, ed. K. L. Ketner and J. E. Cook, 3 vols. plus an index vol. (Lubbock: Texas Tech University Press, 1975–87). References give volume and page.

NEM *The New Elements of Mathematics*, by Charles S. Peirce, ed. C. Eisele, 4 vols. in 5 (The Hague: Mouton, 1976). References give volume and page.

PW *Semiotic and Significs: The Correspondence between Charles S. Peirce and Victoria Lady Welby*, ed. C. S. Hardwick (Bloomington: Indiana University Press, 1977). References are to page numbers.

W *Writings of Charles S. Peirce: A Chronological Edition*, ed. Max H. Fisch et al., 4 vols. to date (Bloomington: Indiana University Press, 1982–). Citations give volume and page.

CHARLES S. PEIRCE
and the Philosophy of Science

Introduction: Charles S. Peirce and the Philosophy of Science

Edward C. Moore

One of the themes in the work of Charles Peirce that has been of greatest interest to students of his thought centers on his view of the philosophy of science and its logic. Peirce was himself a physical scientist.[1] His father, Benjamin Peirce (1809–90), was a distinguished professor of mathematics at Harvard University. Peirce received a bachelor of arts degree from Harvard in 1859, a master of arts in 1862, and a bachelor of science in chemistry in 1863. He worked as an assistant at the Harvard Astronomical Observatory from 1869 to 1872 and made a series of astronomical observations there from 1872 to 1875. Solon I. Bailey (1931:198) says of these observations, which were the basis of Peirce's only published book, *Photometric Researches*, "The first attempt at the Harvard Observatory to determine the form of the Milky Way, or the galactic system, was made by Charles S. Peirce. . . . The investigation was of a pioneer nature, founded on scant data."

Peirce was employed for over thirty years by the United States Coast and Geodetic Survey as a physicist. He did significant work on the determination of the gravitational constant and ultimately left the Coast and Geodetic Survey in a dispute with his superiors about the reliability of the pendulum technique used for making such determinations. His position on this matter has only recently been vindicated, and a whole series of subsequent observations are now considered questionable (Lenzen 1969:5–24).

Peirce made major contributions also in fields as diverse as mathematical logic and psychology. C. I. Lewis has remarked that "the head and font of mathematical logic is found in the calculus of propositional functions as developed by Peirce and Schröder" (Lewis and Langford 1932:21). From 1879 to 1884, when Peirce was a lecturer on logic at Johns Hopkins University, he carried out, with Joseph Jastrow, some fundamental experiments in psychology (see the chapter by Behrens in this volume), and he subsequently invented, almost from whole cloth, semiotics—the science of the meaning of signs. British authors Ogden and Richards (1949:279) say that "by far the most elaborate and determined attempt to give an account of signs and their meaning is that of the American logician C. S. Peirce, from whom William James took the idea and the term Pragmatism, and whose Algebra of Dyadic Relations was developed by Schröder."

In 1963 the Coast and Geodetic Survey commissioned a survey ship, *Peirce CSS 28*, named after Charles Peirce. At that time the director of the Coast and Geodetic Survey, Rear Admiral H. Arnold Karo, wrote me that, "in addition to being a logician and philosopher, Peirce made many important scientific and technical contributions to the Coast and Geodetic survey during his thirty years of service in the bureau."

Aside from small earnings from lectures and book reviews, the income Peirce earned in his lifetime came mainly from his work as a scientist. I think it is appropriate to say, as Max Fisch has said, that Peirce was primarily a scientist and only secondarily a philosopher. This career gives a credibility and a substance to his views about the philosophy of science that they might not otherwise have.

I

But a philosopher of science must know not only science but also philosophy. This side of Peirce's career was determined mainly by two philosophers: Immanuel Kant and Duns Scotus. Immanuel Kant (1724–1804) was the leading philosopher of science in the nineteenth century. Kant asked, How is the synthetic a priori possible? a famous question that was, and is still, the most difficult question in the philosophy of science. How is it possible to predict the future successfully in matters of science? Many of the papers in this present volume address this question either directly or indirectly. Kant had been led to it mainly by David Hume's analysis of causality, which showed the difficulties in asserting any causal scientific statements.

Today we rightly think of Kant as a great philosopher. We forget that, like Peirce, he was a scientist. Kant had a thorough training in mathematics and physics. He received his doctorate from the University of Königsburg, where he became an instructor lecturing mainly in physics. In 1770 he was appointed professor of logic and mathematics, a post he held until his retirement in 1797. B. A. G. Fuller (1955) says of Kant's scientific work:

> His first writings were concerned with physics and astronomy, and his later philosophical works were interspersed with excursions into the field of the physical sciences. Of his scientific work, *A General Natural History and Theory of the Heavens* (1755) is perhaps most important. . . . In it Kant applies Newtonian principles to the fixed stars, develops a mechanical theory of the whole sidereal universe, and suggests for the first time the nebular hypothesis of the origin of planetary systems. His other scientific writings ranged over such subjects as the action of the tides on slowing up the rotation of the earth, the influence of the earth's rotation upon winds, the causes of earthquakes, the different races of man and the beginnings of human history, the volcanoes on the moon, and the influence of the moon upon the weather. (2:217)

Like Peirce, Immanuel Kant was a scientific-philosophical polymath.

Fuller also says, "It is safe to say that no one can understand philosophy since Kant unless he knows his Kant" (ibid.). This was certainly true in Peirce's day. He took Kant seriously. He tells us, "I devoted two hours a day to the study of Kant's *Critic of Pure Reason* for more than three years until I almost knew the whole book by heart and had critically examined every section of it" (*CP* 1.4). He says in another place that he "believed more implicitly in Kant's table of categories than if they had been brought down from Sinai" (*CP* 4.2). He would eventually take the term "pragmatism" from Kant's *Critic of Pure Reason* as the name for his own philosophy (*CP* 5.3). This early study of Kant started him down the path of developing his own philosophy of science. (See the review of this process by C. F. Delaney that opens the present collection.) He was to travel a long road of many years and much thinking to find his own answers.

II

Peirce was to find his answer to the problem of how the synthetic a priori is possible from his study of another great philosopher, John Duns Scotus (1266–1308), an answer that came through Duns Scotus from Aristotle—yet another scientist-philosopher polymath. I have discussed the influence of Scotus on Peirce elsewhere (Moore 1964:401–13). Here I wish to call attention only to what I consider Peirce's principal indebtedness to Scotus: "the scholastic doctrine of realism. . . . The opinion that there are real generals" (*CP* 5.453).

The doctrine that there are real generals allowed Peirce to answer Kant's question by saying that it is possible to know synthetic a priori propositions because there are real generals (laws) in the world, and we are able to know them. It will be helpful to consider this answer.

If we think about what scientists do, we quickly sense that it is not quite what they say they do. Scientists talk about their experiments and their laboratories and their expensive instrumentation as though these were their essential business, but these are only the prelude to doing the essential business of science. What we value science and scientists for is not that they have at such and such a time and place observed such and such a singular fact, but rather that the cumulative effect of experimentation and observation leads them to state general propositions about nature.

These general propositions may be variously described as theories or hypotheses or laws. The important thing about them is that they are not singular. They are general. They do not tell us just about the specific instance which the scientist has observed. Instead they refer to all instances, those not observed as well as those observed, those which are yet to come as well as those which have already passed. Scientists are of such overwhelming value to us because we do indeed believe that they can tell us about the future. They enable us to predict that if such and such is done now, then something else will occur in the future. These predictions enable us to perform actions now which we believe will mold the future world nearer to our desires. We can avoid actions which will lead us to undesirable outcomes, and we can perform those actions which will bring us to desired outcomes.

The network of all such laws and theories and hypotheses is what we call knowledge. It is the base on which we determine our actions. This knowledge base is made of strands coming from many sciences and varying in their validity and reliability. But what is so philosophically interesting about them is that no matter how valid or reliable a particular strand may be, no scientist has ever observed one. No one has ever observed the law of gravitation or the second law of thermodynamics or the principle of relativity. We have observed instances of the laws and theories, but never the laws and theories themselves. Science, which calls itself empirical, relies in its finest hours on the nonempirical, the unexperienced, the never-seen.

I am sure that such language will strike many as exaggerated. What is a law, some will say, but the sum of its instances? What is a theory but a prediction about individual instances that may occur? Important as the observed instances of a law are, however, just as important, and often more important, are the unobserved instances. No matter how much of a hurry we are in, we do not step out of an airplane in flight. That action is an unobserved instance of our belief in the law of gravitation. We did not do something because we believed that if we did do it, then something unfortunate would have occurred, which did not occur. But since we did not do it, and since it did not occur, there is nothing observable about it.

We have here the familiar problem of the so-called hypothetical, contrary-to-fact conditional. We do not step off the curb when cars are coming because we believe the law of inertia that says that they will keep coming and hit us. We do not eat ground glass because we believe that it will kill us. We do not drink hydrochloric acid because we believe that it will destroy our digestive tract. Our days are filled as much with these unobserved and unobservable exemplifications of the laws and theories of science as they are with the observable ones. I refer to them as unobservable not because our failure to act cannot be observed but because the law which we rely on to guide our actions is evidenced as much by these negative actions as by our positive ones; the negative actions, however, cannot be observed, since they do not occur.

If these avoided actions, these refusals to act, are based as much on scientific propositions as are those positive actions we take, those which we actually perform, then it follows that a scientific law is not described just by the instances of it which actually occur but is equally represented in the instances which do not occur. In short, knowledge, as we use it and think of it, means something general, it means a proposition that applies to all cases of a specific sort, whether actualized in experience or not and whether now or in the future.

Peirce put the matter this way: "No collection of facts can constitute a law; for the law goes beyond any accomplished facts and determines how facts that may be, but all of which never can have happened, shall be characterized" (*CP* 1.420). In another place he says, "the will-be's, the actually-is's and the have-beens are not the sum of the reals. They only cover actuality. There are besides would be's and can be's that are real" (*CP* 8.216).

Whatever philosophers may say or hold to be the case, a scientist believes that there are general laws which apply to all cases, present and future, actualized or not. The scientist seeks to discover these general statements about the universe and to give them to us as guides to our actions. That is why we value science and spend so much money on it.

Let us now revert to philosophical terminology. Such general laws and theories are called concepts in our professional epistemological jargon, and the individual instances of them result in percepts. The difficulty here is that our highest flights of human knowledge—our laws of science, our theories about the origin of the universe, our hypotheses about genetic change—rest on these concepts. But what do these concepts rest on? We have seen that it requires more than individual percepts to form a foundation for them, since we do not have percepts of the unobservable instances. If our best knowledge consists of concepts, is there nothing in the physical world that conforms to them? Are they not real? Are they only ideas in our minds? Surely not! If these general laws exist only as concepts in our minds, if they are only our ideas, there is no reason to think that the world will conform to them. On the contrary, what we mean by a false or inaccurate law is one that does not conform to the real world, no matter how firmly we may personally believe in it.

And this, I think, is the dilemma in which Charles Peirce found himself as a scientist. If all our laws are general, and if there is nothing in the physical world that is general and that conforms to them, then they are simply fictions, or, in Peirce's terminology, "figments of our imagination." If they are not fictions, if they are real, it must be because there is something real in the physical world that conforms to them. But these concepts are general, that is, they apply not just to this case or that case but to all cases, observed and unobserved, actualized or only potential. To believe that the laws and concepts of science are real is to believe that there is something general in the physical world that corresponds to them. And to believe that is to be what Peirce called a realist. To be unwilling to believe it is to be a nominalist—to believe that the law is only a name, not itself a reality.

The situation can be put in more modern terms, however. We need not sound Platonic and say that there are real, external ideas, but using contemporary terminology, we may say that there are real general forces in the universe of which we may have ideas and that these forces have a structure. The structure may be determined by other physical forces or by chance space-time encounters with other systems or in a variety of ways. But that there are generalized structures of energy forces that apply to all of a common set of instances was what Peirce meant by a law. And that our ideas may conform to them through the processes of knowledge was what Peirce believed when he described himself as a Scotistic realist. Thus the ideas, the concepts, formed in our minds are real because real counterparts to them exist in the energy systems of the external world.

A realist in the context of a scientist is a person who believes that there are real generals in the real world corresponding more or less precisely to our ideas of them. The world, then, is not limited to particulars. It is also constituted of generals.

The reader may say that this is nothing but Platonism, the view that there are real universals. Not quite. Plato put the real universals in a separate realm—the Realm of Ideas. What Peirce is advocating is Aristotelianism, which says that there are generals, but they are in the real physical world and are open to observation and study. This is the path by which Peirce got to the Aristotelian views of Duns Scotus and is why he called himself a Scotistic realist and held that science was a form of what is known as moderate realism. (See Demetra Sfendoni-Mentzou's chapter in this volume, as well as Sfendoni-Mentzou 1991 and Moore 1968.)

However, it needs to be said that conceptualism is not a way out of this difficulty. The conceptualist denies that generals exist in the real world. The general ideas in our minds are, in this view, combinations formed from many particulars, but the general is only in the mind, not in the physical world. Since there is nothing in the real world corresponding to our general ideas, the ideas are not of real objects. The nominalist goes even further and says the general ideas do not exist even in the mind, since there is nothing in the mind that does not come from experience, and since we only have experience of particulars, there cannot be any generals even in our minds. What we call general ideas are only names, *nomina* (hence "nominalism"). Thus, neither conceptualism nor nominalism allows for real general laws and real general forces. As a scientist, Peirce believed such general laws and forces existed, and so he called himself a realist.

III

The above analysis raises another question: What is the nature of potentiality? This comes up because when we believe that a law will indeed operate in the future to produce a certain result, we are saying that the law (or rather, the force which the law describes) has in it now a potential for providing a specific result in the future. The problem of the potential is: What is there in the force *now* that represents that fact? I do not believe that the potential can be described simply by statements about the future. The potential does not exist in the future but in the present. If the potential is actualized in the future, that actualization will not be a potential. It will be an actualization of a potential. It will be an actuality, not a potentiality.

I am asking, What is the nature of the potential now, when it is unactualized? If we believe in laws, we believe in real potentials. But the potential appears to be unobservable. All we are able to observe is the actualized consequence of the potential, not the potential itself. And so we have another strange phenomenon. We saw above that the forces described in laws and theories were unobservables. The actual instances of them were observable, but the general forces themselves were not observable. Now, here, we have another unobservable. The actual instances of a potentiality may be observed when they are made into actualities, but the potentiality itself is unobservable.

Perhaps these two unobservables are only two ways of saying the same thing. Perhaps what is unobservable in the predictive character of a law or theory is simply the potentiality that is in it. Yet here we all are in the twentieth century, empiricists in epistemology of one stripe or another, but governed in all our actions by unobserved and unexperienced potentialities. No matter how high a degree of resolution our electron microscopes give us, or how far back into time our radio telescopes may peer, we never find anything but actualities. The potential is nowhere to be seen. And so the question, What is the actuality of the potential; what is a potential really? goes unanswered.

We cleverly define our metaphysics so as to avoid this question. Rather than candidly admit that the world contains the potential as well as the actual, we assert that we can define the potential in terms of the actual consequences and then, by a sly reductionism, eliminate the potential—reducing it to the actual. Fortunately for us, the real world continues to have real potentials, even though our metaphysics ignores it. Peirce put it thus, "There is no objection to saying that a law is a general fact, provided it be understood that the general has an admixture of potentiality, so that no congeries of actions here and now can ever make a general fact" (*CP* 1.420).

IV

But these difficulties are not new. The Greeks faced them, the medievals faced them, and we face them more acutely only because we live in an age that has more scientific knowledge. Charles Peirce, at the dawn of our age, also faced them. More honest, perhaps, than we, he admitted an actuality-potentiality dualism into his thought which we have sought to deny. We have sought to avoid these problems by analyzing language, by producing therapeutic theories of meaning which rule these problems to be meaningless.

Peirce tried that route also. His pragmatism is an empirical theory of meaning which defines concepts in terms of percepts. But the percepts have to be tied together. To do that, Peirce identified the notion of thirdness. But thirdness is essentially the notion of a general. Being a general, it contains potentiality. But since percepts, which are of actualities, cannot alone define potentiality, Peirce's own theory of meaning fails to define thirdness, which is the central concept of his metaphysics. His metaphysics and his pragmatism are therefore incompatible. In commenting on my position that pragmatism is incompatible with Peirce's metaphysics, Murray Murphey (1961:169) commented, "Indeed, as Moore has pointed out, if possibility is not actualized, it cannot be cognized; if it is actualized, it is no longer a mere possibility."

The same problem presents itself to every empirical theory of meaning with which I am familiar. They remain halfway measures. Peirce, at least, accepted the contradiction. We, instead, use it to rule metaphysics out of philosophy.

V

Let us conclude by looking briefly at a related question: How does knowledge of these generals get into our minds, if we cannot perceive them? In the first place, we have not yet fathomed the mysteries of the mind to such an extent that we can say with certainty what is in it. And with even less certainty can we speak to how it gets there. I do not wish to propose an epistemology, but I can say that Peirce did not ignore this question but sought to answer it.

In his 1898 lectures "Vitally Important Topics," he argued that humankind has developed a survival characteristic, which he calls an instinct, which gives us "certain tendencies to think truly about physics, on the one hand, and about psychics, on the other" (*CP* 5.591). And why might this not be right? We have developed, in the course of evolution, an excellent set of senses for perceiving actualities. If potentialities are so important to our survival, why may we not have developed some sense that would at least give us a tendency toward recognizing potentialities in the world about us. Such a tendency would obviously have great survival value. According to Peirce:

> Besides, you cannot seriously think that every little chicken, that is hatched, has to rummage through all possible theories until it lights upon the good idea of picking up something and eating it. On the contrary, you think the chicken has an innate idea of doing this; that is to say, that it can think of this, but has no faculty of thinking anything else. The chicken you say pecks by instinct. But if you are going to think every poor chicken endowed with an innate tendency toward a positive truth, why should you think that to man alone this gift is denied? (*CP* 5.591)

In commenting on this passage, Bertrand Russell (1946:xv) said, "This is an important question to which I do not know the answer." He goes on to say of Peirce, "He holds—and I confess that an examination of scientific inference has made me feel the force of this view—that man is adapted, by his congenital constitution, to the apprehension of natural laws which cannot be proved by experience, although experience is in conformity with them."

Hans Reichenbach (1942:94), in speaking of Einstein's discovery of the principle of equivalence for mass and energy, said,

> From the standpoint of logic, one cannot here speak of an inference, for this far-reaching assumption cannot be logically demonstrated by means of the scantily available facts. Rather we have here a typical procedure in physics, that of the formation of a hypothesis. . . . There seems to exist something like an instinct for the hidden intentions of nature; and whoever possesses this instinct, takes the spade to the right place where gold is hidden and thus arrives at deep scientific insights. It must be said that Einstein possesses this instinct to the highest degree.

With such support from two of the most eminent philosophers of science of this century, we must at least give some credit to the possibility that Peirce was right when he wrote, "It cannot any longer be denied that the human intellect is peculiarly adapted to the comprehension of the laws and facts of nature" (*CP* 2.750). And so Kant's question is at last answered. The synthetic a priori is possible because there are general laws and general forces at work in the universe and because the human intellect is adapted to the comprehension of them. The world is general, as Peirce first taught us, and we can know its generality.

But when all of that is said and done, what Peirce has left us are many problems that still need answers. There has been excellent work done on these problems by many Peirce scholars whose studies appear in this volume and by such Peirce scholars as Arthur Burks, Hilary Putnam, and Nicholas Rescher, among others. Peirce's best legacy was to leave the questions involved in an adequate philosophy of science in a little sharper focus than they had been when he found them. And perhaps that is an epitaph of which any philosopher would be proud.

Note

1. I have summarized the critical evaluations of Peirce's scientific work in the introduction to Edward C. Moore, *Charles S. Peirce: The Essential Writings* (Harper and Row: New York, 1972). Much of this introductory section is taken from that account.

References

Bailey, Solon I. 1931. *The History and Work of Harvard Observatory, 1839–1927*. Harvard Observatory Monograph No. 4. New York: McGraw-Hill.

Fuller, B. A. G. 1955. *A History of Philosophy*. 3d ed. New York: Henry Holt.

Lenzen, Victor F. 1969. An Unpublished Scientific Monograph by C. S. Peirce. *Transactions of the Charles S. Peirce Society* 5:5–24.

Lewis, C. I., and Langford, H. I. 1932. *Symbolic Logic*. New York: Century.

Moore, Edward C. 1952. The Scholastic Realism of C. S. Peirce. *Philosophy and Phenomenological Research* 12:406–17.

———. 1964. The Influence of Duns Scotus on Peirce. In *Studies in the Phi-*

losophy of Charles Sanders Peirce: Second Series, ed. Edward C. Moore and Richard S. Robin, 401–13. Amherst: University of Massachusetts Press.

———. 1968. On the World as General. *Transactions of the Charles S. Peirce Society* 4 (2):90–100.

———, ed. 1972. *Charles S. Peirce: The Essential Writings*. New York: Harper and Row.

Murphey, Murray G. 1961. *The Development of Peirce's Philosophy*. Cambridge: Harvard University Press.

Ogden, C. K., and Richards, I. A. 1949. *The Meaning of Meaning*. London: Routledge and Kegan Paul.

Reichenbach, Hans. 1942. *From Copernicus to Einstein*. New York: Philosophical Library.

Russell, Bertrand. 1946. Preface to *An Introduction to Peirce's Philosophy*, by James Feibleman. New York: Harper.

Sfendoni-Mentzou, Demetra. 1991. Towards a Potential Pragmatic Account of Peirce's Theory of Truth. *Transactions of the Charles S. Peirce Society* 27:27–78.

Part 1

Logic and Mathematics

1 Peirce on the Conditions of the Possibility of Science

C. F. Delaney, The University of Notre Dame

One associates the phrase "the conditions of the possibility of science" with Immanuel Kant, and Peirce's commitment to the Kantian project is certainly well documented. He states that, during his formative years as an undergraduate at Harvard, he "devoted two hours a day to the study of Kant's *Critic of Pure Reason* for more than three years until I almost knew the whole book by heart and had critically examined every section of it" (*CP* 1.4). In the early 1860s he singlemindedly devoted himself to a rethinking of the transcendental analytic, which culminated in his 1867 paper "On a New List of Categories," and he then described Kant as "now acknowledged everywhere as the master of philosophy" (*W* 1.241). To invoke an earlier idiom, Kant was, for Peirce, "The Philosopher," and Peirce's own philosophical reflections are so related to his mentor that, in his own words, "I am forced back again to Kant and find myself unable to take a single step until I have defined somewhat the principles upon which his philosophy is founded" (ibid.).

As in most cases of genuine philosophical inspiration or tutelage, however, there is a substantial difference between the master and the pupil, a difference which is best understood along the lines of a critical rethinking of the Kantian project rather than as a misunderstanding or critique of it. The project at issue is that of exhibiting the conditions of the possibility of science. It is in their conceptions of science that Kant and Peirce fundamentally differ, a difference which has its ultimate ground in the centrality of the notions of "history" and "community" in Peirce's overall philosophical orientation. This general orientation disposed Peirce to what I will call a concrete, as opposed to an abstract, conception of science. In contrast to the focal conception of science being that of a static set of propositions (Euclidean geometry or Newtonian mechanics), he conceived of it as a sociohistorical process of inquiry with a specific structure. This fundamental shift of perspective on science gave a quite different tonality to Peirce's project of exhibiting its conditions of possibility.

For Peirce there were two different traditional conceptions of science. The first was the characterization of science primarily as a systematic or organized body of knowledge, and the second was the characterization of it primarily as a method of knowing. The former he viewed as a rather "shallow cut" which captured only "the fossilized remains of science" (*MS* 614:7). The latter he saw as a "deeper cut" which was surely on the right track but which, in its traditional articulations, was compromised by an excessively individualistic and not sufficiently dynamic conception of methodology. He drew on his own experience as a scientist, his knowledge of the history of science, and his expertise as a methodologist of science in an effort to characterize the concrete reality that was living science in contrast to some abstract specification of some feature thereof.

He argued that the focal meaning of the word "science" should be to designate the concrete life of a social group of inquirers, informed by a particular methodological strategy and animated by the desire to discover the truth. His general characterization of scientific methodology as involving abductive, deductive, and inductive phases is well known. The abductive phase of inquiry is concerned with the original generation and recommendation of explanatory hypotheses; the deductive phase has to do with the logical elaboration of a selected hypothesis; and the inductive phase bears on the confirmation of the hypothesis by future experience. The inquiry process itself is construed as thoroughly historical and thoroughly social, and the ampliative phases of the process (abduction and induction) are intrinsically characterized in sociohistorical terms. The justification of particular abductive and inductive procedures is not in terms of their cognitive status for the individual investigator at a given moment but in terms of their contribution to the success of the community of investigators in the long run.

The sociohistorical picture of science that emerges from Peirce's account of scientific methodology receives its ultimate generalization in his construal of science as a "mode of life." Applauding Francis Bacon's vision of science (while demurring at many particulars), Peirce proposes the following definition of the word "science" as a follower of Bacon would use it:

> For him man is nature's interpreter; and in spite of the crudity of some anticipations, the idea of science is in his mind inseparably bound up with that of a life devoted to single-minded inquiry. That is also the way in which every scientific man thinks of science. That is the sense in which the word is to be understood in this chapter. Science is to mean for us a mode of life whose single, animating purpose is to find out the real truth, which pursues this purpose by a well-considered method, founded on thorough acquaintance with such scientific results already ascertained by others as may be available, and which seeks co-operation in the hope that the truth may be found, if not by any of the actual inquirers, yet ultimately by those who come after them and who shall make use of their results. It makes no difference how imperfect a man's knowledge may be, how mixed with error and prejudice; from the moment that he engages in an inquiry in the spirit described, that which occupies him is *science* as the word will here be used. (*CP* 7.54–55)

This concrete characterization he also extends to our conceptions of the particular branches of science. He views a particular science—for example, chemistry—as "no mere word manufactured by some academic pedant but as a real object, being the very concrete life of a social group constituted by real facts of inter-relation" (*CP* 7.52). From Peirce's perspective, then, when we speak of science in general or of some particular science, what we are concretely talking about is a community of inquirers, extended over time, with a unity of purpose and method which enables the product to be much more than the sum of the individual contributors. It is in this spirit that Peirce returns to the question of definition and states, "If we are to define science, not in the sense of stuffing it into an artificial pigeonhole where it may be found again by some insignificant mark, but in the sense of characterizing it *as a living, historical entity*, we must conceive it as that about which men such as I have described busy themselves" (*CP* 1.44).

It is this understanding of science that informs Peirce's project of exhibiting the conditions of the possibility of science. Given this conception of science as a specific mode of inquiry, two points clearly call for some accounting. Science is, from Peirce's own perspective, only one model of cognitive inquiry, and one which has not always been dominant and whose continuance is by no means inevitable. Moreover, it is a mode of inquiry whose objective validity, whose possibility of attaining the truth, has no obvious guarantee. Hence, what are the conditions of the possibility of its continuance and flourishing, on the one hand, and its objective validity, on the other? As I will reconstruct Peirce's project, then, it will have two facets: first, the articulation of certain qualities of inquirers and institutions necessary to sustain the process; second, the articulation of certain features of our world necessary to guarantee its objective validity. Together these will constitute the conditions of the possibility of science as we know it.

The Conditions of the Possibility of the Development and Continuance of Science as a Mode of Inquiry

Since other models of cognitive inquiry (tenacity, authority, and self-evidence) have had historical periods of dominance and continue to vie with science for our allegiance, the development and continuance of science as we know it depends on the persistence of certain interrelated social *practices* which in turn depend on certain *virtues* being embodied in the individual members of the community of investigators. It is to these sustaining virtues that Peirce turns his attention in his initial exploration of the conditions of possibility of the scientific mode of life.

Peirce wants to argue that the continued development of science as we know it is grounded in certain moral dispositions which mark its individual practitioners: "The most vital factors in the method of modern science have not been the following of this or that logical prescription—although these have had their value too—but they have been the moral factors" (*CP* 7.87). The specific moral factors on which he focused are *the love of truth*, *the sense of community*, and *the sense of confidence*.

The first of what Peirce calls moral factors seems initially to be completely uncontroversial: "The first of these has been the genuine love of truth and the conviction that nothing else could long endure" (*CP* 7.87). This apparently straightforward claim, however, masks some complexities. Peirce is not claiming that an inquiry qua scientific is either disinterested or presuppositionless. On the contrary, he is well aware of the many interests that can motivate a given line of scientific inquiry. His claim is that for the truly scientific mind the search for truth is the dominant one. One of the contrasts he is alluding to is that between the mind of the inquirer and the mind of the pedagogue. He sees the scientist as the one whose dominant, driving interest is in the search for truth wherever it may lead, whereas he sees the teacher (or preacher) as one whose dominant interest is in organizing and communicating what he or she already knows. At times he characterizes this distinction as that between the laboratory mind and the seminary mind. This contrast between the quest for truth and the elaboration and dissemination of belief runs deep into the human character, and it is a contrast that Peirce sees "writ large" in the difference between the spirit of modern science and that of the Middle Ages.

Nor does Peirce think that scientific inquiry is presuppositionless. Any given scientific inquiry is conducted not only against a background of the "established scientific verities" of the moment but also against the background of more general metaphysical assumptions which guide our orientation to the world (*CP* 7.82). It is our attitude toward these presuppositions that can be either scientific or not. If the quest for truth is dominant, these background presuppositions are never regarded as beyond question. Although, as entrenched, the presumption is in their favor, if the results of the inquiry seem to call for their revision, then such a revision must be regarded as a real option in the interest of truth.

The second moral factor, the sense of community, is more complicated and attracts more of Peirce's attention. In addition to making the obvious points about the requirement of intersubjectivity of evidence imposed by the social character of scientific investigation, he goes on to explore the deeper commitments of self-sacrifice involved in the enterprise which is science.

> The method of modern science is social in respect to the solidarity of its efforts. The scientific world is like a colony of insects in that the individual strives to produce that which he cannot himself hope to enjoy. One generation collects premises in order that a distant generation may discover what they mean. When a problem comes before the scientific world, a hundred men immediately set all their energies to work on it. One contributes this, another that. Another company, standing on the shoulders of the first, strikes a little higher, until at last the parapet is attained. (*CP* 7.87)

Mixing his metaphors between a colony of insects and a company of troops, Peirce makes the point that the life of science is essentially that of a historical community that is teleological in structure. The development and continuance of this life depends on the social sense becoming supreme, such that the individual investigators will subordinate their own satisfaction to the long-range goals of the community.

For Peirce this sense of community is not merely an extrinsic support of the life of science but is essentially tied to the very logic of scientific method. The rationality of both the abductive and the inductive phases of inquiry requires one to look beyond the particular inference in question to its role in the overall inquiry process: "It can be shown that no inference of any individual can be thoroughly logical without certain *determinations of his mind* which do not concern any one inference immediately" (*CP* 5.354). And these "determinations of mind" involve the individual's viewing his or her particular inferences not just as part of the larger set of that person's own inferences but in terms of their role in that ongoing inquiry, the proper logical subject of which is the historical community. With this in mind Peirce articulates what he calls the three logical sentiments: "namely, interest in an indefinite community, recognition of the possibility of this interest being made supreme, and hope in the unlimited continuance of intellectual activity as indispensable requirements of logic" (*CP* 2.655). The life of science demands the transcendence of both selfishness and skepticism through the active hope that rational cooperative effort will, in the end, prevail.

This leads to the third moral factor undergirding the development and continuation of science, namely, the sense of confidence. Peirce thinks that a sense of confidence is particularly crucial in an inquiry that proceeds by the method of conjecture and refutation. With regard to our specific, proposed explanations, we are clearly going to be wrong more often than we are right, so it is important that we continue to view our proximate failures in terms of their contribution to the long-range effort. He sees this confidence as characteristic of scientists: "Modern science has never faltered in its confidence that it would ultimately find out the truth concerning any question to which it could apply the check of experiment" (*CP* 7.87). This attitude implies both the correctness and completeness of science and takes the form of the action-guiding hope that the indefinite application of scientific methodology will lead to the attainment of truth in the long run. The ground of this particular hope will be explored in the next section of this chapter.

Given Peirce's pragmatism, it should not be surprising that he saw these moral factors not as private, internal states but as embodied in a coordinated set of practices constitutive of the scientific community. In addition to the constraint to make experiment replicatable and evidence intersubjectively available, Peirce points to the scientists' "unreserved discussions with one another" and their "availing themselves of that neighbor's results," which practices constitute a given science as "a real object, being the concrete life of a social group constituted by real facts of inter-relation" (*CP* 7.52). This *institution* so constructed not only is the judge and repository of past results but is dynamically characterized by criteria of evaluation of present programs and prognoses of future ones, which give it an identity over time that transcends any individual or group of practitioners. In summary, the reality of these virtues, practices, and institutions is the condition of the possibility of the development and continuation of science as a mode of inquiry.

Conditions of the Possibility of the Objective Validity of Science as a Mode of Inquiry

What must reality be like such that science so construed may be cognitively successful? Peirce now turns to the dynamic variant on the classical Kantian question, and by this move the project undergoes two important modifications. First, the historical dimension is introduced in that the notion of validity *here and now* is replaced by the notion of validity *in the long run*. The question is no longer, What must the world be like such that these specific claims of science be true? but rather, What must the world be like such that this scientific mode of inquiry will ultimately attain the truth? Second, Peirce's characteristic fallibilism is introduced, severing the tie between "necessarily true" and "objectively valid." Objective validity will be tied to the process of inquiry rather than merely to a priori structures. It is in this spirit that Peirce displaces the priority of the Kantian question How are synthetic a priori judgments possible? in favor of the more general question How are synthetic judgments *in general* possible? (*CP* 5.348)

Since the distinctive features of science as a cognitive mode of life were seen to be the regulative principles embodied in the abductive and inductive phases of scientific inquiry, it seems reasonable to focus on these if we are to get clear about the objective validity of science. What explanation can be given for our ability to come up with antecedently plausible explanations of the world? And what justification can we give for our confidence that the continued application of confirmation procedures will lead to truth in the long run? Both foci, that of discovery and that of confirmation, seem to involve evidential leaps which call for some kind of metaphysical grounding if our confidence in the objective validity of science is to be justified.

In his more fine-grained account of the abductive phase of scientific inquiry, Peirce distinguishes two moments. The first moment bears on discovery properly so-called, namely, the origination of those conjectures (hypotheses) which will make up the list of possible explanations of the phenomena under consideration. Since, absolutely considered, he thinks that the number of possible explanations for any finite set of phenomena is infinite, Peirce puzzles about humankind's ability to select from that potentially infinite set a subset of possible explanations that will contain the ultimately adequate account. In the end Peirce posits a "natural instinct for guessing right," not reducible to strict formulas or rules of procedure, as part of any account of the de facto historical success of science. This instinct is "strong enough not to be overwhelmingly more often wrong than right" (*CP* 5.173), and given the repetition of the process, this should be enough to get the best explanations on the list to be considered.

The second moment of the abductive process is, to a degree, rule governed and bears on the task of singling out, from our general list of possible explanations, those we are to take as serious candidates for investigation and the order in which we are to take them. With regard to this second moment of abduction Peirce develops methodological criteria such as "simplicity" and those bearing on "the economics of research," but even at this second level there is an irreducible role for our "natural instinct for guessing right" in our preference ordering of those hypotheses that should inform our research programs. (*CP* 7.220)

What is the metaphysical ground of this cognitively crucial natural instinct? Given the obvious presence of survival instincts in the rest of the animal kingdom, Peirce's first thought is that surely it is not unreasonable to believe that we too have those instincts necessary for the effective continuance of our distinctive mode of life (*CP* 6.476). The details of the metaphysical picture are broadly evolutionary in nature.

Peirce sees this ability to guess right as having obvious adaptive value and hence as being a clear candidate for natural selection. In particular, he views a rudimentary grasp of certain fundamental principles of mechanics as crucially important to certain organic practices necessary for survival.

> The great utility and indispensableness of the conceptions of time, space and force, even to the lowest intelligence, are such as to suggest that they are the results of natural selection. Without something like geometrical, kinetical, and mechanical conceptions, no animal could seize his food or do anything which might be necessary for the preservation of the species. . . . But, as that animal would have an immense advantage in the struggle for life whose mechanical conceptions did not break down in a novel situation (such as development must bring about), there would be a constant selection in favor of more and more correct ideas of these matters. (*CP* 6.418)

It is important to note that Peirce is here talking about the ability to *guess* right (the impetus toward "more and more correct ideas of these matters"); he thinks that only beliefs that are on the right track will have staying power and developmental fecundity, given ever-changing circumstances. Since it is correct beliefs that will have survival value, it should not be at all surprising that we, the survivors, should have this ability to guess correctly to a considerable degree.

Can anything more be said about how we came to have this ability in the first place? If we are operating on the Cartesian picture of humankind's relation to nature such that the human mind is, as it were, outside nature, looking in, trying to guess at the laws which describe its structure, then its apparent high success rate would indeed seem mysterious. But if we assume that the human mind is constituted by nature's evolving development, its affinity with its object becomes less mysterious. "If the universe conforms, with any approach to accuracy, to certain highly pervasive laws, and if man's mind has been developed under the influence of those laws, it is to be expected that he should have a *natural light*, or *light of nature* or *instinctive insight* or genius, tending to make him guess those laws aright or nearly aright" (*CP* 5.604).

Quite paradoxically, Peirce finds the ultimate ground for this Cartesian notion of a natural light in a decidedly anti-Cartesian picture of mind: "Our minds having been formed under the influence of phenomena governed by the laws of mechanics, certain conceptions entering into these laws become implanted in our minds so that we readily guess at what the laws are" (*CP* 6.10). Being nature's products, we have ready access to her secrets.

Peirce now turns to the inductive phase of the inquiry process to explore the question of its metaphysical grounding. Is our confidence that the continued application of confirmation procedures will lead to the attainment of truth in the long run merely an unfounded hope, or can it be shown to have some objective foundation? Can our metaphysical picture of the world be extended to render intelligible the success of this self-monitoring feature of scientific inquiry and thereby the long-run objectivity of science as a whole? Specifically, this is a question about the self-corrective feature of scientific inquiry, while more generally it is the question of the justification of induction. If the confirmation procedures of science are going to be viewed as playing this crucial role in the movement toward truth over time, then we must have reason to believe that the inductive sampling that functions in the confirmation stage is not destined to be misleading and can function as a reliable guide to the structure of the real.

That Peirce believes that inductive procedures have this feature is quite clear. He views induction as a procedure which "if steadily persisted in must lead to true knowledge in the long run of cases of its application whether to the existing world or any imaginable world whatsoever" (*CP* 7.207). It is self-monitoring in that its continued use will uncover the mistakes in its earlier uses, and by a process of elimination, truth will eventually be attained. He sees this as quite independent of any particular features of the world. In fact, he cannot even imagine a world in which such procedures would not be reliable.

> If men were not able to learn from induction it might be because, as a general rule, when they have made an induction the order of things would then undergo a revolution. But this general rule would itself be capable of being discovered by induction; and so it must be a law of such a universe that when this rule was discovered it would cease to operate. But this second law would itself be capable of discovery. And so in such a universe there would be nothing which would not sooner or later be known; and it would have an order capable of discovery by a sufficiently long course of reasoning. (*CP* 5.352)

If even in such a demonically contrived universe inductive procedures would be reliable, surely they would be reliable in any ordinary universe—ours in particular.

Peirce's reasons for thinking this are tied to his construal of the monitoring phase (induction/confirmation) of scientific inquiry, specifically in terms of statistical inductions. Specific predictions as to the character of our world are derived from the hypothesis under investigation, and then our world is checked for this character. Samples are drawn from our world to see if it has the characteristics we suppose it to have. Obviously there is some initial evidence for this character, but if controlled sampling belies its validity, continued sampling should reveal which of the other proposed characterizations really map onto our world. Why this should be the case brings us to the heart of Peirce's pragmatic realism.

> [An] endless series must have some character; and it would be absurd to say that experience has a character which is never manifested. But there is no other way in which the character of that series can manifest itself than while the endless series is still incomplete. Therefore, if the character manifested by the series up to a certain point is not that character which the entire series possesses, still, as the series goes on, it must certainly tend, however irregularly, toward becoming so; and all the rest of the reasoner's life will be a continuation of this inferential process. This inference does not depend on any assumption that the series will be endless, or that the future will be like the past, or that nature is uniform, or any other material assumption whatsoever. (*CP* 2.784)

Peirce's idea is that inductive inference is basically an inference from part to whole, and its validity depends simply on the fact that parts do make up and constitute the whole. In confirmation we are basically involved in drawing samples from a population, and if the frequency with which some relevant property is distributed over the individuals of the sample does not correspond to its frequency of distribution over the population, the discrepancy is sure to become apparent as the sampling process is extended over the long run. To resist this line of thought is to entertain a conception of the population of the whole which will never manifest itself in the samples or the parts. But to entertain this is to conceive of truth as transcendent, reality as possibly incognizable, as a thing-in-itself, views which Peirce thinks he has good reason to reject.

Having thus given some kind of metaphysical accounting for the evidential leaps involved in abduction (instinct to guess right) and induction (logic of sampling), Peirce feels he has given us good reason to accept the objective validity of that mode of inquiry which he has designated scientific. The continual monitoring of the inductive phase occasions conceptual revision through the abductive phase, with the total process of inquiry resulting, over time, in an increasingly adequate picture of the world.

Conclusion

This, in broad outline, is the general structure of Peirce's attempt to delineate the conditions of the possibility of science. As generally characterized, the project is unmistakably Kantian, but in its execution, it takes a decidedly pragmatic turn. We have seen the principal difference between the two variants to be traceable to differing conceptions of the focal meaning of the word "science," the meaning shifting from that of a given body of propositions to that of a sociohistorical process of inquiry. Two transpositions in the Kantian project are effected by this shift. First, the grounds of the necessary validity of science is still the object, but the focus shifts from particular propositions to the inferential procedures through which these propositions are attained. The focus is no longer the specific claims of Euclidean geometry and Newtonian mechanics but the long-range stability and objectivity of a specific kind of sociohistorical inquiry. Second, speaking metaphorically, the function of the phenomenal/noumenal distinction can be seen to be transposed from the spatial to the temporal axis. There is still a functional distinction between how things are represented by us and how they really are in themselves, but the contrast now is between our present, imperfect representations and the ultimately adequate ones of the scientific community in the long run. Fallibilism and realism are seen to be facets of one and the same overall picture of science.

2 Peirce's Realistic Approach to Mathematics: Or, Can One Be a Realist without Being a Platonist?

Claudine Engel-Tiercelin, Université de Paris (Pantheon-Sorbonne)

Peirce's position on the problem of universals is that of a sophisticated realism inherited from the Avicennian-Scotistic tradition, which may be briefly characterized by its opposition to metaphysical realism (or Platonism) and various forms of nominalism. Against the idea that there could be either a reality totally independent of our ways of knowing or verifying (an incognizable *ding an sich*) or realities that could be reduced to singular existents (sense-data, concepts, determinate individuals), Peirce's realism is a unique and ambitious analysis of all the aspects—logical, physical, and metaphysical—of the problem of universals, combining scholastic and categorical elements with scientific, pragmatistic, and straightforward metaphysical concerns.

In this chapter I consider how Peirce's realism fits his approach to mathematics, which is often presented as a somewhat incoherent mixture of Platonistic and conceptualistic elements. Without denying these, I shall attempt to claim that the subtlety of Peirce's position not only helps to clear up some of these so-called inconsistencies but offers many insights for contemporary ways of dealing with the mathematical aspects of the problem of universals.

Peirce's Definition of Mathematics: Or, How to Apply Realism to Mathematics

Briefly one could say that Peirce's definition of mathematics has two main characteristics. First, it does not cover a special domain of entities.

It is not defined by the specificity of its objects (space, time, quantity) or by the nature of its propositions (analytical, a priori) or by the kinds of truths it can exhibit. Against Hamilton and De Morgan, Peirce denied any dependence of mathematics on space, time, or any form of "intuition" (*CP* 3.556). As to the analytic or synthetic nature of mathematical propositions, he said almost nothing, convinced that the real issues were elsewhere and that they had to be thought through and formulated in new terms (and especially through the distinction of corollarial and theorematic forms of deductive reasoning). If mathematics has nothing to say about truth, it is because it is not—contrary to logic—a science of facts but a science of hypotheses and abstractions.

But for the son of Benjamin Peirce, a second characteristic of mathematics was that it was also a science of reasoning, more specifically, "the science which draws necessary conclusions" (*CP* 3.558; 4.229). This is a very wide definition of mathematics, since not only is all mathematical reasoning diagrammatic, but all necessary reasoning is mathematical reasoning, no matter how simple it may be (*NEM* 4:47). In that sense, when Peirce affirmed the fundamentally iconic, observational, and experimental character of deduction, he not only defined mathematical deduction as such, he accounted for all kinds of deduction, thus reviving the whole conception of logical necessity.

Such a definition presents obvious difficulties for any kind of realism. First, how is one to adjust the idea of mathematics as a purely formal and ideal system to realism? What is the status of these *entia rationis*? Are they pure conventions, arbitrarily chosen, which never refer to reality? Are they simple tautological or analytical statements, incapable of being qualified as true except as concerns the meaning of the expressions they involve? Then why insist on the practical side of mathematics? Why take so seriously the problem of its application? If one accepts the notion of applied mathematics as something which is needed by all sciences, what is to warrant that such idealizations have the objective validity which justifies their being used by these other sciences?

Second, if Peirce's realism is a realism of indeterminacy—which implies, on the one hand, that it should be possible to think without contradiction not only about reals but also about possibles and, on the other hand, that indeterminacy renounces any idea of absolute or infallible necessity and exactitude—how is such indeterminacy going to be taken care of by mathematics? According to Susan Haack, one can measure the importance of the issue by noticing Peirce's many hesitations about answering the question of whether or not fallibilism should be extended to mathematics. For Haack (1979:37), Peirce never really made up his mind, at times declaring that the necessity of mathematics prevents our being wrong about our mathematical beliefs and that we are only fallible as far as our factual beliefs are concerned (*CP* 1.149, written c. 1897); at other times he emphasized that fallibilism does extend to mathematics (*CP* 7.108, 1892; 1.248, 1902) and that mathematical inferences are only probable after all. Clearly enough, for Peirce, mathematical necessity was perfectly compatible with the notion of an ideal system in which one reasons about possibles (hence indeterminates) and not about real cases. Not only does mathematics allow for individuals that are not perfectly determinate in all respects (*NEM* 4:xiii), but also one can think about the infinite. In that sense, the basic principles of Peirce's realism that are attached to indeterminacy (possibility and generality, or firstness and thirdness) seem to be obeyed. But can one be satisfied with such a definition of necessity?

Peirce's Conventionalism and Conceptualism

Before trying to make sense of Peirce's realism in mathematics, we have to ask if it makes sense to speak of Peirce's realism at all. Indeed, a number of instances seem to suggest that he adopted a conceptualistic, or nominalistic, position in which pure mathematics is the domain of hypotheses and ideal creations. Moreover, its necessity is apparently not due to any characteristics of its objects or to any particular nature of its propositions that would afford a specific objectivity.

Conceptualism always has realism in its background. Thus, after defining a pure diagram as that which is "designed to represent and to render intelligible, the Form of Relation, merely" and asserting that "an intelligible relation, that is a relation for thought, is created only by the act of representing it" (*NEM* 4:316 n. 1), Peirce added that if we should some day find out the metaphysical nature of relation, that would not mean that it would thereby be created, for the intelligible relation surely existed before, in thought or in the way God represented the universe. But this does not necessarily conflict with conceptualism. As David Wiggins (1980:139) says:

> Conceptualism properly conceived must not entail that before we got for ourselves these concepts, their extensions could not exist autonomously, i.e., independently of whether or not the concepts were destined to be fashioned and their compliants to be discovered. What conceptualism entails is only that, although horses, leaves, sun and stars are not inventions or artifacts, still, in order to single out these things, we have to deploy upon experience a conceptual scheme which has itself been fashioned or formed in such a way as to make it *possible* to single them out.

And this is, indeed, the spirit that animates such Peircean assertions as the following: "All necessary reasoning is reasoning from pure hypothesis in this sense, that if the premise has any truth for the real world, that is an accident totally irrelevant to the relation of the conclusion to the premise, while in the kinds of reasoning that are more peculiarly topics of logical [rather than mathematical] discussion, it has all the relevancy in the world" (*NEM* 4:164; cf. *CP* 4.233; *NEM* 4:270).

If mathematics is "the study of pure hypothesis regardless of any analogies they may have in our universe" (*NEM* 4:149; *CP* 3.560), which is particularly clear in the case of arithmetic (*NEM* 4:xv–xvi), and if "it certainly never would do to embrace pragmatism in any sense in which it should conflict with this great fact" (*NEM* 4:157), it would seem that mathematical necessity was derived, not from some necessity in things, but merely from the link of logical consequence between premises and conclusion (*CP* 4.232) and from the hypotheses, conventions, and rules which the mathematician has chosen to adopt (cf. defs. 32 and 33 in MS 94 [*NEM* 2:251]).[1]

So mathematical systems are purely formal. The meaning of the terms appearing in the postulates, hypotheses, and theorems is totally irrelevant as such: "A proposition is not a statement of perfectly pure mathematics until it is devoid of all definite meaning, and comes to this—that a property of a certain icon is pointed out and is declared to belong to anything like it, of which instances are given" (*CP* 5.567). For example, the only definitions that have to be retained as conforming to the "dignified meaninglessness of pure algebra" (*CP* 4.314) are those implicit definitions that postulates impose on their terms (see *CP* 3.20). In their turn, the postulates will be considered as implicitly defining the object(s) to which they apply, in the exact sense in which Riemann declared that the axioms of geometry provide a definition of space (see Murphey 1961:235).

No doubt all this contributes to qualifying Peirce's position as *antirealistic* in the sense given by Dummett (1987) that mathematical propositions have apparently no predeterminate meaning or truth, or such that it should suffice to discover them. Their meaning is such as is postulated, then demonstrated. In mathematics, there are no propositions which could benefit from truth conditions that were utterly independent of our capacity to recognize them as such, or whose truth conditions could be realized without our being able to recognize that they are so. In other words, one should always be able to define the truth conditions of any mathematical statement in terms of its conditions of assertibility. This is clearly stated in *MS* 94: "The *meaning* of any speech, writing, or other sign is its translation into a sign more convenient for the purposes of thought; for all thinking is in signs. The meaning of a mathematical term or sign is its expression in the kind of signs in the imaginary or other manifestation of which the mathematical reasoning consists. For geometry, this [expression] is [in] a geometrical diagram" (*NEM* 2:251).

This is also why one must ordinarily attach great importance to the mathematical procedures of demonstration, to the modus operandi (*CP* 4.429; *NEM* 2:10–11). That is, meaning is not given from the start but, on the contrary, is determined by the demonstration. To reason is not to use meanings; it is to construct them, to manipulate them in order to determine them. And the analysis of the way these constructions work may help in clarifying, if not in constituting, such a determination of meaning (*CP* 3.363). Hence the great importance of the iconicity of reasoning (*CP* 2.279; *NEM* 4:47–48) but not only of that (Engel-Tiercelin 1991), for indeed, there are two other essential procedures in the determination of the meaning of mathematical statements. These are *abstraction* and *generalization*, both of which allow a better grasp of the status of the *entia rationis* the mathematician works upon. Hypostatic abstraction has a decisive role (*NEM* 4:49) because it is that operation by which something "denoted by a noun substantive, something having a name," which belongs to the category of substance as such, is transformed into an assertion, and the *reality* of which "can mean nothing except the *truth* of statements in which the real thing is asserted" (*NEM* 4:161–62; see *CP* 4.234). Thus, to say that numbers are real is not to reduce them to some singular existing entities but merely to indicate that there are statements in which numerical expressions are used to describe classes adjectively (*CP* 4.154–55). Thanks to such abstractions as numbers, lines, or collections, "it thus becomes possible to study their relations and to apply to these relations discoveries already made respecting analogous relations" (*CP* 3.642). It thus becomes unnecessary to assume some kind of objects preexisting in some kind of mathematical universe.

All this tends to suggest the image of a conceptualist and conventionalist who is more eager to present mathematics, not as a realm of objects to be discovered or of independent truths describing already-given facts and transcending all possibility of verification or refutation, but as a body of rules, practices, mental constructions, procedures of decision, and methods of demonstration from which mathematics derives its instrumental value and the ground of its necessity.

Arithmetic and the Temptation of Platonism

It would be an exaggeration, however, to say that Peirce's position is utterly clear on all these points. It seems indeed difficult to deny that some of his analyses, mostly in arithmetic but not only there, do not sound conceptualistic at all, but realistic, in the most traditional or Platonistic sense of the word.

Thus Peirce did not hesitate to speak of the "innate" propositions of mathematics, preferring that rather than such a term as "a priori," which "involves the mistaken notion that the operations of demonstrative reasoning are nothing but applications of plain rules to plain cases" (*CP* 4.92). It is, of course, in arithmetic that the temptation of Platonism is the strongest, numbers being at times qualified as "ideas" belonging as such to a different "universe of experience" from that of facts and laws;

the "Platonistic world of pure forms" (*CP* 4.118), or that "Inner World" (*CP* 41.161) in which these eternal, abstract, and "airy nothings" (*CP* 6.455) are not "absolutely created by the mathematician" (*CP* 4.161).[2] In that respect, we would not be very far from Frege's universe of true "thoughts," from "laws of Pure Being," or thoughts independent of the senses and of the empirical world. Indeed, it is in a way that reminds us very much of Frege inasmuch as the development of the whole theory of numbers is described by Peirce as arising from a small number of first and primitive propositions (*CP* 2.361; 6.595). For the same reasons, we could understand some of Peirce's assertions as signifying that the truth conditions of any mathematical proposition are transcendent in respect to their conditions of verification and that arithmetic, at least, is an independent domain of entities (cf. *CP* 4.114).

The Real Nature of Peirce's Realistic Solution in Mathematics

I would like to claim, however, that these inconsistencies between conceptualistic and Platonistic elements are a wrong way to look at the matter. In fact, they hide the real nature of Peirce's way of dealing with the problem of universals in mathematics and the typical realistic solution he proposes. Let us start with the problem of conceptualistic conventionalism.

Hypotheses and Definitions

First of all, it is clear enough that Peirce's conventionalism is never so absolute as that of, say, a Poincaré, (who is at times criticized and viewed harshly, though wrongly, as such). No doubt, Peirce would not be ready to reduce mathematics to an ideal science of hypotheses that would consist in nothing more than a simple game of abstract, arbitrary, and convenient formulas.

If mathematics is a matter of creations, these are never totally arbitrary (*NEM* 4:xii), first, because most often the mathematician's hypothesis is provoked by some real problem which arises in other sciences (*NEM* 4:xv).[3] This is why the distinction between pure and applied mathematics is not so decisive after all (*NEM* 2:vi).

But the arbitrariness of hypotheses is weakened in another way. Contrary to the poet's hypotheses (*CP* 4.238; *NEM* 4:68) the mathematician's hypotheses, which indeed require the highest intellectual capacity to be found, are subject to the rules of deduction. A mathematician is interested in hypotheses only for the forms of inference that can be drawn from them (*NEM* 4:268). Before going any further, let me add that it is surely not enough to invoke the ideal character of hypotheses, especially in geometry, to solve entirely the question of the nature of their content: namely, that topics is the sole part of geometry that is really pure and ideal, because it deals more with pure continua. A lot could be said here about the interfaces between mathematics and straightforward metaphysics (cf. Engel-Tiercelin 1986 and 1990).

Rules of Deduction and Habits of Reasoning: Our "Hereditary Metaphysics"

If one wishes to define mathematics as a system that is ideal, and yet not arbitrary, one must first be able to *verify* whether its creations are relevant or just simply not meaningless. This explains why we do not describe surfaces or lines as mere ad hoc constructions. "We here use the traditional phraseology which speaks as if lines and surfaces were something we make. This is not strictly correct. The lines and surfaces are *places* which are *there*, whether we think of them or not. They are quite ideal, it is true. But they are there, in this sense that we can think of them as being there without being drawn into any absurdity" (*NEM* 2:387).

But the lack of arbitrariness is also due, in good part, to the character of logical necessity, that is, to the fact that the rules we follow in our demonstrations do nothing except exhibit the habits we have acquired through reasoning (*CP* 5.367). The leading principles of inference are but the linguistically codified formulation (*logica docens*) (*CP* 1.417) of these habits of reasoning under the form of habits of inference, which must not be confused with some psychological constraint (in the mode of Sigwart or Schröder (*CP* 2.52, 209; 3.432). Even if it is true, in the end, that it is not the constraint exerted by the rule but rather the fact that the conclusion is true when the rules of transformation and substitution are correctly applied (*CP* 2.153; 5.365) that determines the objective validity of the reasoning, one must not underrate the important role played in inference by habits. It explains why logical necessity is something like a fact which is felt as such and which, for that reason, hardly needs any justification. Here Peirce is very near Wittgenstein: necessity is present in our acts and practices; we stumble against something whose explanation could hardly be anything other than "we use it so."[4]

Because logical necessity exerts itself as a constraint difficult to explain and to justify, the teacher in mathematics often has difficulties in trying "to make another person feel the force of that demonstration who does not do so already" (*NEM* 4:xiv). Peirce's realism comes to the fore here: logical necessity is one of those habit-facts which are absolutely real and yet irreducibly vague, before which no further explanation seems appropriate or even required (*CP* 2.173). Something like a sort of constraint of necessity should be granted here. One can wonder if it is enough to warrant the self-evident character of mathematics (*CP* 2.191). But obviously, Peirce would have answered that in such a case, it was precisely the hopeless attempt at asking for some kind of justification that would need to be justified.

But such habits of reasoning also have the strength of general rules (and this is part of Peirce's realism too)—on the one hand, because they express collective and not subjective practices and presuppose some agreement among the members of the community (which is a decisive constituent of the kind of exactitude that can be reached in mathematics) (*CP* 5.577) and, on the other hand, because they result from a controlled use ruled by intelligence (or thirdness). Here lies all the difference there is between the mere thinking *in* images (which is more often a handicap than a help) and the experimenting *on* images, that is, icons or abstract schemes. Because intelligence rules practice, it is of no consequence in mathematics if one uses formulas or notations which are mere *flatus vocis*. In fact, the more they are so, the better it is, for it will facilitate experimentation. Hence, the utility of children's rhymes like "eeny, meeny, miney-mo" in which one counts words, not things. What is essential is what one does with them, the way one thinks in this notation, for "one secret of the art of reasoning is to *think*" (*NEM* 1:136).[5]

Peirce's realism insists as much on the irreducible vagueness of our habits of reasoning as on the fact that thirdness is the category of generalization, of abstraction, of all the operations of reasoning for which self-control or criticism is always on the lookout. In that sense, mathematics does not constitute an exception *in principle* to fallibilism. Rather it is, most often, an exception *in fact*. But this again does not warrant any absolute or fundamental value being assigned to mathematical necessity: "Mathematical certainty is not absolute certainty. For the greatest mathematicians sometimes blunder, and therefore it is *possible*—barely possible—that all have blundered every time they added two and two" (*CP* 4.478).

Realism as an Alternative to Platonism in Arithmetic

What about Peirce's Platonistic claims in arithmetic? I think they are counterbalanced by other kinds of claims. First, paradoxically, such concepts as "number," "zero," and "successor" are counted not as primitive concepts (Russell) but as mere variables (Peano) (Murphey 1961:238ff.). Peirce tried to construct several systems of pure number (see *CP* 4.160ff., 677–81; 3.562ff.), giving only an implicit definition of its primitive terms, thus allowing a perfectly formalistic interpretation of his system (Murphey 1961:244–45).

Second, the system of pure number was for him but a particular case of ordinals,[6] which were the primitive pure numbers (*CP* 3.628ff.; 4.332, 657–59, 673ff., etc.) not only because they expressed a relative place but because they illustrated it, so that Peirce believed they exemplified the pure serial relation which was instantiated in all the series (Murphey 1961:273–74): "But the highest and last lesson which the numbers whisper in our ear is that of the supremacy of the forms of relation for which their tawdry outside is the mere shell of the casket" (*CP* 4.681).

But there are reasons other than Peirce's relational realism that explain his preferences for ordinals, and these are taken either from sensible experience (*CP* 4.154) or from straightforward pragmatical considerations.[7]

It would be as wrong to overrate any so-called supremacy of arithmetic above the other sciences for it is as vain to hold to the idea of some compartmentalization of mathematics (*CP* 4.247) as to underrate the pragmatic considerations that are part of Peirce's way of dealing with numbers. This is true, even though he was careful to distinguish between pure, or "scientific," arithmetic (*CP* 3.562A), which "considers only the numbers themselves and not the application of them to counting," and practical arithmetic (*CP* 4.163). Peirce spent a lot of time achieving a whole pedagogy for arithmetic. What does this signify, except that in arithmetic, too, what is important is not only the type of objects or propositions[8] but consideration of the system, in which the learning of the rules is decisive?

The learning of arithmetic, and the role played in it by iconic representations, was taken by Peirce to be essential and yet responsible for so many errors that he kept reconstructing the steps that ought to be performed in the techniques of education and of learning.[9] This is not to say that Peirce would reduce meaning to use (this is why, for example, instinct should not decide for the right interpretation of a system). One should manage to teach this at the pedagogical level. But we can go further than this: "Some children learn by first acquiring the use of a word, or phrase, and then, long after, getting some glimmer of what it means" (*NEM* 1:213–14). So it is surely true that a distinction should be maintained between pure arithmetic, which is "the knowledge of numbers," and practical arithmetic, which is "the knowledge of how to use numbers" (*NEM* 1:107). But how are we to understand, in the writings of this so-called avowed Platonist in arithmetic, such claims as the following one: "The way to teach a child what number means is to teach him to count. It is by studying the counting process that the philosopher must learn what the essence of number is" (*NEM* 1:214)? Does this not say that, in a certain way, the reality of numbers is to be found as much in the rules determining their meaning as in any primitive predeterminate meaning which would only need to be discovered in some Realm of Ideas?

Conclusions

First, it is not easy to determine what Peirce's final position is concerning the mathematical aspects of the problem of universals. However, I have tried to show that we can reconcile the obvious conceptualistic and conventionalistic claims with an equally obvious Platonism in other respects.

I do not think that Peirce's waverings express merely an ill-assumed tension between the metaphysically inclined logician, more eager to respect the principles and conclusions of his categorical and realistic analysis, and a kind of natural reflex of the mathematical practitioner who is easily led toward Platonism. Of course, we could always say, to the detriment of Peirce, that Platonism is not easy to avoid, even if, as B. Van Fraassen says (1975:40), Platonists win Pyrrhic victories because the conventionalists fail to provide the good arguments.

But such a conclusion is unsatisfactory. In fact, as I have shown, I am not at all convinced about the seriousness of Peirce's Platonism, which, should we be content with it, would constitute a severe objection to his whole theoretical project.

Second, a number of Peirce's vacillations merely reflect the difficulty of a coherent position on the subject; they do not always result in confusions but, on the contrary, succeed in stressing the immense complexity of the problem. In that respect, one of the paradoxes and characteristics of Peirce's reflections on mathematics is, by his wide definition of the domain of mathematics, to have pointed out its specificity, namely, that in mathematics, what is most difficult is not the solution of problems but the fact that there are problems to be solved and that consequently acute analysis is required as much on the methods and on reasoning as on the nature of mathematical objects and propositions.

Third, even if Peirce did not always avoid inconsistencies, because he openly adopted positions which came closer either to conventionalism or to Platonism, he was also perfectly aware of the impossible alternative that would consist in believing that one accounts for the specificity of the mathematician's work and of mathematical invention in terms of an opposition between a mathematics of discovery and a mathematics of invention, and he tried to find a third way to understand why there can be real problems in mathematics to be solved. I have tried to show that Peirce's idiosyncratic realism is such a third way, a very interesting attempt to answer the obvious major difficulty for a solution to the problem of universals in its mathematical form. A problem which he himself formulated quite clearly is, how is one to explain that "although mathematics deals with ideas and not with the world of sensible experience, its discoveries are not arbitrary dreams but something to which our minds are forced and which were unforeseen" (*NEM* 2:346)?

Fourth, Peirce obviously felt himself facing the following difficulty: on the one hand, his anti-Platonism, which was a natural consequence of his belief in the realism of indeterminacy, forced him to consider mathematics as simple meanings to be displayed or to "use" in practice, and not as truths to be discovered. But on the other hand, realism, or simply good sense, forbade him to adopt some strict form of verificationistic constructivism and to take mathematical demonstrations as pure determinations of meaning, free products of arbitrary creations and constructions.

The stress put on the ideal and hypothetical character of mathematics, the definition of it as the science of pure reasoning, and the fact that questions of method, practice, procedure, and demonstration are at least as important as questions bearing on the nature of objects or propositions all tend to show Peirce's awareness of the fact that a purely realistic answer, in the Platonist sense, cannot constitute a sufficient warrant for the necessity and objectivity of mathematics. For that reason, it is clear that Peirce's view cannot be compared to Frege's type of solution.

Again, if one of the Platonist's arguments consists in assuming a universe of objects, entities, and truths not only independent of, but transcending, our capacity to recognize them, Peirce's version cannot be reduced to such a position. First, because mathematics was not for him a science of truths (for reasons different from Wittgenstein, who, in somewhat related terms, thought that mathematics consisted not of true or false propositions but of autonomous rules of grammar), Peirce did not want mathematical statements, which are hypotheses to be taken as true or false, as describing any kinds of facts whatsoever. Second, Peirce's view cannot be reduced to Platonism because he thought that the meaning of mathematical statements could not be given independently of any demonstration: in that sense, although Peirce's pragmatistic realism of indeterminacy prevented him from reducing the meaning of a proposition to its conditions of verification, or reducing meaning to use, Peirce never separated the meaning of any mathematical proposition from its conditions of assertibility.

Such a view prevails, even when Peirce seems to show some form of Platonism. For example, although he admits that mathematics has a certain autonomy as an ideal system and goes so far as to talk of it as a universe ruled by dichotomies and truth, whose reality lies in its entities subsisting, even when no one is thinking about them or trying to know them, he also adds that if they can be called real ideas, it is because, one day or another, they will be "capable of getting thought," and that is but a question of time (*CP* 3.527; 6.455). We are far from the Platonist definition of a completely independent and transcendent universe. Faithful to the principle of the impossibility of incognizables, Peirce never defined mathematics as a universe totally independent of our possibility of knowing it, nor did he assume some completely given and preestablished meaning of mathematical statements which was waiting to be discovered, without any construction.

Indeed, Peirce was so interested in "pure numbers" that he tried to construct several systems of pure number. The parallel with the intuitionists seems obvious. In that respect, Murphey has shown (1961:286–87) that the way Brouwer defines a set—as a law according to which the elements of the set may be constructed and which is not a finished totality or any particular element of it necessarily a finished totality—is very near to Peirce's analysis, by its generating relation, of an infinite collection. But we have here all the differences between the antimetaphysical verificationism of the intuitionists and Peirce's realism. For example, Peirce would admit with Heyting (1966:15) that mathematical objects must have a consistency which in a way renders them independent of the acts of thought which aim at them, and at the same time he would consider that it makes no sense to think of an existence of these objects independently of any relation to human thought (Heyting 1964:42). But Peirce would find truistic the intuitionist's confusion between the fact of an object to be actually thought by an individual and the fact of an object to be dependent upon some form of general or possible thought, which is

Peirce's definition of reality. For Peirce, then, intuitionism would be a form of nominalistic Platonism, born from a misunderstanding of the real issues in the problem of universals. No wonder if the intuitionists finally drew antimetaphysical conclusions, for fear of the Platonism in which they fell after all (Heyting 1966:3), and decided to adopt a purely historical and constructivist definition of mathematics by considering only *actual* mental constructions (ibid.:8).

Even if Peirce used intuitionistic distinctions (e.g., between enumerate, denumerable, and nondenumerable collections), since his realism forced him to admit that one could, without contradiction, talk about possibilities, he could not be identified with intuitionism, for at least two reasons. First, we do not have to adopt strict verificationism (*CP* 6.455, 4.114) or "actually construct the correspondences" (*CP* 4.178); second, we can always treat possibilities as forming collections and extend the operations of classical logic (including the law of excluded middle) to such collections (*CP* 6.185ff.), even if Peirce also considered borderline cases and multivalued systems. Thus, we can think about the infinite, and Peirce believed that it was by calling up collections of possibilities that the paradoxes of the theory of sets could be avoided. As Murphey said (1961:287), nothing could be more opposed to intuitionism.

Fifth, Peirce has also shown, in defining numbers, collections, and multitudes as *entia rationis*, that for him they were inseparable from their conditions of assertability. In that respect, stressing as he did the operations of mathematical reasoning, Peirce also understood that it is necessary to remove the problems from the metaphysical ground and to concentrate first on their elaboration, on the problem of their meaning conditions. As Dummett wrote (1987:2), there is perhaps no hope of settling the argument between those who favor mathematical realism, who hold that we discover mathematical objects, and those who favor an idealist position, for whom mathematical objects are creations or conventions of our mind. The only way to clarify the issue is therefore to place it first at the level of meaning: Peirce's way of dealing with problems in mathematics seems perfectly appropriate to that kind of recommendation.

Finally, it is nonetheless true that Peirce would never dream of renouncing the metaphysical ground either. In that respect his treatment of the continuum problem is typical. If there is any reality in mathematics, it is perhaps much less in some particular nature of its objects than in that fundamental idea of the synechistic metaphysics, according to which there is, after all, no difference in nature between the "Inner Universe" (the universe of our representations) and the "Outer Universe" (or reality). One may consider either that such a position is fatal to any serious treatment of the mathematical aspects of the problem of universals or that a Dummettian way of dealing with realism leaves the ontological questions undecided, which in some cases may also be fatal to a grasp of some mathematical issues (think of the debates over the continuum problem). Even if we do not decide in favor of the second solution, I think Peirce's idiosyncratic realism is full of insights that may help clarify the issues and contribute to their ultimate solution.

Notes

1. It is indeed easier for arithmetic to reach such a formal ideal of purity, but even if it is difficult, as space is very often a matter of experience (*NEM* 4:xv), it is necessary (as Riemann showed) to come to a purely formal conception of geometry too (namely, one dealing only with pure continua), so that no distinction could finally be made between geometry and pure arithmetic, both containing analytical propositions, that is, deductions derivable from definitions in a purely logical way with no consideration of their possible empirical validity. Hence the question of some real correspondence between mathematical and real space or between the hypotheses or axioms of geometry and their empirical validity has no real value (*NEM* 2:251–52). The reason is that, even if it existed, it could not be demonstrated; real space is a *ding an sich*; we cannot apply our cognitive devices to it. And even if it were meaningful to say that we do apply them, we should have to consider their fallible character and possible margin of error. All of this explains why we are justified in adopting constructions that are extremely far from the properties of real space, insofar as they are practical and convenient.

2. As Murphey (1961: 239) rightly points out, we can see how faithful Peirce was to Platonism even after 1885, since he titled the fourth volume of his "Principles of Philosophy" (a project in twelve volumes) "Plato's World: An Elucidation of the Ideas of Modern Mathematics" (1893).

3. It is, indeed, because we find ourselves in such complicated situations

that it is impossible to determine with exactitude what the consequences could be that one calls for the help of the mathematician (*NEM* 2:9). So it is most often from such a practical suggestion that the mathematician will "frame a supposition of an ideal state of things," then "study that ideal state of things and find out what would be true in such a case," before generalizing to a third stage from that state of things, namely, by "considering other ideal states of things differing in definite respects from the first" (*NEM* 2:10). In so doing, "he not only finds out, but also produces a rule by which other similar questions may be answered" (ibid.).

4. Compare Wittgenstein 1974:2:74 and Peirce *NEM* 4:59: "It is idle to seek any justification of what is evident. It cannot be rendered more than evident."

5. This is why "the student must learn to use notations to think *in*, but he must not try to make the notation think for him, if he wishes to push his reasonings far. Thinking is done by experimenting in the imagination. Notations are excellent things to experiment with; but still experimentation requires intelligent supervision to come to much" (in Eisele 1979:186).

6. Contrary to most authors (among whom is Cantor), Peirce thought that ordinal and not cardinal numbers were the primitive pure numbers. A cardinal number, "though confounded with multitude by Cantor, is in fact one of a series of vocables the prime purpose of which, quite unlike any other words, is to serve as an instrument in the performance of the experiment of counting" (*CP* 3.628). In consequence of which, "The doctrine of the so-called ordinal numbers is a doctrine of pure mathematics; the doctrine of cardinal numbers, or rather, of multitude, is a doctrine of mathematics applied to logic" (*CP* 3.630). For that reason, Peirce thought that Dedekind could have gone even further when he and others considered "the pure abstract integers to be ordinal. . . . [They] might extend the assertion to all real numbers" (*CP* 4.633). Ordinals express a relative position in a simple sparse sequence (*CP* 4.337) and not, as in Cantor, the sequences themselves (Murphey 1961:247, 255). Ordinals only name "places" which are relative characters, which determine classes of members of these sequences. Hence, being classes, ordinals are more general than their members, among which are the collections to which multitudes are attributed.

7. Indeed, if ordinals are more primitive than cardinals, it is also because "the essence of anything lies in what it is intended to do." Now what are numbers? "Simply vocables used in counting. In order to subserve that purpose best, their sequence should stick in the memory, while the less signification they carry the better." The children are quite right in counting with nonsense rhymes, but these are always purely ordinals. Second, "The ultimate utility of counting is to aid reasoning. In order to do that, it must carry a *form* akin to that

of reasoning. Now the inseparable form of reasoning is that of proceeding *from* a starting-point *through* something else, *to* a result. This is an ordinal, not a collective idea" (*CP* 4.658–59).

8. For example, Peirce was not at all convinced by the superiority in arithmetic of any one system above another; in particular, he did not believe that the decimal system should be more "natural" than, say, the "secundal," or binary, numerical system which he proposed. He went so far as to say that if the ten fingers account for the almost universal use of base ten among all races of humankind, then the decimal system is a monument to human stupidity (*NEM* 1:237, 241). On the contrary, base six would seem extremely advantageous, even for counting on fingers and toes, although he thought that much the prettiest of the Aryan systems is the secundal. It is, to say the least, extremely convenient in logic, especially in the logic of denumerable and abnumeral series (*MS* 1121), its major merit lying in "its having several different methods of performing each operation, from which one can at sight select the one most convenient for the case in hand." But, as Carolyn Eisele has pointed out, although aware of its merits, Peirce did not believe "that any propaganda would ever move the world, because there is nothing in secundals to excite the emotional nature" (*MS* 1, in Eisele 1979:205).

9. See "Teaching Numeration" (*MS* 179; *NEM* 1:212ff.). "Imagination, concentration, generalization"—such are the qualities that have to be educated. In a seventeen-page manuscript entitled "Practical Arithmetic" (*NEM* 1:107ff.), we have a sketch of the maxims of a work to help acquire exactitude and agility in the use of numbers. Most individuals cannot help thinking about an abstract number without accompanying it with colors and forms having no intrinsic connection with the number. One has to take account of such "phantasms" and try to prevent "the formation of associations so unfavorable to arithmetical facility" (*NEM* 1:213). Again, "The teacher must not fail in his teaching to show the child, at once, how numbers can serve his immediate wishes. The schoolroom clock should strike; and he must count the strokes to know when he will be free. He should count all stairs he goes up. In school recess, playthings should be counted out to him; and the number required of him. This is to teach the ethical side of arithmetic" (ibid.).

References

Dummett, M. 1987. [An interview on] Philosophical Doubts and Religious Certainties. *Cogito* 2 (1):1–3.

Eisele, Carolyn. 1979. *Studies in the Scientific and Mathematical Philosophy of Charles S. Peirce.* Ed. R. M. Martin. The Hague: Mouton.

Engel-Tiercelin, C. 1986. Le vague est-il réel? Sur le realisme de C. S. Peirce. *Philosophie* 10:69–96.

———. 1990. Le problème des universaux chez C. S. Peirce. Doctoral diss., University of Paris.

———. 1991. The Semiotic Version of the Semantic Tradition in Formal Logic. In *New Inquiries into Meaning and Truth*, ed. N. Cooper and P. Engel, 187–213. London: Simon and Schuster.

Haack, Susan. 1979. Fallibilism and Necessity. *Synthese* 41:37–63.

Heyting, A. 1964. The Intuitionist Foundations of Mathematics. 1931. In *Philosophy of Mathematics: Selected Readings*, ed. P. Benacerraf and H. Putnam, 52–61. Oxford: B. Blackwell. 2d ed., 1983.

———. 1966. *Intuitionism: An Introduction.* 2d ed. Amsterdam: North-Holland Publishing. 1st ed., 1956.

Murphey, Murray G. 1961. *The Development of Peirce's Philosophy.* Cambridge: Harvard University Press.

Van Fraassen, Bas C. 1975. Platonism's Pyrrhic Victory. In *The Logical Enterprise*, ed. A. R. Anderson, M. R. Barcan, and R. M. Martin, 38–47. New Haven: Yale University Press.

Wiggins, David. 1980. *Sameness and Substance.* Oxford: B. Blackwell.

Wittgenstein, L. 1974. *Bermerkungen über die Grundlagen der Mathematik.* Frankfurt: Suhrkamp.

3 Peirce as Philosophical Topologist

R. Valentine Dusek, University of New Hampshire

"There is nothing of the nature of 'philosophy' in my book on topology" (*NEM* 2:xxi), Peirce wrote to his prospective publisher, Ginn and Company. Clearly Peirce was responding to the publisher's incomprehension of a work that was at least seventy years ahead of its time in education about mathematics, a work that introduced high school students to a field that had barely been born. Peirce meant simply that topology was genuine mathematics.

Yet Peirce was clearly a philosophical topologist in several senses. He was a philosopher uniquely knowledgeable about topology for his day, or even for our day. Since topology is the study of continuity, or the abstraction from mathematical structures of all but that which is strictly necessary for continuity, topology plays a highly central role in Peirce's philosophy of "synechism."

I wish to suggest that Peirce was a philosophical topologist in an even stronger sense, a sense in which he did *not* describe himself. Peirce used topology as a philosophical canon in the way in which the philosophical logician uses logic.

Peirce always described himself as primarily a logician. He compared himself, not boastfully but accurately, to Leibniz with respect to his strictly logical powers (Fisch 1986:250). Yet Peirce in his later philosophical work frequently appeals to "geometrical topics," what today is called combinatorial or algebraic topology. Carolyn Eisele has fostered a revolution in Peirce studies in the last decade by emphasizing the role of mathematical exactitude in Peirce's philosophy. It is not unreasonable that a philosopher who, in Hartshorne's words (1929:521), treats continuity as "the form of forms" should use the most advanced mathematical results concerning continuity to present his views.

Peirce was a precursor, among many other things, of mathematics without foundations, that is, a rejection of logicism and a recognition of the autonomy of mathematical thought, even with respect to the mathematical aspect of logic. For instance, Peirce wrote, "Mathematics is almost, if not quite, the only science which stands in need of no aid from the science of logic" (*CP* 2.81), and "Logic can be of no avail to mathematics; but mathematics lays the foundation on which logic builds" (*CP* 2.197). Therefore, it is reasonable to treat his use of topology independently of his use of logic, even the "logic of continuity."

Because twentieth-century philosophers have been mesmerized by Frege and Russell's account of logic as the foundation of mathematics and by the foundational account of Russell's "logic as the essence of philosophy" (Russell 1960:33), they have tended, even in discussing Peirce, to subordinate the intuitive geometrical aspect of topology to its set-theoretical "foundation."

Perhaps this claim can be clarified by a consideration of the last pages of Murray Murphey's pioneering book. Murphey (1961:405, 405n) remarked in concluding that "the topology which Peirce possessed was hopelessly inadequate to the demands he made on it; the topology Peirce needed was still a thing of the future, and when it was discovered it was developed on principles diametrically opposite to his. . . . It was precisely Cantor's theory of continuity which made possible the set-theoretical reconstruction of topology."

This conclusion errs in several respects. One respect has been extensively discussed by Dauben, by Potter and Shields, and by others. That is Peirce's relation to and knowledge of Cantor's set theory. It was hardly ignorance of, or even misunderstanding of, Cantor which led Peirce to develop his alternative account of the set-theoretical structure of the continuum.

But the passage is misleading in another respect on which this paper focuses. Murphey seems to confuse two areas of inquiry which later developed into the two major branches of topology. These were not distinguished during the 1890s and were only retrospectively so classified. These are the so-called point-set topology and the so-called combinatorial topology (later called algebraic topology). Peirce concerned himself with both. His work on the logic of number and the continuum were often concerned with what was to become point-set topology. His work on Listing's census theorem and the classification of surfaces and spaces was part of what later was to be called combinatorial, or algebraic, topology. The latter is my concern here (the former having been masterfully treated by Dauben).

Although the set-theoretical reconstruction of topology often modified the exposition and enriched various areas of investigation in algebraic topology, it is hardly true, as Murphey's remark suggests, that algebraic topology (whose rudimentary form was topics, the topology that Peirce possessed) was somehow intrinsically inadequate to the sort of inquiries with which Peirce concerned himself, such as the structure of real space for the purposes of cosmology. During the period of 1895–1906 Henri Poincaré was laying the foundations of twentieth-century combinatorial topology. Unfortunately, it appears that Peirce was not acquainted with Poincaré's path-breaking papers in this area, although he was familiar with Poincaré's philosophy of geometry (*NEM* 3:979) and with parts of his celestial mechanics, at least as they influenced the American G. W. Hill (*NEM* 3:1050). However, Poincaré not only did *not* found his work on Cantor's transfinite set theory but instead, in an article of 1906, rejected it (Poincaré n.d.:195). Recall that Poincaré was a precursor of intuitionism and of what would later be called constructive analysis and predicative set theory. An interesting aside on this is that L. E. J. Brouwer, who made the next major advances in topology in 1910–11, was also *the* intuitionist who wished to expel us from Cantor's paradise of transfinite set theory altogether. The case of Brouwer is more complex than that of Poincaré, however, in that Brouwer did make use of point-set-theoretical methods at the same time that he was founding intuitionism, and much later attempted to reconstruct as much of his topology as he could on an intuitionistic basis.

Poincaré and Brouwer, the founders of twentieth-century geometrical topology, were both highly guided by spatial intuition. But the "intuitionism" that resulted from Brouwer's work rejected the very spatial intuition that guided topology, in part because of non-Euclidean geometry, and rejected the infinite as well. Ketner (1982:330) is surely correct that "Peirce's diagrammatic thought might deal more effectively with some issues addressed by mathematical intuitionism."

Further evidence for the relative independence of topics or combinatorial topology from point-set considerations are the following. First, the work on triangulation of surfaces related to homology theory could be carried out, and was carried out, largely independently of point-set topological considerations. Second, in the early twentieth century a large number of combinatorial techniques and results were put in algorithmic form, and later logical investigations showed that for three dimensions or less the whole field could be developed independently of considerations of point-set structure of the continuum (Stillwell 1980:viii). (For four- and higher-dimensional surfaces, undecidability is the result, for the word problems in group theory show that this algorithmic approach is impossible for attaining a complete classification, so that Murphey's point may be correct here.) Third, the revival of geometrical intuition in higher mathematics during the 1960s led to the publication of numerous textbooks on algebraic topology in the 1970s which, for simplicity of exposition and richness of visualizable heuristic content, develop algebraic topology independently of point-set considerations (Giblin 1977:v). Certainly this is even more the case with graph theory (the theory of networks, not to be confused with Peirce's "graphics" or "perspective" [*CP* 2.197], which is now called projective geometry).

Max Fisch (1986:385) has written, "What Peirce called topology, topical geometry or topics, was something very different from what topology became. Is there any reason for trying with Peirce the road not taken, as non-standard analysis has returned from the doctrine of limits to that of infinitesimals?" I believe, at the level of undergraduate education, that return has to some extent been made. Many texts of the last decade have returned to the concrete, intuitive geometrical approach to algebraic topology. After six decades of formalizing and presenting in terms of abstract algebra the intuitive spatial insights of Poincaré, there has been a return by pedagogues to Peirce's insights that "topology, or connective geometry, is the best field for the growth of imagination, demanding little logic and not very much demonstration" (Eisele 1979:179).

However, I do not think that on the research level the parallel with nonstandard analysis is quite correct. Much that happened in algebraic topology after Peirce was not simply the replacement of analytic for synthetic methods, as Fisch suggests, but simply the tremendous development of theorems and results in the field. Point-set topology and the more abstract, general topology became separate fields from algebraic topology. The roots of point-set topology are different from those of algebraic topology. Point-set topology arose out of considerations of Fourier analysis specifically, or of analysis generally (Mannheim 1964).

Algebraic topology, in contrast, arose in Poincaré's work from the consideration of qualitative behavior of differential equations in physics and astronomy. It took mathematicians four decades to recapture in precise and rigorous form some of Poincaré's intuitions, and only afterward, from the 1950s on, was there a widespread return to the physical and geometrical considerations which had originally motivated the birth of algebraic topology. In this respect, at least, there has been a return to the intuitive and visual content of early topics. Perhaps it is in differential topology, which has developed greatly since the late 1950s, and in modern differential geometry and global analysis, with their use of the methods of both algebraic and differential topology, that the strongly intuitive study of spatial structures and the concern with spaces related to physics and cosmology, which was of interest to Peirce, has returned. Fields which have made a splash in the popular press, such as catastrophe theory in the 1970s and chaos theory in the 1980s, make extensive use of topological methods to deal with scientific problems. These theories of bifurcation, catastrophe, and chaos focus on the generation of singularities from smooth, continuous manifolds. Here Peirce's metaphysical claims about singular points representing individuality as opposed to the generality and possibility of the continuum are very germane (*NEM* 2:xiii; *MS* 159). The singularities of smooth mappings are precisely what are identified with the observable physical structures in catastrophe theory, for instance.

I might throw out one speculation, however, about the connection of Peirce's role as precursor of nonstandard analysis and his work on topics, or the topology of surfaces. Peirce contrasted his richer notion of the continuum with Cantor's "Continuity of the Theory of Functions which I

[Peirce] call Pseudo-Continuity" (*NEM* 4:325). Peirce several times referred to this full continuity as "true topical continuity" (*NEM* 2:483) and said that "pseudo-continuity does not answer the purposes of topics" (*NEM* 2:532). If Peirce meant that the theorems of combinatorial topology cannot be proven using merely Cantorian continuity, he would seem to be just wrong. Indeed the major theorems for low-dimensional topology (less than four dimensions) can be proven without using any set-theoretic notion of continuity at all (Stillwell 1980:vii). However, if one puts together Peirce's defense of real differentials with his rejection of Weierstrass's continuous but nondifferentiable (nonsmooth, saw-tooth) curve (*NEM* 2:579), perhaps what Peirce wanted was not just continuity but smoothness, the province of differential topology, and here his extended continuum, as a means of defining differentials as real objects, would be necessary to do calculus the way Peirce wished to do it. This addition of smoothness to continuity would require modifications in one's metaphysical extrapolations from topology.

To return to Peirce's philosophical topology, it both has a prehistory and is itself the forerunner of a number of contemporary developments which are not as systematic and self-conscious in their employment of topology to make philosophical points as Peirce was. Certainly the prehistory of philosophical topology begins with Anaximander's infinite, with Zeno's paradoxes, and with the discovery of irrational magnitudes by the Pythagoreans, perhaps by Hippasus of Metapontum, and by the qualitative continuum of Anaxagoras. Zeno rejected continuous motion (and motion as such) as contradictory. The Pythagoreans allegedly did not welcome—indeed, were horrified by—the discovery of the irrational magnitude (as testified to by Hippasus's alleged fate). It was really with Aristotle that the analysis of continuity in a positive and systematic manner begins. It has its origins in books 3 and 4 of Aristotle's *Physics*. The discussion in late antiquity and the Middle Ages developed from Aristotle.

Leibniz, with both his emphasis on the principle of continuity and his proposal of the field of mathematics called *analysis situs*, the precursor of geometric topology, marks the beginning both of topology and of its philosophical use.

In a less explicit way, Kant was a major precursor of philosophical topology. Metrical geometry is not introduced by Kant in the first *Critique* until the category of quantity is introduced in the axioms of intuition in the analytic. The "Transcendental Aesthetic" discusses geometry without a metric or distance, and it is misleading to discuss the "geometry" of the aesthetic as if it is the metrical analytic geometry of physics, something which can be developed only much later in the *Critique*. Despite the well-known story, told by the logical positivists and others, of the overthrow of Kant's "Aesthetic" by non-Euclidean geometry, most of the discussion in the "Aesthetic" was of what we should today call topological characteristics of space and time. For every sentence in Kant about parallels, three-dimensionality, or right angles there are hundreds about the continuity and connectedness of space. This claim for Kant as topologist manqué, or *malgré lui*, may be controversial, and a topic for another paper, but there is Kant's explicit concern with infinity and infinite divisibility and with Zeno and others in the "Antinomies," particularly the "Second Antinomy."

Although Peirce does not explicitly detail the influence of Kant's "Aesthetic" or "Antinomies" on his excursions into philosophical topology, we know of his intensive and exhaustive study of the first *Critique*. We also note his reference to Kant (along with Aristotle) in his terminology concerning continuity (Potter and Shields 1977:25). Peirce's use of "Aristotelicity" and "Kanticity" (*CP* 6.166) to designate features of continuity—"closure" (with respect to limit points) and "denseness" respectively in modern terminology—shows his debt to Kant in his work on continuity. Peirce made a second intellectual tribute to Kant on continuity, using Kant's definition of continuity as "that, all of whose parts have parts of the same kind" (*CP* 6.168).

Despite this prehistory and these precursors, it is only with Peirce that philosophical topology comes into its own. Here mathematical discoveries within topology, such as Listing's census theorem, Riemann surfaces, and the nineteenth-century work on graph theory and the four-color theorem are mathematically understood and extended, as well as applied philosophically.

The major areas of these philosophical applications are well known to students of Peirce's general philosophy, although usually not grouped under the rubric of philosophical topology. They include (1) Peirce's metaphysics of synechism, with all its multiform manifestations within his writings; (2) the mathematical aspects of Peirce's cosmology, especially with regard to the structure, curvature, and twist of space; and (3) the existential graphs in their various forms and applications. These areas of Peirce's metaphysics, cosmology, logic, epistemology, and psychology have been discussed by numerous writers, who have not, for the most part, seen them as benefiting from the application of Peirce's topological studies.

There are several areas of philosophy in which Peirce the philosophical topologist was precursor. In each of these areas he was more systematic and self-conscious than the contemporary figures who have unsystematically applied topological ideas to philosophy. In the light of twentieth-century advances in topology, it is unfortunate that Peirce's magnificent beginnings were not systematically followed up by others.

One area of contemporary philosophy wherein philosophical topology might be claimed to be pursued is in the philosophy of space-time. Peirce was very much an inspired precursor here. Randall Dipert in his article "Peirce on Mach and Absolute Space" notes (1973:84) the similarity of Earman's defense of the intelligibility of absolute space to that of Peirce. It is not surprising that Peirce, as the major capable defender of absolute space in his day, as well as one to use topology early for purposes of philosophical cosmology, should be a precursor of the philosophical work on space-time theory of the last two decades. His work shows a shift from the relational account of space and time to an account of space-time as general-relativistic but given structure, the "substantualist" view of space-time. The relationist conception of space of Mach, the early Einstein, and the logical positivists was dominant during the first half of our century. But in the last decades writers such as Sklar, Earman, Michael Friedmann, and a number of others severely criticized the relational view of space and space-time. Scientific realism inclined a number of philosophers to the substantial view of space-time. In this, Peirce, swimming against the tide at the beginning of the century, is an eminent precursor.

In recent decades, philosophers of space-time have appealed to topological considerations as a canon for evaluating traditional arguments and positions in the philosophy of space and time. Topology has been used to clarify the presentation of philosophical theses. Various topologies, some basic and some pathological, have been used as counterexamples to philosophical claims. The writers who do this can be seen as philosophical topologists, although, for the most part, they do not make explicit their manner and method of the use of topology in argument in the same way in which traditional analytical philosophers have been explicit and reflective about their use of logic. In this, Peirce is not merely a precursor but a forerunner in advance of much contemporary work.

Another area in which Peirce was in advance, not only of his own time but of ours, is what might be called the topology of logic or, more informally, the geometry of logic.

The history of mathematics, and of the philosophy stimulated by mathematical advances, is punctuated by the forging of major connections between previously disparate fields of mathematics. The systematic presentation of geometry in logico-deductive form by Euclid's predecessors influenced Aristotle. Euclid influenced philosophers from ancient times through Spinoza until the rise of non-Euclidean geometries and the nineteenth-century rigorization of geometry. Descartes's unification of algebra and geometry in analytic geometry influenced both Descartes's optimism concerning the future of scientific knowledge and his epistemological turn in philosophy. The nineteenth-century non-Euclidean geometries, axiomatizations of geometry, and rigorization of analysis led to a separation of pictorial intuition from logical syntax and a discrediting of graphical intuition (certainly not a final one), which strongly influenced the logical positivists and their model of science, and which sharply separated uninterpreted formal systems from interpretation. Frege, Russell, and Whitehead's founding of arithmetic on logic influenced much of twentieth-century philosophy, both in the major shift from Descartes's epistemological turn to the post-Fregean linguistic turn and in supplying a model for philosophical analysis in general.

From Euclid to Hilbert the logical formulation of geometry was developed, which, in the Euclid formulation, served as a model for scholastics and rationalists and, in the Hilbert formulation, served as a model for logical positivists. Descartes unified algebra and geometry and centered philosophy on epistemology. Frege logicized arithmetic and shifted the focus from epistemology to language. Gödel turned the tables and arithmetized logic, reversing the trend from Weierstrass's arithmetization of analysis to Frege's logicization of arithmetic. The full philosophical depths of Gödel's move are still being plumbed.

What is missing from this story is the geometrization of logic. After the rather spotty and scattered results of Aristotle's square of opposition and the diagrams of Euler and of Venn (and the self-contained lattice theory of propositional logics), the major systematic contribution is that of Peirce's existential graphs. Peirce's beta graphs present a complete and consistent system of predicate logic with identity, with an unsurpassed economy of rules. As mentioned, not only are the graphs evidently topological, but Peirce related them to topological considerations. In at least one long passage Peirce explicitly applies the terminology and results of Listing's census theorem in topology to his graphs (*NEM* 4:323).

Peirce's existential graphs represent a major system of the topology of logic. Unfortunately, they have not yet become part of the mainstream of contemporary logic. Hilary Putnam (1982:292) correctly refers to them as "what is or has become esoteric in Peirce's logical work." However, the potential field of the topology of logic has not become a systematized field since Peirce's time. There is a scattering of topological methods in logic today. Examples are the compactness theorem, topological completeness proofs, the topological interpretation of intuitionistic logic, the theory of *topoi* based on category theory, and fixed point theorems in recursive function theory. These are sometimes isolated techniques or tricks used for certain purposes. Perhaps there is a general field of the topology of logic, of which Peirce's existential graphs are our as yet most comprehensive portion, but which are a precursor of developments yet to come.

References

Biggs, N. L.; Lloyd, E. K.; and Wilson, R. J. 1976. *Graph Theory, 1736–1936*. Oxford: Clarendon Press.

Croom, Fred H. 1978. *Basic Concepts of Algebraic Topology*. New York: Springer-Verlag.

Dauben, Joseph W. 1977. C. S. Peirce's Philosophy of Infinite Sets. *Mathematics Magazine* 50:123–35.

———. 1982. Peirce's Place in Mathematics. *Historia Mathematica* 9:311–25.

Dipert, Randall R. 1973. Peirce on Mach and Absolute Space. *Transactions of the Charles S. Peirce Society* 9:79–94.

Eisele, Carolyn. 1979. *Studies in the Scientific and Mathematical Philosophy of Charles S. Peirce*. Ed. R. M. Martin. The Hague: Mouton.

Fisch, Max H. 1986. *Peirce, Semeiotic, and Pragmatism.* Bloomington: Indiana University Press.

Giblin, P. J. 1977. *Graphs, Surfaces, and Homology.* London: Chapman and Hall.

Hartshorne, Charles. 1929. Continuity, the Form of Forms, in Charles Peirce. *Monist* 39:521–34.

Ketner, Kenneth. 1982. Carolyn Eisele's Place in Peirce Studies. *Historia Mathematica* 9:326–32.

Manheim, Jerome H. 1964. *The Genesis of Point-Set Topology.* Oxford: Pergamon Press.

Murphey, Murray G. 1961. *The Development of Peirce's Philosophy.* Cambridge: Harvard University Press.

Poincaré, Henri. n.d. *Science and Method.* New York: Dover Publications.

Potter, Vincent G., and Shields, Paul B. 1977. Peirce's Definitions of Continuity. *Transactions of the Charles S. Peirce Society* 13:20–34.

Putnam, Hilary. 1982. Peirce the Logician. *Historia Mathematica* 9:290–301.

Russell, Bertrand. 1960. *Our Knowledge of the External World.* New York: New American Library.

Stillwell, John. 1980. *Classical Topology and Combinatorial Group Theory.* New York: Springer-Verlag.

4 Peirce and Propensities

James H. Fetzer, The University of Minnesota/Duluth

In his "Notes on the Doctrine of Chances" (*CP* 2.661) and related reflections, Charles S. Peirce advanced a conception of probabilities according to which a die and tossing device, for example, possesses "would-be's" for its various possible outcomes, where these would-be's are intensional, dispositional, directly related to the long run, and indirectly related to singular events. Among the most influential contemporary accounts—the frequency, the personal, and the propensity—the most promising, the propensity theory, provides an account according to which probabilities are intensional, dispositional, directly related to singular events, and indirectly related to the long run. Thus, in his general conception, if not its specific details, Peirce appears to have anticipated what seems to be the most adequate solution to one of the most difficult problems in the theory of science.

During his lifetime, Peirce shifted from the conception of probabilities as long-run frequencies to the conception of probabilities as long-run dispositions, that is, as tendencies to produce long-run frequencies. Section 1 below outlines Peirce's conception of probabilities as long-run dispositions. Section 2 sketches the superiority of this view over its long-run frequency and personality probability alternatives. Section 3 explains the necessity to displace the conception of probabilities as long-run dispositions by one of probabilities as single-case dispositions. And section 4 suggests how this successor to Peirce's account can contribute to the solution of contemporary scientific problems. In passing, we shall consider how Peirce's earlier views are related to his later views. Although a thinker's later views do not always improve upon his earlier opinions, in this case Peirce's later views turn out to be more adequate.

I

If the theory of probability had its origin in "games of chance" involving tosses of coins, throws of dice, draws of cards, and the like, then it is entirely appropriate that Peirce illustrated his conception by using this example:

> I am, then, to define the meanings of the statement that the *probability*, that if a die be thrown from a dice box it will turn up a number divisible by three, is one-third. The statement means that the die has a certain "would-be"; and to say that a die has a "would-be" is to say that it has a property, quite analogous to any *habit* that a man might have. Only the "would-be" of the die is presumably as much simpler and more definite than the man's habit as the die's homogeneous composition and cubical shape is simpler than the nature of the man's nervous system and soul. (*CP* 2.664)

In this passage, Peirce characterizes probability as a dispositional property of a specific type of physical arrangement (such as a die and tossing device), explicitly invoking the subjunctive mood concerning "what would happen if." Since this disposition is probabilistic, its effects are complex.

> Now in order that the full effect of the die's "would-be" may find expression, it is necessary that the die should undergo an endless series of throws from the dice box, the result of no throw having the slightest influence upon the result of any other throw. . . . It will be no objection to our considering the consequences of the supposition that the die is thrown an endless succession of times . . . that such an endless series of events is impossible, for the reason that the impossibility is merely a physical . . . impossibility. (*CP* 2.665–66)

Having already established that probabilities are supposed to be dispositional properties whose nature requires the subjunctive mood for its characterization, Peirce relates this intensional conception to its long-run displays.

> The fact [is] that the probability of the die turning up a three or a six is not *sure* to produce any determination [of] the run of numbers thrown in any *finite* series of throws. It is only when the series is endless that we can be *sure* that it will have a particular character. Even when there is an endless series of throws, there is no syllogistic certainty, no "mathematical" certainty . . . that the die will not turn up a six obstinately at every throw. . . . It sanely *would not*, however, unless a miracle were performed; and moreover if such a miracle *were* worked, I should say (since it is my use of the term "probability" that we have supposed to be in question) that during this . . . series the die took on an abnormal, a miraculous, habit. (*CP* 2.667)

The normal, nonmiraculous habit that a probability represents, Peirce explains, displays the outcome of interest with a limiting frequency which equals its generating probability when subjected to an endless sequence of trials. The relation involved here, moreover, is not logical but causal, insofar as these probabilities are displayed by corresponding limiting frequencies but are not defined by means of them. Indeed, as Peirce emphasizes, a deviation from this correspondence could occur, but this is merely a logical and not a physical possibility: "I say it *might*, in the sense that it would not violate the principle of contradiction if it did" (ibid.). Peirce thus endorses a conception of probabilities according to which these properties are dispositional, intensional, and directly related to their long-run displays.

II

The frequency conception of probability itself draws a strong connection between probabilities and their long-run displays, but it is precisely the definitional relationship that Peirce is prudent to deny. Indeed, while the theory of probability as an abstract domain of pure mathematics can be successfully developed by means of the frequency definition, it has a serious defect in application to physical sequences. For any case in which an endless series (of tosses, of throws, of draws, or of whatever) does not occur during the world's history (including every finite sequence), there are no physical probabilities that satisfy the frequency conception. Although it may initially appear as though the advantage lies on the other side, with respect to its applicability to physical domains, the frequency definition offers no competition. (For an introduction to the frequency view, see, for example, Salmon 1966.)

Indeed, the magnitude of this difficulty for the frequency conception can be appreciated in relation to the problem of assigning appropriate values to the occurrence of singular events, such as the next toss of this coin, the next throw with that die, and the next draw from this deck. Any property whose presence or absence makes a difference to the limiting frequency with which corresponding outcomes occur presumably ought to be taken into account in assigning specific cases to proper reference sequences, since otherwise their assigned values would not be appropriate. Since every singular event has to have properties that distinguish it from every other event, it becomes increasingly difficult to satisfy the implied desiderata of the existence of infinite sequences of repetitions of such events under increasingly specific descriptions. (Cf., e.g., Reichenbach 1949, Salmon 1966, and Fetzer 1981:chap. 4.)

The personal conception, by contrast, falls prey to a different but related problem. According to this interpretation, probabilities are subjective properties, not of chance arrangements (like dice and tossing devices) but of persons and their beliefs. Thus, these probabilities are construed as reflecting the degrees of belief specific agents have in specific propositions, where a belief might acquire a probability of 1 as a function of its indubitability or a probability of 0 as a function of its incredibility. Most propositions would fall somewhere in between as neither personally necessary nor personally impossible. The problem with this approach, however, is that "degrees of belief" could exist as properties of a specific agent's beliefs, even in the absence of probabilistic properties of any chance arrangements in the world. (For an introduction to the personal view, see, for example, Skyrms 1986.)

This situation is exemplified by "pull tab" lotteries, in which the purchase of a ticket permits the participant to remove the covering from a description of a possible prize. If pulling the tab discloses the description of a prize, the participant wins that prize, but otherwise not. Such "lotteries" are "games of chance" in the sense that the purchase of a ticket creates the possibility of a prize, and only a fixed number of prizes can be won. These "lotteries," however, are not "games of chance" in the sense that they assume the existence of probabilistic properties or of indeterministic causation in the physical world. They are altogether compatible with causal determinism and are only "probabilistic" in relation to their participants' personal convictions. (For various views, see Savage 1954, de Finetti 1964, and Jeffrey 1965.)

III

If Peirce's conception confronts difficulties of its own, it does not suffer in comparison with these alternatives for the purpose of understanding the occurrence of events in the physical world. His account improves upon the frequency conception because its applicability does not presuppose the existence of infinite sequences of individual trials. It improves upon the personal conception because it applies to events in the world rather than to degrees of belief that we might have about them. In both of these respects, Peirce's views appear to be theoretically superior to those reflected by the alternative conceptions. Nevertheless, because it is directly related to the long run, it may not be entirely satisfactory in the solution that it affords for the single case.

That Peirce was aware of this difficulty is not in doubt. The problem is vividly expressed in his reflections on the nature of probabilistic inference.

> An individual inference must be either true or false, and can show no effect "l" of probability; and, therefore, in reference to a single case considered in itself, probability can have no meaning. Yet if a man had to choose between drawing a card from a pack containing twenty-five red cards and a black one, or from a pack containing twenty-five black cards and a red one, and if the drawing of a red card were destined to transport

him to eternal felicity, and that of a black one to everlasting woe, it would be folly to deny that he ought to prefer the pack containing the larger proportion of red cards, although, from the nature of the risk, it could not be repeated. (*CP* 2.652)

Thus, if probability can have no meaning relative to any single case considered in itself, it becomes very difficult to understand how probability is to serve its role as a "guide in life" with respect to the explanation and the prediction of immediate events. As Niiniluoto (1989) notes, Peirce sought to resolve this problem within the context of "A Theory of Probable Inference" by introducing an argument form called *Simple Probable Deduction*.

(1) The proportion p of M's are P's.
S is an M.
It follows, with probability p, that S is a P. (*CP* 2.696)

Here Peirce relates relative frequencies of events with truth frequencies of conclusions, which he exemplifies by means of the following illustration:

(2) About two percent of persons wounded in the liver recover.
This man has been wounded in the liver.
Therefore, there are two chances out of a hundred that he will recover. (*CP* 2.694)

(For related discussion, cf. Niiniluoto 1981, 1982, and now especially 1989.)

The rationale Peirce supplies for deriving such an inference, moreover, has two aspects. On the one hand, he maintains that "to say . . . that a proposition has the probability p means that to infer it to be true would be to follow an argument such as would carry truth with it in the ratio of frequency p" (*CP* 2.697). On the other hand, he maintains that "it is requisite, not merely that S should *be* an M, but also that it should be an instance drawn *at random* from among the M's," where "randomness" is intended to be understood as a function of our personal belief that the case under consideration should not be treated as a special, atypical case (*CP* 2.696).

It is not difficult to see that Peirce's account here is not entirely satisfactory. Even if truth frequencies of conclusions correspond to relative frequencies of events over the long run, that does not establish their relevance for the occurrence of singular events. Moreover, our belief that *S* is a typical instance of *M* surely does not make it one, no matter the degree of conviction with which that belief happens to be held. Insofar as Peirce intended *Simple Probable Deduction* as a pattern of inference for both explanation and prediction, it cannot succeed unless it is utilized on the foundation of an objective conception of physical probability that can resolve the problem of the single case without subjective entanglements.

An adequate resolution of this difficulty requires a conceptual revision according to which probabilities are now to be understood as characterizing the strength of the causal tendency for a single trial with a physical arrangement to bring about one or another of its various possible outcomes. The propensity for obtaining the outcome of eternal felicity by drawing a card from a deck would equal 25/26 if that were the strength of the tendency for that outcome to be brought about on a single trial of that arrangement. These single-case propensities, in turn, tend to generate relative frequencies as expectable outcomes of repetitious trials of similar arrangements, an approach that affords a suitable foundation for improving on Peirce's account. (See, for example, Popper 1957, 1959 and Fetzer 1971, 1981.)

IV

Probabilities as "propensities" are understood as intensional, dispositional, directly related to singular events, and indirectly related to the long run. In order to suggest the contemporary significance of this successor to Peirce's conception, consider the illumination that it may shed upon problems from some of the most important areas of current research in physics, in biology, and in psychology. The first of these concerns the difficulties posed by the apparent paradoxes of quantum mechanics, which can be illustrated by the Schrödinger "cat paradox." The second concerns the definition of fitness as it occurs within evolutionary biology, especially in relation to the notions of survival and reproduction. And the third concerns the nature of the architecture of the mind in relation to the connectionist conception of the brain.

Schrödinger's cat paradox envisions an arrangement consisting of a live cat within an opaque chamber connected to an electrical device activated by a Geiger counter linked to a radioactive substance. If decay then occurs and is registered by the Geiger counter (with probability 1/2, let us say), then an impulse is activated that electrocutes the cat. From the perspective known as the Copenhagen interpretation, until the chamber is actually opened and it is discovered to be dead or alive, the cat is presumed to be neither dead nor alive but rather somewhere in between, which is represented by a superposition of psi-functions for both of these events prior to this observation being made. The process of taking a look and discovering, say, that the cat is dead is thus supposed to effectuate "the collapse of the wave packet," whereby in effect the result of being dead or of being alive is brought about.

From the perspective of the propensity conception, the Schrödinger cat paradox appears to be generated by the personal point of view. It is only possible for a quantum outcome to be assigned a personal probability value between 0 and 1 when that specific outcome, such as a cat's being dead, remains unknown. As soon as an agent becomes aware that the cat is dead, the agent's degree of belief that that is the case has to change to 1! This result, moreover, is not peculiar to quantum contexts but is commonplace with respect to games of chance in general. The propensity interpretation takes the mystery out of situations like these, therefore, by eliminating the need for a "collapse of the wave packet" as reflecting a confusion between changes in degrees of belief, on the one hand, and in strengths of causal propensities, on the other. (See, for example, especially Popper 1967, 1982, and Fetzer 1983.)

The notion of fitness as it occurs in evolutionary biology is an especially important explanatory conception. The principal benefit of the analysis of fitness as a propensity is that it identifies fitness with a tendency to survive and reproduce rather than with its actual attainment. It thus becomes possible for comparative theoretical determinations of fitness to differ from actual survival and reproduction, where "the more fit" might have fewer offspring than "the less fit" as a permissible probabilistic outcome "by chance." As a propensity, of course, fitness is relative to whatever factors affect the prospects for an individual's survival and reproduction, which will normally include not only its physical attributes but also each element of its environment that contributes to its chance of surviving and reproducing offspring.

Moreover, various measures of fitness can be constructed on the foundation that the propensity conception provides, some of which introduce reliance upon averages or "means" within heterogeneous populations or across divergent environments. The problems that arise with frequency-dependent causation, for example, where the adaptability of a trait can vary with the number of members of a species that possess it, should be properly understood as manifestations of the context-dependence of propensity values generally, rather than as problematic special cases. And the discovery that "mean propensities" may turn out to be the appropriate measures of evolutionary prediction in specific situations enhances the prospects for biology to catch up with physics in its reliance upon sophisticated average values. (Cf. Mills and Beatty 1979, Fetzer 1986, but also Beatty and Finsen 1989.)

Connectionism, perhaps the most exciting development within the domain of cognitive science, exhibits a similar potential for benefiting from the propensity approach. Connectionism models the brain as a neural network of numerous nodes that are capable of activation. These nodes can be connected to other nodes where, depending on their levels of activation, they may bring about increases or decreases in the activation of those other nodes. These patterns of activation, in turn, can function as signs for the larger systems of which they are otherwise meaningless elements by coming to stand for other things as "signs" for those systems. These might include features of their internal states or of their external environments as a consequence of the ways these things can function as signs for those systems.

Since the connections that are established between these various subsymbolic neural networks depend not only upon their levels of activation but also upon their predispositions to connect with other elements, it may prove exceptionally promising to entertain these nodes as endowed with propensities. From this point of view, the brain turns out to be a complex arrangement of neural propensities to establish connections under various conditions of system stimulation, where patterns of activation themselves can function as causal antecedents of human behavior. Indeed, a theory of the mind which appears ideally suited to complement such a conception of the brain—according to which minds are semiotic, or "sign using," systems—can be developed on the basis of Peirce's theory of signs. The problems that might benefit from conceptions stimulated by his work seem almost endless. (See Rumelhart et al. 1986, Smolensky 1988, and Fetzer 1988b, 1990, 1991.)

The notion that Peirce advanced thus anticipated what appears to be the most adequate solution to a wide variety of difficult problems in the theory of science. The fertility of the propensity approach, moreover, promises to penetrate other perplexing conundra, including the deeper differences between determinism and indeterminism. The emergence of chaos theory as a branch of mathematical inquiry, for example, invites clarification on the basis of the consideration that the strength of causal tendencies depends on, and varies with, the presence or absence of every property whose presence or absence makes a difference to their outcomes, no matter how minute. It even appears as though chaotic phenomena and normal games of chance are deterministic rather than indeterministic, as true propensities require—an enlightening but nevertheless ironic fate for Peirce's account to endure. (See Hacking 1980, Crutchfield et al. 1986, Gleick 1988, and Fetzer 1988a.)[1]

Notes

I am grateful to Paul Humphreys and Cheryl Misak for critical comments.

1. Peirce's paper "A Theory of Probable Inference" was published in 1883, while "Notes on the Doctrine of Chances" appeared in 1910. For further analysis of issues related to the fashion in which Peirce's later conception of probability and its successor satisfy the general features of his earlier theory of probable inference, cf. Fetzer 1981 and Niiniluoto 1981, 1982, and especially 1989.

References

Beatty, J., and Finsen, S. 1989. Rethinking the Propensity Interpretation: A Peek inside Pandora's Box. In *What the Philosophy of Biology Is: Essays for David Hull*, ed. M. Ruse, 17–31. Dordrecht: Kluwer.

Crutchfield, J. P., et al. 1986. Chaos. *Scientific American* 255:46–57.

de Finetti, B. 1964. Foresight: Its Logical Laws, Its Subjective Sources. In *Studies in Subjective Probability*, ed. H. Kyburg and H. Smokler, 93–158. New York: John Wiley and Sons.

Fetzer, J. H. 1971. Dispositional Probabilities. In *PSA 1970*, ed. R. Buck and R. Cohen, 473–82. Dordrecht: D. Reidel.

———. 1981. *Scientific Knowledge*. Dordrecht: D. Reidel.

———. 1983. Probability and Objectivity in Deterministic and Indeterministic Situations. *Synthese* 57:367–86.

———. 1986. Methodological Individualism: Singular Causal Systems and Their Population Manifestations. *Synthese* 68:99–128.

———. 1988a. Probabilistic Metaphysics. In *Probability and Causality*, ed. J. Fetzer, 109–32. Dordrecht: D. Reidel.

———. 1988b. Signs and Minds: An Introduction to the Theory of Semiotic Systems. In *Aspects of Artificial Intelligence*, ed. J. Fetzer, 133–61. Dordrecht: Kluwer.

———. 1990. *Artificial Intelligence: Its Scope and Limits*. Dordrecht: Kluwer.

———. 1991. *Philosophy and Cognitive Science*. New York: Paragon.

Gleick, J. 1988. *Chaos: Making a New Science*. New York: Penguin Books.

Hacking, I. 1980. Grounding Probabilities from Below. In *PSA 1980*, ed. Peter D. Asquith and R. N. Giere, 110–16. East Lansing, Mich.: Philosophy of Science Association.

Jeffrey, R. 1965. *The Logic of Decision*. New York: McGraw-Hill.

Mills, S., and Beatty, J. 1979. The Propensity Interpretation of Fitness. *Philosophy of Science* 46:263–86.

Niiniluoto, I. 1981. Statistical Explanation Reconsidered. *Synthese* 48:437–72.

———. 1982. Statistical Explanation. In *Contemporary Philosophy: A New Survey*, ed. G. Fløistad, 2:157–87. The Hague: Martinus Nijhoff.

———. 1989. Peirce's Theory of Statistical Explanation. Reprinted in the present volume.

Popper, Karl R. 1957. The Propensity Interpretation of the Calculus of Probability, and the Quantum Theory. In *Observation and Interpretation in the Philosophy of Physics*, ed. S. Korner, 65–70. New York: Dover Publications.

———. 1959. The Propensity Interpretation of Probability. *British Journal for the Philosophy of Science* 10:25–42.

———. 1967. Quantum Theory without "The Observer." In *Quantum Theory and Reality*, ed. M. Bunge, 7–44. New York: Springer-Verlag.

———. 1982. *Quantum Theory and the Schism in Physics*. Totowa, N.J.: Rowman and Littlefield.

Reichenbach, H. 1949. *The Theory of Probability*. Berkeley: University of California Press.

Rumelhart, D. E.; McClelland, J. L.; and the PDP Research Group. 1986. *Parallel Distributed Processing*. Cambridge: MIT Press, Bradford Books.

Salmon, W. C. 1966. *The Foundations of Scientific Inference*. Pittsburgh: University of Pittsburgh Press.

Savage, L. 1954. *The Foundations of Statistics*. New York: Dover Publications.

Skyrms, B. 1986. *Choice and Chance: An Introduction to Inductive Logic*. 3d ed. Belmont, Calif.: Dickenson Publishing.

Smolensky, P. 1988. On the Proper Treatment of Connectionism. *Behavioral and Brain Sciences* 11:1–23.

5 Induction and the Evolution of Conceptual Spaces

Peter Gärdenfors, Lund University, Sweden

Peirce's Constraint on Inductive Inference

Humans do not perform inductions in an arbitrary manner. At the beginning of his discussion of inductive inferences, Peirce notes that there are certain forms of constraints that delimit the vast class of possible inferences. As he puts it:

> Nature is a far vaster and less clearly arranged repertory of facts than a census report; and if men had not come to it with special aptitudes for guessing right, it may well be doubted whether in the ten or twenty thousand years that they may have existed their greatest mind would have attained the amount of knowledge which is actually possessed by the lowest idiot. But, in point of fact, not man merely, but all animals derive by inheritance (presumably by natural selection) two classes of ideas which adapt them to their environment. In the first place, they all have from birth some notions, however crude and concrete, of force, matter, space, and time; and, in the next place, they have some notion of what sort of objects their fellow-beings are, and how they will act on given occasions. (*CP* 2.753)

Here, Peirce hints at an evolutionary explanation of why "the human intellect is peculiarly adapted to the comprehension of the laws and facts of nature" (*CP* 2.750).

My aim in this chapter is to elaborate some of his ideas about innate constraints for a theory of induction. I focus on the problem of *projectability*, that is, the problem of which predicates may be used in inductive inferences. My approach to this problem is a combination of Peirce's evolutionary perspective and some ideas from recent cognitive science. In particular, I introduce the notion of conceptual spaces as a framework for selecting projectable properties and outline how this notion can throw light on the so-called prototype theory of concept formation.

Logical Positivism and the Problems of Induction

The most ambitious project for analyzing inductive inferences during this century has been that of the logical positivists, where Carnap was the prominent figure. As a background to an analysis of Peirce's constraints on inductive inferences, I start by commenting briefly on this effort.

Inductive inferences were important for the logical positivists, since such inferences were necessary for their verificationist aims. The basic objects of study for them were sentences in some more or less regimented language. Ideally, the language was a version of first-order logic, where the atomic predicates represented observational properties. These observational predicates were taken as primitive, unanalyzable notions. The main tool used in studying the linguistic expressions was logical analysis. In its pure form, logical positivism allowed only this tool. A consequence of this methodology was that all observational predicates were treated in the same way, since there were no *logical* reasons to differentiate between them. For example, Carnap (1950:sec. 18B) requires that the primitive predicates of a language be logically independent of each other.

However, it became apparent that the methodology of the positivists led to serious problems. The most famous ones are Hempel's (1965) "paradox of confirmation" and Goodman's (1955) "riddle of induction." I will not repeat these well-known difficulties but will only state my diagnosis of what causes them (for a more detailed account, see Gärdenfors 1990). What I see as the root of the troublesome cases is that if we use logical relations alone to determine which inductions are valid, the fact that all predicates are treated equally induces symmetries which are not preserved by our understanding of the inductions: "Raven" is treated the same as "nonraven," "green" as "grue" (the hypothetical combination of green and blue), and so forth. What we need is a nonlogical way of distinguishing those predicates that may be used in inductive inferences from those that may not.

There are several suggestions for such a distinction in the literature. One idea is that some predicates denote "natural kinds" or "natural properties," while others do not, and only the former may be used in inductions. Natural kinds are normally interpreted realistically, following the Aristotelian tradition, and thus are assumed to represent something that exists in reality independently of human cognition. However, when it comes to inductive inferences, it is not sufficient that the properties exist out there somewhere; we need to be able to grasp the natural kinds by our minds. In other words, what is needed to understand induction, as performed by humans, is a *conceptualistic*, or *cognitive*, analysis of natural properties. One of the aims of the present chapter is to outline such an analysis.

Another relevant notion here is that of "similarity." Quine discusses this notion and its relation to that of a natural kind. He notes (1969:117) that similarity "is immediately definable in terms of kind; for things are similar when they are two of a kind." Furthermore, he says about similarity that we

> cannot easily imagine a more familiar or fundamental notion than this, or a notion more ubiquitous in its application. On this score it is like the notions of logic: like identity, negation, alternation, and the rest. And yet, strangely, there is something logically repugnant about it. For we are baffled when we try to relate the general notion of similarity significantly to logical terms. (ibid.)

To substantiate this claim he discusses, and rejects, several attempts to define similarity or natural kinds in logical or set-theoretic terms. Again, what is needed for the present project is not a realistic analysis of similarity (like Lewis's [1973] similarity between possible worlds) but a conceptualistic, or cognitive, analysis. I return to this kind of similarity in relation to my discussion of prototype theory.

Conceptual Spaces

As mentioned above, Peirce says that humans and animals "have from birth some notions . . . of force, matter, space, and time" (*CP* 2.753). I propose to call these notions *quality dimensions* (see Gärdenfors 1988, 1990, 1991). Apart from those mentioned by Peirce, one can mention other perceptual qualities such as color, temperature, and weight. When we talk and think about the qualities of objects, it is in relation to such dimensions. A *conceptual space* consists of a class of quality dimensions (see Quine's [1960] notion of "quality space").

The dimensions should be given a conceptualistic, or cognitive, interpretation. Some of the dimensions are closely related to what is produced by our sensory receptors, but there are also quality dimensions that are of a more abstract character. Furthermore, the dimensions are prelinguistic in the sense that we (and other animals) can think about the qualities of objects, for example when planning an action, without presuming a language in which these thoughts can be expressed.

The notion of a dimension should be taken seriously. Each of the quality dimensions is endowed with a certain topological or sometimes even a metrical structure. For example, we normally think of the time dimension as being isomorphic to the line of real numbers, and the weight dimension as isomorphic with the positive real numbers (there are no negative weights). Since the dimensions are cognitive entities, their topological or metrical structure should not be determined by scientific theories which attempt to give a "realistic" description of the world, but by *psychophysical* measurements which determine the structure of how our perceptions are represented. For instance, as regards the hue of colors, the psychological, or cognitive, representation is given by the well-known color circle (see fig. 1), which thus determines the basic topology of this quality dimension, while a more "scientific" approach would use the wave length of the light as the underlying dimension.

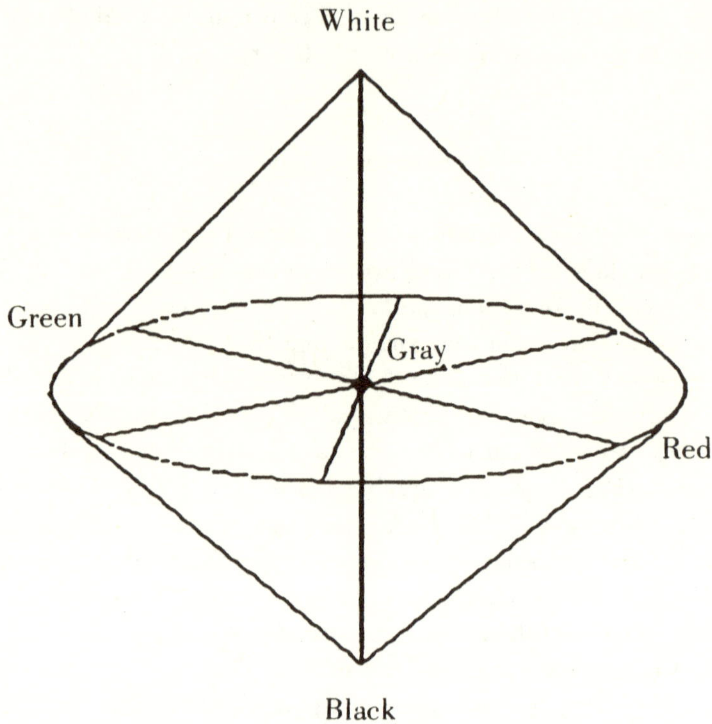

Figure 1. The three-dimensional color space

In general, representing information about the world in terms of quality dimensions involves using geometric or vectorial notions. In contrast, the program of the logical positivists (and much of current artificial intelligence work) is based on symbolic or logical modes of representation. (See Sloman's [1971] distinction between "analogical" and "Fregean" representation and Churchland's [1986] criticism of the "sentential paradigm.") The move to a new form of representation is, I believe, necessary; the problems that plagued the logical positivists indicate that we have to go beyond logic and language to find a solution to the problem of projectability in inductive reasoning. Quine (1969:123) notes that something like a conceptual space is needed to make learning possible.

> Without some such prior spacing of qualities, we could never acquire a habit; all stimuli would be equally alike and equally different. These spacings of qualities, on the part of men and other animals, can be explored and mapped in the laboratory by experiments in conditioning and extinction. Needed as they are for all learning, these distinctive spacings cannot themselves all be learned; some must be innate.

This concludes my general presentation of conceptual spaces. I believe it can be seen as a generalization of the state-space approach, advocated, among others, by P. M. Churchland (1986) and P. S. Churchland (1986), and of the vector function theories of Foss (1988). To some extent, conceptual spaces also function like the domains in Langacker's (1986) semantic theory. The theory of conceptual spaces is a theory for representing information and not primarily an empirical psychological or neurological theory. I believe the theory can be applied to a number of philosophical problems in epistemology and semantics. Here, my aim is to show its viability as a foundation for an analysis of projectable properties.

Natural Properties

With the notion of conceptual space, it is quite easy to give an analysis in cognitive terms of "natural properties" and "similarity" (for a more detailed presentation, see Gärdenfors 1990). The key idea is that a natural property can be identified with a convex region of a given conceptual space. By way of the notion of "convexity," this utilizes the topological properties of the quality dimensions. A convex region is characterized by the criterion that for every pair o_1 and o_2 of points in the region, all points between o_1 and o_2 are also in the region. This definition presumes that the notion of "between" is meaningful for the relevant dimensions. This is, however, a rather weak assumption, which demands very little of the underlying topological structure.

As an application of the criterion of a natural property, I conjecture that all color terms in natural languages express natural properties with respect to the psychophysical representation of the quality dimensions of colors. In other words, the conjecture predicts that if some object o_1 is described by the color term *C* in a given language and another object o_2 is also said to have the color *C*, then any object o_3 with a color which lies between the color of o_1 and the color of o_2 will also be described by the color term *C*. It is well known that different languages carve up the color circle in different ways, but all carvings seem to be done in terms of convex sets. Strong support for this conjecture can be gained from Berlin and Kay (1969), although they do not treat color terms in general but concentrate on basic color terms.

Using the present criterion for natural properties, it is now possible to formulate a more precise version of a constraint on induction, which I believe is consistent with the quotation from Peirce presented at the beginning.

(C) Only properties corresponding to a convex region of the underlying conceptual space may be used in inductive inferences.

It is proposed only that convexity is a necessary condition, but perhaps not a sufficient one, for a property to count as natural and thus projectable. In any case, the criterion is strong enough to avoid many Goodman-type problems. The motivation for why convexity is an appropriate condition will become clearer in the next section.

Prototype Theory and Similarity

Apart from giving intuitively plausible solutions to the old riddles of induction (see Gärdenfors 1990), the definition of natural properties in terms of convex regions of conceptual spaces derives independent support from the prototype theory of categorization developed by Rosch and her collaborators (Rosch 1975, 1978; Mervis and Rosch 1981; Lakoff 1987). This theory can also be used to provide a reasonable explication of the notion of similarity that is presupposed in inductive inferences. (Quine [1969:119–20] argues that "natural kind" is definable in terms of "similarity," and he actually proposes a precursor to prototype theory.)

The main idea of prototype theory is that within a category of objects, such as those instantiating a property, certain members are judged to be more representative of the category than other members. For example, robins are judged to be more representative of the category *bird* than are ravens, penguins, or emus; and desk chairs are more typical instances of the category *chair* than rocking chairs, deck chairs, or beanbag chairs. The most representative members of a category are called prototypical members. It is well known that some properties, like "red" and "bald" have no sharp boundaries, and for these it is perhaps not surprising that one finds prototypical effects. However, these effects have been found for most properties, including those with comparatively clear boundaries like "bird" and "chair."

With the traditional definition of a property, it is very difficult to explain such prototype effects. Either an object is a member of the class assigned to a property or it is not, and all members of the class have equal status as category members. Rosch's work has been aimed at showing asymmetries among category members and asymmetric structures within categories. Since the traditional definition of a property does not predict such asymmetries, something else must be going on.

In contrast, if natural properties are defined as convex regions of a conceptual space, prototype effects are indeed to be expected. In a convex region one can describe positions as being more or less central. For example, if color properties are identified with convex subsets of the color space, the central points of these regions would be the most prototypical examples of the color. Rosch has, in a series of experiments, been able to demonstrate the psychological reality of such "focal" colors.

For more complex categories, like "bird," it is perhaps more difficult to describe the underlying conceptual space. However, if something like Marr and Nishihara's (1978) analysis of shapes is adopted, we can begin to see what such a space would look like. Their scheme for describing biologic forms uses hierarchies of cylinder-like modeling primitives (see fig. 2).

Each cylinder is described by two coordinates (length and width). Cylinders are combined by determining the angle between the dominating cylinder and the added one (two polar coordinates) and the position of the added cylinder in relation to the dominating one (two coordinates).

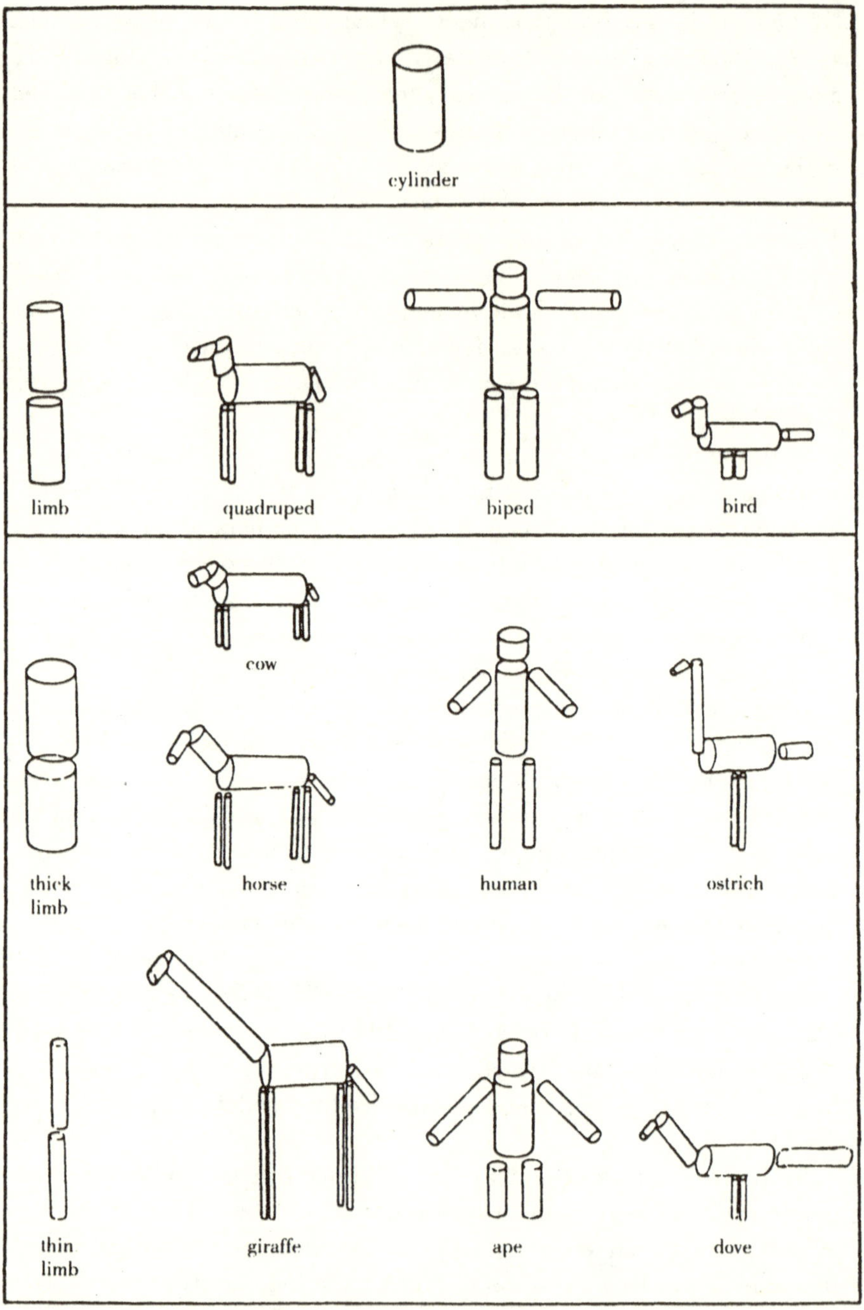

Figure 2. Hierarchical representation of animal shapes, using cylinders as modeling primitives (from Marr 1982)

The details of the representation are not important in the present context, but it is worth noting that on each level of the hierarchy an object is described by a comparatively small number of coordinates on the basis of lengths and angles. Thus the object can be identified as a hierarchically structured vector in a (higher-order) conceptual space.

It should be noted that, even if different members of a category are judged to be more or less prototypical, it does not follow that some of the existing members must represent "the prototype." (Rosch has been misunderstood on this point; see Lakoff 1987:chap. 2.) If a category is viewed as a convex region of a conceptual space, this is easily explained, since the central member of the region specifies a possible individual but need not be among the existing members of the category.

It is possible to argue in the converse direction also, showing that if prototype theory is adopted, then the representation of properties as convex regions is to be expected. Assume that some quality dimensions of a conceptual space are given, for example, the dimensions of color space, and that we want to partition it into a number of categories, for example, color categories. If we start from a set of prototypes $p_1, \ldots, p_n$ of the categories, for example, the focal colors, then these should be the central points in the categories they represent. One way of using this information is to assume that for every point p in the space, one can measure the distance from p to each of the p_i's. If we now stipulate that p belongs to the same category as the closest prototype p_i, it can be shown that this rule will generate a partitioning of the space that consists of convex areas (convexity is here defined in terms of the assumed distance measure). This is the so-called Voronoi tessellation (see fig. 3).

Thus, assuming that a metric is defined on the subspace that is subject to categorization, a set of prototypes will, by this method, generate a unique partitioning of the subspace into convex regions. Hence, there is an intimate link between prototype theory and the above analysis where properties are defined as convex regions in a conceptual space. Furthermore, such a metric is an obvious candidate for a measure of similarity between different objects. In this way, the Voronoi tessellation gives an explicit answer to how a similarity measure determines a set of natural kinds.

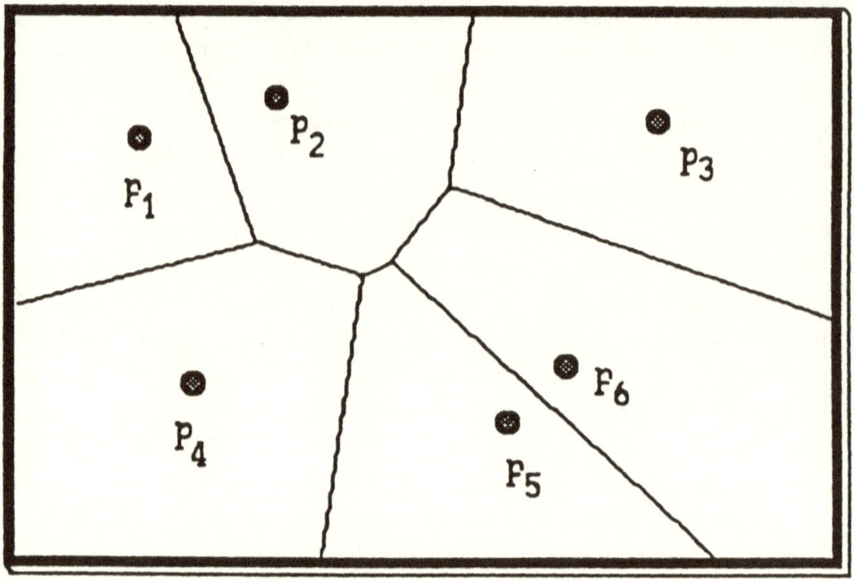

Figure 3. Voronoi tessellation of the plane into convex sets

It is interesting to note that Quine (1969:119–20) formulates a precursor to this kind of prototype theory.

> One may be tempted to picture a kind, suitable to a comparative similarity relation, as any set which is "qualitatively spherical" in this sense: it takes in exactly the things that differ less than so-and-so much from some central norm. If without serious loss of accuracy we can assume that there are one or more actual things (*paradigm cases*) that nicely exemplify the desired norm, and one or more actual things (*foils*) that deviate just barely too much to be counted into the desired kind at all, then our definition is easy: *the kind with paradigm a and foil b* is the set of all things to which *a* is more similar than *a* is to *b*. More generally, then, a set may be said to be a *kind* if and only if there are *a* and *b*, known or unknown, such that the set is the kind with paradigm *a* and foil *b*.

Quine notes that, as it stands, this definition of a kind is not satisfactory.

> Thus take red. Let us grant that a central shade of red can be picked as norm. The trouble is that the paradigm cases, objects in just that shade of red, can come in all sorts of shapes, weights, sizes, and smells. Mere degree of overall similarity to any such paradigm case will afford little evidence of degree of redness, since it will depend also on shape, weight, and the rest. If our assumed relation of comparative similarity were just comparative chromatic similarity, then our paradigm-and-foil definition of kind would indeed accommodate redkind. What the definition will not do is distill purely chromatic kinds from mixed similarity. (ibid.:120)

The problem for Quine is that he does not assume anything like a conceptual space to help in structuring the relations of similarity. However, if such a structure is given, we need not rely on actual objects as paradigm cases but can use focal points on a particular quality dimension, like the color dimension, as a basis for comparing chromatic similarity. The shape, weight, and the rest of the qualities of objects will simply not be relevant for such comparisons.

The Origin of Quality Dimensions

How do we know that the inductions generated from the natural properties determined by a conceptual space will be successful? In order to answer this question we must first consider the origins of the quality dimensions we use in classifying kinds and judging similarities.

There does not seem to be a unique origin of our quality dimensions. Some of the dimensions are presumably innate and to some extent hardwired in our nervous system, as, for example, color, pitch, and probably also ordinary space. These subspaces are obviously extremely important for basic activities like finding food and getting around in the environment.

Other dimensions are probably learned. Learning new concepts often involves expanding one's conceptual space with new quality dimensions. "Volume" is an example here. According to Piaget's "conservation" experiments with five-year-old children, small children do not make a distinction between the height of a liquid and its volume. The conservativity of volume, which is part of its conceptual structure, is something that must be learned. Another example is provided by the functional properties which are used for describing artifacts. Such properties are characterized by a set of associated actions (see, for example, Vaina's [1983] analysis of functional representation). Even if we do not know very much about the conceptual dimensions underlying actions, it is quite obvious that there is some such nontrivial structure.

Still other dimensions may be culturally dependent. "Time" is a good example: in contrast to our linear conception of time, some cultures conceive of time as circular, so that the world keeps returning to the same point in time, and in other cultures it is hardly meaningful at all to speak of time as a dimension. A sophisticated time dimension is needed for advanced forms of planning and coordination with other individuals but is not necessary for the most basic activities of an organism. Although Peirce, in the quotation given earlier, counts some concepts related to our fellow beings to be among the innate ideas, I believe that many social dimensions are culturally dependent. For example, kinship relations, which are fundamental to social concepts, are specified differently in different cultures.

Finally, some quality dimensions are introduced by science. For example, the distinction between temperature and heat is central for thermodynamics but has no correspondence in human perception. Human perception of heat is basically determined by the amount of heat transferred from an object to the skin, rather than by the temperature of the object.

The Relativism of Inductive Inferences

Natural properties have here been defined in relation to a given conceptual space. But is not the choice of a conceptual space arbitrary? Since a conceptual space may seem like a Kuhnian paradigm, are we not thereby stuck with an inescapable relativism? Since we can pick our own conceptual space, we can, in this way, make our favorite properties come out as natural in that space. For example, it is not difficult to construct a conceptual space where "grue" would correspond to a convex region and thus be a natural property in that space.

I argue that freedom in choosing a conceptual space is rather limited, and thus that the relativism inherent in this choice is not as problematic as it may at first seem. As was noted in the previous section, many quality dimensions are innate. This is the main reason why human beings, to a remarkable extent, agree on which properties are projectable —in particular, if the properties are closely connected to what is provided by the senses. For instance, different cultures show high agreement in identifying species of animals and plants, at least in what concerns basic categories.

But even if such an agreement exists, we must answer the question of why our way of identifying natural properties accords so well with the external world as to make our inductions tend to come out right. Peirce notes that "it can no longer be denied that the human intellect is peculiarly adapted to the comprehension of the laws and facts of nature" (*CP* 2.750).

Quine (1969:126) formulates the problem in the following way: "Why does our innate subjective spacing of qualities accord so well with the functionally relevant groupings in nature as to make our inductions tend to come out right? Why should our subjective spacing of qualities have a special purchase on nature and a lien on the future?"

The answer, it seems to me, comes from evolutionary theory. As is indicated by the earlier quotation, Peirce (*CP* 2.753) seems to take the same line. Natural selection has made us all develop a conceptual space that results in inductions that are valid most of the time and thus promote survival. It is not that "there is a general tendency toward uniformity in the universe" (*CP* 2.749) but that our quality dimensions are what they are because they have been selected to fit the surrounding world. In Peirce's words, "There is a special adaptation of the mind to the universe, so that we are more apt to make true theories than we otherwise should be" (ibid.). As indicated in the previous section, not all of our quality dimensions have a genetic origin, but the agreement between the conceptual spaces of different individuals of our species is high enough to produce very similar conceptions of, at least, the perceptual categories of the world. The upshot is that there will exist rather limited freedom for humans in choosing a conceptual space as a basis for inductive inferences. The relativism that will occur will apply only to the more advanced, learned, and culturally dependent quality dimensions.

However, a far-reaching consequence of this evolutionary account of conceptual spaces is that *our inductive capacities will be dependent on the ecological circumstances under which they have evolved.* Human conceptual spaces are attuned to the environment of thousands of years of hunting and gathering. Consequently, we cannot expect our intuitions about which properties are projectable to be successful in environments that widely diverge from those present during our evolutionary history.

Conclusion: The Role of Science

The conclusion to be drawn is that we should trust our capacities for inductive reasoning only in situations which are "ecologically valid," to borrow a concept from J. J. Gibson. What should we do about the increasing proportion of artificial objects and situations that surround us?

This is where science enters the scene. By introducing theoretically precise, nonpsychological quality dimensions, a scientific theory may help us find new inductive inferences that would not be possible on the basis of our subjective conceptual spaces alone. A scientific breakthrough is often made when a new quality dimension is introduced in a theory. Witness, for example, the distinction between temperature and heat that was mentioned earlier. This distinction was of crucial importance for the development of thermodynamics. Or witness Newton's distinction between weight and mass. This is the very first definition introduced in the *Principia*; without it, his celestial mechanics would have been impossible.

The quality dimensions of scientific theories and their associated measurement procedures help us in producing new, successful inductive conclusions in environments which are completely different from that of our evolutionary cradle. Furthermore, the precise metrics of the scientific quality dimensions make it possible to formulate functional laws which enable us to compute more sophisticated predictions. Once a scientific conceptual space has been established, the formulation of such laws is a kind of inductive inference. Clearly, such laws supersede inductions of the form "All *F*'s are *G*'s," which are the only possible inductive inferences on quality dimensions which have only a crude or rudimentary metric. As Quine (1969:137–38) notes, it is a sign of mature science that notions of similarity become less and less important, being replaced by theoretically more sophisticated concepts.

In this way, science builds upon our more or less evolutionarily determined conceptual spaces, but in its most mature form becomes independent of them. It seems appropriate to conclude with the following quotation from Peirce which supports this general thesis: "Side by side, then, with the well established proposition that all knowledge is based on experience, and that science is only advanced by the experimental verifications of theories, we have to place this other equally important truth, that all human knowledge, up to the highest flights of science, is but the development of our inborn animal instincts" (*CP* 2.754).

Note

Research for this paper has been supported by the Swedish Council for Research in the Humanities and Social Sciences and by the Erik Philip-Sörensen Foundation. The article draws heavily on material in Gärdenfors 1990, and some of the text on prototype theory is borrowed from Gärdenfors 1991.

References

Berlin, B., and Kay, P. 1969. *Basic Color Terms: Their Universality and Evolution.* Berkeley: University of California Press.

Carnap, Rudolph. 1950. *Logical Foundations of Probability.* Chicago: University of Chicago Press.

Churchland, Patricia S. 1986. *Neurophilosophy: Toward a Unified Science of the Mind/Brain.* Cambridge: MIT Press, Bradford Books.

Churchland, Paul M. 1986. Some Reductive Strategies in Cognitive Neurobiology. *Mind* 95 (379):279–309.

Foss, Jeff. 1988. The Percept and Vector Function Theories of the Brain. *Philosophy of Science* 55:511–37.

Gärdenfors, Peter. 1988. Semantics, Conceptual Spaces, and the Dimensions of Music. In *Essays on the Philosophy of Music*, ed. V. Rantala, L. Rowell, and E. Tarasti, 9–27. *Acta Philosophica Fennica*, vol. 43. Helsinki.

———. 1990. Induction, Conceptual Spaces, and AI. *Philosophy of Science* 57:78–95.

———. 1991. Frameworks for Properties: Possible Worlds vs. Conceptual Spaces. In *Language, Knowledge, and Intentionality*, ed. L. Haaparanta, M. Kusch, and I. Niiniluoto, 383–407. *Acta Philosophica Fennica*, vol. 49. Helsinki.

Hempel, Carl G. 1965. *Aspects of Scientific Explanation, and Other Essays in the Philosophy of Science*. New York: Free Press.

Goodman, Nelson. 1955. *Fact, Fiction, and Forecast*. Cambridge: Harvard University Press.

Lakoff, George. 1987. *Women, Fire, and Dangerous Things*. Chicago: University of Chicago Press.

Langacker, Ronald W. 1986. *Foundations of Cognitive Grammar*. Vol. 1. Stanford, Calif.: Stanford University Press.

Lewis, David K. 1973. *Counterfactuals*. Oxford: B. Blackwell.

Marr, David. 1982. *Vision*. San Francisco: Freeman.

Marr, David, and Nishihara, H. K. 1978. Representation and Recognition of the Spatial Organization of Three-dimensional Shapes. *Proceedings of the Royal Society in London*, vol. B 200, pp. 269–94.

Mervis, Carolyn, and Rosch, Eleanor. 1981. Categorization of Natural Objects. *Annual Review of Psychology* 32:89–115.

Quine, Willard V. O. 1960. *Word and Object*. Cambridge: MIT Press.

———. 1969. Natural Kinds. In *Ontological Relativity and Other Essays*, 114–38. New York: Columbia University Press.

Rosch, Eleanor. 1975. Cognitive Representations of Semantic Categories. *Journal of Experimental Psychology: General* 104:192–233.

———. 1978. Prototype Classification and Logical Classification: The Two Systems. In *New Trends in Cognitive Representation: Challenges to Piaget's Theory*, ed. E. Scholnik, 73–86. Hillsdale, N.J.: Lawrence Erlbaum Associates.

Sloman, Aaron. 1971. Interactions between Philosophy and AI—the Role of Intuition and Non-logical Reasoning in Intelligence. *Artificial Intelligence* 2:209–25.

Vaina, Lucia. 1983. From Shapes and Movements to Objects and Actions. *Synthese* 54:3–36.

6 Abduction, Justification, and Realism

Anthony J. Graybosch, California State University at Chico

Abduction is the key concept in Peirce's epistemology, so important to Peirce that he claimed the question of pragmatism is nothing else than the question of the logic of abduction (*CP* 1.196). The pragmatic approach to both meaning and truth is to deduce the consequences of hypotheses in order to subject them to inductive tests. Pragmatism is the logic of abduction, to the extent it has a logic, because it determines whether an abductive hypothesis is admissible (Frankfurt 1958:597). Abduction is more extensive than pragmatism, since it includes considerations of economy in selecting hypotheses for further inquiry from those which are pragmatically admissible (*CP* 2.776).

Peirce divided inquiry into three irreducible stages: abduction, deduction, and induction (*CP* 2.807). We will be concerned here with Peirce's later views of abduction. But he consistently distinguished two forms of abduction—perceptual and interrogatory—on the basis of their possible controllability by reason. Perceptual abduction provides us with the data to perform inductive tests, and interrogatory abduction provides the hypotheses we are testing.

I consider the two types of abduction identified by Peirce, the degree of justification each confers on a belief, Peirce's justification of abduction, and two contemporary attempts to base scientific realism on abduction. My major original claim is that a legalistic theory of epistemic justification can be built upon Peirce's views. Just what such an epistemological theory is will become clearer presently, but its exposition will involve drawing some philosophical connections between the views of Wittgenstein and Peirce. It is my view that the key notions of a legalistic epistemology—suspicion, warrant, beyond a reasonable doubt—can be abstracted from the practice of a community of inquirers. I offer this argument somewhat hesitantly in light of Peirce's remark that logic "puts its heel upon" the suggestion "of our atrocious legal procedure" that hot and partisan debate is the way to investigate (*CP* 2.635).

However, Peirce's concern was with the methodological principles adopted in legal practice; ours is with the concepts of evidence found in legal practice. Legal notions are helpful in fully characterizing interrogatory and perceptual abductions, and other concepts borrowed from legal practice help illuminate the kind of collective inquiry Peirce advocated. By suggesting a connection between an area of social science and Peirce's theory of inquiry, I do not mean to criticize Peirce's view as a theory of science. Rather, I wish to suggest that inquiry in science, common sense, and law rely on similar epistemic concepts. Certainly epistemology and law have at least one common concern: determining how to regulate belief. And perhaps they have a common historical root in the practice of the Sophists.

Knowledge, Certainty, and Truth

Peirce did not claim that certainty was impossible; rather, he observed that he was unable to find any truth which seemed to him to be free from doubt (*MS* 829). Peirce cannot mean that he is able to find some theoretical reason to doubt every belief; his critique of Cartesianism makes that an unacceptable interpretation. Rather, he must mean that every one of his supposed certainties has been called into doubt in some concrete way.

Now if we accept the traditional notion of knowledge as involving the truth of what is known, then we will have to deny that Peirce has an epistemology. The fixation of belief is brought about by practicing the scientific method to remove doubt induced by a surprising experience. The removal of the doubt does not, however, establish the truth of the belief we are able to fix upon. Such belief does involve a form of subjective certainty, though, in that we temporarily have no reason to doubt. So Peirce's theory of knowledge, if we are going to call it that, is one in which knowledge does not involve truth but does involve a form of temporary certainty.

Wittgenstein (1972:8,18) suggests a useful way of looking at the relationship between knowledge and non-Cartesian certainty.

> The difference between the concept of "knowing" and the concept of "being certain" isn't of any great importance at all, except where "I know" is meant to mean: I can't be wrong. In a law-court, for example, "I am certain" could replace "I know" in every piece of testimony. We might even imagine it being forbidden to say "I know" there. . . . "I know" often means: I have the proper grounds for my statement.

One might want to argue that Wittgenstein was not dealing with the concept of knowledge, but with the conditions under which one is warranted in claiming to have knowledge. But Wittgenstein is looking at language games in which knowledge figures as an element. And this is his means to determine its meaning—the right way, I believe, as well as being very amenable to pragmatism. In the language game played in the law courts, knowledge and certainty are identical concepts: both are used where one feels properly grounded to make an assertion. Testimony is not so much proved true or false as its grounds are established or impugned. The difference in the concepts is that in some language games knowledge is taken as involving a claim that one cannot be wrong. But that type of claim is not made by a fallibilist.

Peirce will be properly seen as having a legalistic epistemology if his use of the concept of knowledge is analogous to that found in court. There a report or an explanation stands if it meets community-based criteria of evidence which give it warrant to a degree of certainty less than infallibility and incorrigibility—the claim that one cannot be wrong. A full-fledged legalistic epistemology involves applying concepts borrowed from that language game in characterizing the degree of warrant possessed by different beliefs. It is an attempt to create a new language game for epistemology which is essentially historicist.

In a passage similar in import to that cited from Wittgenstein, Peirce noted that to say that a proposition may be false is to say that it will perhaps be found out to be false, and to say that a proposition is certainly true means that it can never be found out to be false (*MS* 312). True and false were unpacked in epistemic terms and involve prediction of the extent of evidential investigation the propositions would sustain. There is some evidence in Peirce's writing that he at least entertained such an approach to knowledge. For instance: "Truth is that strength, that virtue in an idea by which it is bound to triumph in the struggle for existence. There is a higher type of truth which fact has got to conform itself to, and there is the lower kind of truth that conforms itself to fact" (Peirce in Whitney 1889).[1]

Here, the lower type of truth is that supplied by perceptual abductions. They are admissible simply because they cannot be doubted, at least in the moment of their occurrence. And this lack of control establishes our lack of negligence in believing them. The higher type of truth, I presume, includes those interrogatory abductions which survive the process of inquiry, which maintain their standing when tested by the best available methods. Truth is the name of the virtue that such ideas possess. Peirce also described assent as the enunciation of a proposition which renders the subject "responsible in some measure for its truth, so at least his reputation will be endangered in case it turns out not to be true" (in Whitney 1889). Here he speaks the language of epistemic responsibility, not that of apprehension of objective truth. Peirce's evolutionary metaphysics requires such a dynamic vision of truth. And the virtuousness of those temporary resting places called habits or beliefs is unpacked pragmatically as a lack of negligence in their acquisition and their well-groundedness when called into question. Here we have evidence of the beginnings of a legalistic epistemology and kinship with the later Wittgenstein.

Fallibilism, Perceptual Abductions, and the Evidence Base

Any epistemology, even fallibilism, must deal with the question of the source of evidence. Hans Albert, a contemporary fallibilist, notes that if every belief requires evidence to justify our assenting to it, the Munchausen trilemma arises.[2] Epistemologists must then either commit themselves to an infinite regress, accept a logical circle, or arbitrarily break off the process of justification at some point (Albert 1985:16–17). Albert urges epistemologists to avoid the trilemma by replacing their commitment to the principle of sufficient reason with critical examination of all beliefs. All beliefs should be regarded as hypotheses subject to further examination (ibid.:46–47). Another advocate of the fallibilist approach to knowledge offers a similar pragmatist-inspired approach to epistemology. Isaac Levi (1980:1) writes, "There are no immaculate preconceptions. Where all origins are dark, preoccupation with pedigree is self-defeating. We ought to look forward rather than backward and avoid fixation on origins." The particularly pragmatist orientation to fallibilism is the prioritizing of the future usefulness of even the sources of evidence in considering their justificatory status.

Peirce is a fallibilist who holds that inquiry is prompted only by doubt and that doubt cannot be feigned (*CP* 5.375). Beliefs are accepted until reason for doubting them is promoted by a surprising phenomenon (*MS* 309). But Peirce does not abandon the search for justification of naive beliefs. Rather, he argues on naturalistic grounds for the admissibility of perceptual abductions as starting points of inquiry. "Direct experience is neither certain nor uncertain. . . . it testifies to nothing but its own appearance" (*CP* 1.145). His view is that direct experience is not known in itself but grasped as a matter of fact. Sensation and conception place these first impressions into a unity required by the human mind. And the process is an abductive inference performed by the mind, out of our control, and outside of consciousness. "These [sensation and conception] are nothing else than predicates which the mind affixes by virtue of a hypothetical inference in order to understand the data presented to it. A hypothetical predicate is one which is affixed to a thing which has not been experienced as possessing it in order to bring the manifold in the experienced data to unity" (*MS* 357). Peirce sees a difference between the type of information given by sensation and conception. Sensation merely names a datum to give unity to the manifold; it is abduction in form only. Conception adds to the experience a genuine predicate not in the data. The expression of a sensory state in propositional form is a real abduction. All knowledge is abduction refined to some degree in the inductive stage of inquiry (*MS* 692).

The justificatory status of perceptual abductions is that assent to them is not something for which an epistemic agent can be blamed. They are outside one's control. Here concepts borrowed from law are helpful in understanding Peirce's position. In civil law negligence is doing what a reasonable and prudent person would not have done, or not doing what such a person would have done. Negligence involves taking risks inappropriate to a context, or not taking proper precautions.

The legal system also interprets "reasonable" in terms of community standards. The test of reasonableness is supposed to be the conduct of the standard person in a given context, where each jury member consults their common knowledge of reasonable behavior (Terry 1915:41–47). It does not accept mere consistency with other beliefs as good grounds for belief. This concept of reasonableness allows us to make sense of the epistemic intuition that internal and external standards of evidence need to be combined. Persons are held responsible for knowledge which they may not possess but reasonable members of the community do possess. And persons can be held negligent, even when they believed their actions were not negligent. Negligence can also occur even when a person's state of mind involves the desire not to be negligent. The analogy between negligent belief and negligent action is particularly appropriate for pragmatism because of the emphasis pragmatism places upon thought as action. It is not the subjective attitude toward beliefs which matters to their warrant so much as their efficacy in action. Correct conduct can even occur where a negligent attitude is present.

In the case of perceptual abductions, negligence cannot occur because the unconscious "inference" is common to the reasonable members of the community. It is not subject to an attitude of negligence because it cannot be controlled. Assent to perceptual abductions is blameless, since they cannot be controlled.[3] Logic parallels law and morals for Peirce; the absence of the faculty of self-criticism eliminates our ability to be responsible and suspends blame (*CP* 1.606). What seems especially problematic in this position, however, is that we can and do become conscious of our ability to control perceptual abductions and the need to follow the guidelines of some experts in the community in making observations. Science, the practice of the best available methods of inquiry, replaces not just the interrogatory abductions of common sense but the perceptual with those supplied by instruments. Peirce's remark on the two types of truth given above suggests that he at times recognized the controllability of perceptual abductions.

Peirce claimed a type of continuity between common sense and science on the basis of two types of natural gifts. These gifts are the ability to understand the things that an animal must have some understanding of in order to support life, such as the idea of mechanical force, and the natural gift to understand the workings of another's mind (*MS* 755). But it is difficult to see why the common presuppositions of common sense and science should be taken to establish a continuity of method. The breaks in inquiry between common sense and science seem more radical than Peirce allows.

In some places Peirce does note this discontinuity. Perceptual abductions are facts in the world, the way human organisms automatically represent the stimulations of the environment on the nervous systems. They are true for the shared world of human experience. "Man . . . lives in a world that is partly unreal, or rather is relative to his faculties, but is so completely limited to that world, that he can be said to live in it; and what he is obliged to by his nature, as a man, to believe becomes the truth, relatively [*sic*] to that universe of his human thoughts to which his conception of the Real is limited" (*MS* 755).[4] Still, there is tension in this world which leads to the search for the higher kind of truth to which fact must conform. We begin to see the data of perception as rival reports competing for assent in the court of epistemic inquiry, leading us to develop a theory of admissibility first in *logica utens* and then full-fledged theories of knowledge in the sciences or the law courts (*MS* 311). Peirce's theory of interrogatory abductions is, then, an attempt to complete this search for a method.

We observed above that Peirce attempts to supply a naturalistic justification of perceptual abductions. What we might expect is that the warrant of the spontaneous perceptual reports would be established by scientific investigation, that perceptual abductions are vindicated by science. Peirce thought that science had clearly discovered true theories

(*CP* 5.172). And he held that this success can be explained only if evolution had somehow equipped us to make good guesses when formulating hypotheses. Peirce argues persuasively that the probabilities of hitting on true theories by trial and error are infinitesimal. Success through trial and error would be impossible, given the infinite number of hypotheses which would have to have been tried in the finite period of time science has established for human existence (*CP* 5.591).

But the success of science does not help establish the justificatory status of perceptual abductions, because Peirce bases the success of the guessing instinct in science on a natural gift found in perception. So his naturalistic justification of abduction runs in a direction different from the one we expected, from common sense to science. Psychology is relevant to the logic of abduction at least to the extent that we cannot explain the success of the guessing instinct in science without considering the justificatory status of the natural instinct.[5] This natural instinct is characterized as groundless, ubiquitous, and trustworthy (*MS* 692).

Peirce claimed that humans have an insight into the general elements of nature, more often right than wrong. One thing that seems clear about this instinct is that it is not the result of critical thought (*CP* 5.173). Peirce observed that there may be mysterious agencies in ideas but that this sort of answer does not deal with cognitive goodness (*MS* 315).

It is difficult to understand why Peirce felt a justification of the natural instinct was needed, or even a justification of perceptual abductions. As a pragmatist, he should see its vindication in its use. The goodness of abduction seems to be that its deductive consequences can be inductively tested. This amounts to determining that there is no reason at present to believe they are false. Perceptual judgments which cannot be subjected to this process are by definition not good abductions. Only perceptions which take risks, which have implications for future experience, which are more than mere naming, can be good abductions. This seems a perfectly adequate means of justifying abductions, but Peirce wanted to justify the form of inference itself and not just its instances. And this desire seems to have been rooted in his belief that all inquiry relies on the basic abduction that the real is capable of rational comprehension. And since science cannot justify the process of abduction, his concern turns him toward metaphysics. This is a connection I do not think he needed to make for his epistemology.

Why would Peirce not claim that perceptual abductions provide a foundation for knowledge? I suggest it is because they are abductions, and abductions have low provisional warrant. Perceptions may be out of our control when they are made, but they can be revoked in light of subsequent information (*MS* 309). Perhaps, then, Peirce would allow that they could be overruled or replaced by scientific instruments as the process of inquiry moves toward the higher form of truth. Here is another similarity to the law, where eyewitness testimony can be set aside by more reliable evidence. The continuity between science and common sense lies not in a common content but in the natural instinct which supposedly justifies abductions.

Interrogatory Abductions

Doubt cannot be feigned, since it cannot be voluntarily induced. Doubt arises for concrete reasons and has an important occurrent psychological aspect; "experience invariably teaches by means of surprises" (*MS* 309). And so experience in the form of additional perceptual abductions is the motivating force in our attempt to realize that higher truth to which fact must conform. It leads us to formulating hypotheses of greater scope than the momentary presentations of sense covered by perceptual abductions. This form of reasoning is lower in security than perceptual abduction and takes greater risks. Still, it does confer warrant on a hypothesis. The warrant is little more than mathematical possibility, a slight anticipation that a hypothesis might be true. But Peirce argues that an argument is still logical, even if it is weak, provided it does not pretend to be strong. And occurrent interrogatory abductions can be reconstructed in the form of an argument.

Guessing, interrogatory abduction, is both insight and inference. It is controllable reasoning from data to an explanatory hypothesis, it must be at least amenable to the logical form of abduction which Peirce originally based in the syllogism, and it bestows a slight degree of warrant on the hypothesis. The logical form of abduction, however, is not selective enough to guarantee arriving at truth.

Peirce did suggest some methodological considerations for interrogatory abduction. For instance, he remarked that there is no positive sin (negligent action?) against inquiry in trying any theory which comes into our heads provided it permits inquiry to go on unimpeded (*CP* 1.136). Besides being amenable to reconstruction in the correct logical form, a good abduction must have deductive consequences which can all be tested inductively (*CP* 5.189). The hypothesis must be distinctly put, failures and successes honestly noted, and resemblances taken at random (*CP* 2.634).

It might seem that background beliefs would be important in guiding abductive inferences, that the continuity Peirce sees between common sense and science would lead him to accept gradualism and epistemic conservatism. But Peirce's position was more contextual and naturalistic than it was conservative. He emphasized solving a current problem as opposed to preserving accepted theory. This does cohere well with his position that there are no beliefs in science, since there is no risk. Here I think he means that theoretical "beliefs" are not beliefs, since they can be changed without significant change in behavior and sensory experience. Peirce wrote that revolutions in belief which lead to eliminating surprises do not overturn scientific beliefs; scientists have no beliefs connected to actions anyway (*CP* 5.60). In other words, Peirce held an early version of the undeterminacy theory, which led him to regard observations and theories as having different epistemic status: beliefs or hypotheses respectively.

Perhaps Peirce's most metaphoric description of interrogatory abduction is that it is:

> when the mind drinks in the facts, be they observable, testified to, or merely dreamed, drinks them in, breathes their atmosphere, soaks itself in them, becomes an organ of them, in its earnest endeavor to interpret them, and recognize the idea that is beneath them, or behind them, or that pervades them. . . . suddenly a new light illuminates the scene, the facts . . . all at once rearranged themselves in a new figure. (*MS* 756)

The basic justification of interrogatory abduction is that it reduces a confusing tangle of given facts to order (*MS* 831). Peirce remarked that "reason is inferior to instinct in several respects. It is less subtle, less ready, less unerring. The one respect in which it is superior is in being controlled, checked, criticized" (*MS* 832). This emphasis on control as a means of delineating two types of abduction cannot be sustained. Perceptual abductions can be controlled at least to a degree, and Peirce's description of interrogatory abduction suggests that it may not be a controllable process. It is reasoning because we pass from what we know directly to indirect knowledge (*MS* 756). But that also occurs in perceptual abduction when we pass from sense to a judgment. Abduction is simply more continuous than Peirce allowed, and he should have noted that, given the common instinctual basis of perceptual and interrogatory abductions in our natural insight into the workings of nature and other minds.

In any case, the role of background beliefs in selecting among interrogatory hypotheses is minimized. Peirce described abduction as selecting among equal explanations and remarked on the importance that this selection not be made on the basis of antecedent knowledge (*CP* 6.525). Explanations are not restricted to candidates which are consistent with past beliefs. Rather, they are selected as they rise to consciousness from instinctual mind and give new meaning to facts. Selecting an explanation decides the arrangement and characterization of the facts. The hypothesis must explain the facts and make them likely, but it need not take care of past facts also (*CP* 7.202).[6] This likelihood has nothing to do with antecedent probability (*CP* 5.599). In fact, even simplicity is given a subjective, instinctual interpretation. Peirce argued that the simplest hypothesis is the one suggested most immediately to the reasoner by the natural light, citing testimony from Galileo to support this view (*CP* 6.477).

There is similarity between the shaping of data performed in abduction and the global inferences juries perform in weighing competing stories in determining the facts. In both cases a global inference structures data into facts, there are competing stories available, and there is an appeal to instinctual mind or practical knowledge as one's ultimate justification. Judgment is also guided by principles of admissibility in selecting the right arrangement of the facts. In court, the most prominent principles deal with the presumption of innocence. Peirce's appeal to Galileo suggests that the evolving admissibility standards for epistemology be taken from the history of scientific success.

We have seen how Peirce emphasized the continuity in knowledge through the guessing instinct. Here he seems to have been wrong. Psychological tests find opposition between naive inference and the regulative principles of choice of hypotheses in science and probability theory (Stitch 1985:256). Peirce's position is curious, since he did note that different types of survival value were at issue in science and naive practice, supporting different notions of belief. But his position was perhaps unavoidable, given the desire to find a metaphysical justification of abductive inference proper. We can accept his notion of a community of inquirers guided by special legalistic values, but we must reject his core rationality about guessing.

A Contemporary Application: Abduction and Scientific Realism

Peirce held that the conception of the real is discovered when we correct ourselves from an error. Error allows us to distinguish a perception which was true relative to our private determination from a judgment which will stand in the long run. The real is that which reasoning and information will lead to in the long run and involves the notion of an ideal community capable of an increase of knowledge (*CP* 5.311). Such a community must always be in the making. The idea of the real would be lost if it were realized (*MS* 315). It also must be guided by sound principles of procedure if inquiry is to progress. The way Peirce presents realism makes it difficult to determine whether he is referring to real entities or to the linguistic embodiment of the real which survives inquiry. Given Peirce's view of matter as habitual mind, this is not an important distinction for him. Peirce's epistemology can survive his metaphysics; we can take the set of survivable interrogatory abductions as the ideal goal of inquiry. And, more relevant to this section, these two ways of presenting realism are mirrored in contemporary discussions.

Scientific realism is made up of both theory realism and entity realism. The former asserts that theoretical sentences make literal assertions about deep structure and that humans are capable of arriving at such true descriptions. It thus is a view which ordinarily makes ontological claims about theoretical entities (they must exist if true statements can be made about deep structure) and about human intellectual capacities. Entity realism asserts the existence of theoretical entities but does not claim that any of the theories which mention these entities are true. The former position is advocated by Richard Boyd, the latter by Ian Hacking.[7]

One of the more popular applications of abduction in contemporary philosophy is as a defense of scientific realism. For instance, Boyd argues that if a theory provides the best available explanation for some important phenomenon, that is a justification for believing that the theory is at least approximately true. If it is not, "it is hard to see how intellectual inquiry could proceed" (Boyd 1984:67).

But it seems quite clear how inquiry could proceed and even profit from rejecting the connection between explanation and truth. Theories which offer good explanations would be accepted provisionally and subjected to rigorous inductive testing. Inquiry would perhaps progress more rapidly in the fallibilist spirit. Also, Boyd is worried about the truth of scientific theory and the reality of unobservables. But the reality of theoretical unobservables cannot be tested any more rigorously than theories are tested. The abductive inference that says that science is more than just a synthesizing of human perceptual abductions into a neater bundle cannot be tested in a way that warrants it more than rivals such as constructive empiricism and instrumentalism. Realism cannot be established by appeals to how inquiry could proceed. How could inquiry proceed if we do not recognize that explanations are often wrong and stand in the way of inquiry? Also, Boyd (1984:72) remarks that rejection of abductive inferences about unobservables will lead to the rejection of inductive inferences about observables. But rejecting scientific realism does not amount to rejecting abduction. Abduction is accepted as a means of selecting initially plausible hypotheses for further testing, giving them standing in the court of epistemic inquiry. Rejection of the connection of abduction with truth does not leave one speechless, or without an epistemology. The emphasis Peirce puts on the tentativeness of hypotheses is what undercuts the use of abduction in arguing for realism. If we make the abductive inference to realism, there are no unique consequences which follow from the abduction to test it. Abduction is not itself strong enough to warrant belief in scientific realism, although it does support scientific method.

Hacking agrees with Peirce that abduction is a weak evidencing relationship and remarks (1983:52–53) that although we may feel a great gain from the understanding provided by a new explanatory hypothesis, that feeling is not grounds for believing the hypothesis true. But Hacking attributes too little evidencing power to abduction. Certainly his position ignores Peirce's argument that if humans did not have some insight into the workings of nature, they would not have eventually selected correctly from the infinite number of possible theories.

Hacking argues that theoretical, unobservable entities which are used to interfere in nature to create new phenomena are in that way shown to be real. The fact that we can use these entities to create phenomena testifies to their reality. But here the tentativeness of perceptual abduction is ignored. Even though they are not the objects within perceptual abductions, commitment to such entities is still tentative. There is good reason to believe that we may soon use other entities in their place or that our perception of their role in creating new phenomena will be withdrawn. So we find abduction misused in contemporary theory realism, and its fallibilistic dimension ignored in entity realism. Abduction cannot establish theory realism, and its consideration would undermine entity realism in favor of instrumentalism. The better view is Peirce's: the whole fabric of belief is a web of hypotheses, all of which may be subject to revision. In closing, we will take a brief look at Peirce's connection of abduction and realism.

Abduction makes inquiry impossible. So abduction is epistemologically justified because it is the only alternative to blocked inquiry (*CP* 5.171). At times Peirce claimed that abduction had inductive support, but perhaps he meant to speak of individual abductions. At other times he says it has nothing to do with probability. This seems to be the correct view, since he thinks all inductions rely upon abduction.

A second justification of abduction is metaphysical. Here Peirce argued that the real contains genuine chance and that the real is taking on concrete habits, becoming rational. This rationality of the real is a basic abduction, but also it is necessary for inquiry to be worth pursuing (*CP* 5.219). We must hope that a logical process of evolution occurs in nature and can be discovered by inquiry (*CP* 7.480).

But Peirce also observed that different types of minds are more suited to specific disciplines, that all persons are blind to reason in some areas of experience, and even entertained the view that the end of the world may be like its beginning: a state of confused feeling (*MS* 652). There is no guarantee in a world of freedom that convergence of opinion or of nature will occur. The ultimate justification of abduction, then, must be in the type of creatures it allows us to be, reasonable ones who by being reasonable participate in and foster the creation of a reasonable universe (*CP* 1.615). Perhaps that should be the basic hope of all normative disciplines, including epistemology.

Notes

I wish to thank the Institute for Studies in Pragmaticism, Texas Tech University, for access to copies of Peirce's manuscripts. This research was supported in part by a grant from the Oklahoma Foundation for the Humanities and the National Endowment for the Humanities. Findings, opinions, and conclusions do not necessarily represent the views of the foundation or of the NEH.

1. This quotation is from Peirce's interleaved copy of Whitney 1889, from a handwritten entry associated with the entry "conjecture."

2. Named after the Baron of Robert Musil's novel *The Man without Qualities* and Terry Gilliam's recent film.

3. Burks (1946:304) argued that this lack of control means that perception is not a form of inference. But surely the difference is that an inference occurs but is not performed.

4. This manuscript appears to have been dictated to Peirce's wife.

5. See Burks 1946:302–3 for a different view.

6. *CP* 7.37 conflicts with this view.

7. See Graybosch 1990 for a discussion of entity and theory realism.

References

Albert, Hans. 1985. *Treatise on Critical Reason.* Princeton: Princeton University Press.

Boyd, Richard. 1984. The Current Status of Scientific Realism. In *Scientific Realism* ed. Jarrett Leplin, 41–82. Berkeley: University of California Press.

Burks, Arthur W. 1946. Peirce's Theory of Abduction. *Philosophy of Science* 13:301–6.

Frankfurt, Harry G. 1958. Peirce's Notion of Abduction. *Journal of Philosophy* 55:593–97.

Graybosch, Anthony J. 1990. Observation, Instrumentalism, and Constructive Empiricism. *Southwest Philosophy Review* 6 (2):1–17.

Hacking, Ian. 1983. *Representing and Intervening.* Cambridge: Cambridge University Press.

Levi, Isaac. 1980. *The Enterprise of Knowledge.* Cambridge: MIT Press.

Sebeok, Thomas, and Umiker-Sebeok, Jean. 1980. *"You Know My Method."* Bloomington, Ind.: Gaslight Publications.

Skagestad, Peter. 1983. Peirce on Evolution and Progress. In *The Relevance of Charles Peirce*, ed. Eugene Freeman, 348–72. La Salle, Ill.: Hegler Institute.

Stitch, Stephen. 1985. Could Man Be an Irrational Animal? In *Naturalizing Epistemology*, ed. Hilary Kornblith, 249–67. Cambridge: MIT Press.

Terry, Henry. 1915. Negligence. *Harvard Law Review* 28:41–47.

Whitney, William Dwight. 1889. *The Century Dictionary.* New York: Century.

Wittgenstein, Ludwig. 1972. *On Certainty.* New York: Harper Torchbooks.

7 Peirce and the Logic of Logical Discovery

Leila Haaparanta, University of Helsinki

The logic which is labeled modern, and distinguished from both traditional and contemporary logic, can be traced back to Leibniz's philosophy. It was developed by logicians such as Bolzano, De Morgan, Boole, Frege, Peirce, Schröder, Whitehead, and Russell. It can be characterized by two basic features. First, modern logicians tried to construct an artificial language which was meant to overcome natural languages in certain respects: the new language was planned to be unambiguous and hence a more accurate mirror of the world, or of the forms of thought, than natural languages. Second, modern logic was characterized by the idea of a calculus. Modern logicians made an attempt to formulate the rules of logical inference explicitly by presenting logical and nonlogical vocabulary, formation rules, and transformation rules.

The nineteenth century saw a breakthrough of the two ideas. Frege emphasized that he did not want to put forward, in Leibniz's terms, only a *calculus ratiocinator*, by which he primarily meant the rules of logical inference. He argued that his conceptual notation was to be a *lingua characterica*, which was the term that he used for Leibniz's *lingua characteristica*. That is, it was to be a proper language which speaks about all that there is.[1] Frege raised criticisms against Boole because, in his view, Boole merely focused on developing a Leibnizian calculus in his logical works. This was not, however, what Boole himself thought of his project, for Boole (1958:11; 1965:5) included the idea of logic as a mental or philosophical language in his philosophical remarks on logic.

It has been argued in recent literature that since different logicians emphasized different sides of the Leibnizian project, they finally came to advocate conflicting views of the basic nature of logic. It has been claimed that Boole, Peirce, and Schröder, for example, were inclined to stress the importance of developing a calculus, whereas Frege and Russell were among those who laid more emphasis on the idea of logic as a universal language. The systematic consequences of the two views have been under careful scrutiny by a number of authors (van Heijenoort 1967; Hintikka 1979, 1981; Goldfarb 1979). According to these studies, those who stressed the idea of logic as language thought that language speaks about one single world. This was, indeed, the position to which Frege was committed (Frege 1903:56–65). In contrast, those who supported the view that logic is a calculus were ready to give various interpretations or models for their formal systems. This seems to have been Boole's standpoint (Boole 1965:3–4). Obviously, if Frege wanted to develop both a language and a calculus, he could not consistently support both of these implications.

The twentieth-century perspective has given even more content to the two views. It has been claimed that those who support the idea of logic as language tend to think that they cannot step beyond the limits of language and that this prevents them from developing a proper semantic theory for their language. Those, however, who endorse the view of logic as calculus tend to think that it is possible to look at a formal system, as it were, from the outside and to develop a semantic theory for it.

Even if Frege made the distinction between language and calculus on the basis of his interpretation of Leibniz's project, he was not quite conscious of all the implications which have been detected in recent literature. Hence, there are three different, though closely connected, stories to be told, as far as the ideas of universal language and calculus are concerned. There is the content which Leibniz gave to his project, the story which Frege and Boole told about their projects, and the story which is told from the twentieth-century perspective and which tries to capture the far-reaching implications of the original distinction. Some skeptical voices have also been heard. Hans Sluga (1987) has pointed out that the crucial difference of opinion between the pioneers of modern logic does not lie in their allegedly different attitudes toward Leibniz's project of language and calculus, but rather it can be found in their methodological commitments. In what follows, I will discuss Frege's and Peirce's methodological views in more detail. I will also suggest that Frege's idea of a universal language and his methodological choice can both be traced back to the same philosophical source.

What Is the Logic of Logical Discovery?

Even if Frege and Peirce could be located on different sides of this broad division, which ensues from Leibniz's ideas of *lingua characteristica* and *calculus ratiocinator*, there was one common, crucially important discovery which they made independently of each other. Whether the logical vocabulary was meant to constitute a universal language or to belong to the vocabulary of a calculus, both Frege and Peirce rejected the grammatical, subject-predicate analysis of judgments and suggested a new logical language which was characterized by the distinction between arguments, or indices, and functions, or relative terms, and by the idea of quantification.

In order to find out how a radically new logic came into being in the nineteenth century, we may attempt a methodological reconstruction of the work of nineteenth-century logicians. But what is meant by a method of logic? If logic is a science among other sciences, we may speak about its logic (or the method) of discovery and the logic (or the method) of justification. What, then, are the curious "entities" that a logician tries to discover or justify? Good candidates for the objects of discovery are the axioms of logic, the theorems of logic, and the rules of logical inference. The theorems of logic and some of the rules of inference naturally raise the problems of justification.

In the preface of his *Begriffsschrift*, Frege distinguished between the discovery and the justification of logical laws and excluded the problems of discovery from the field of logical studies. He thought that it was the task of psychologists to ponder the questions of discovery, whereas the problems of justification were within the proper interests of logicians (Frege 1964b:ix–x). However, there was one problem of discovery in which Frege was interested within the limits of logic. This problem was the discovery of the structure of a logical language. In his view, the discovery of the formula language of pure thought was a proper interest of logic itself (Frege 1964c:101; 1969a:273).

In the preface of the *Begriffsschrift*, Frege (1964b:xi) also claimed that all great scientific improvements of modern times have their origin in the improvement of method. The new method which he proposed for science was precisely his new logical language. However, there was an even deeper truth in his statement in view of the above discussion. In order to improve the science of logic by discovering a new logical language, Frege needed a new method of logical discovery. When he clarified his project of creating a Leibnizian universal language, he stressed that, unlike Aristotle, Leibniz, and Boole, he did not start from concepts but from judgments (Frege 1964c:101; 1969a:273). He believed that his method differed from that of Boole's, for he analyzed judgments, whereas Boole started with concepts. In Frege's day, there was also one logician who used the method of analyzing judgments in order to justify the basic concepts of traditional logic: Edmund Husserl (Husserl 1913:244–45; 1929:184; 1948:70).

In this chapter, I will argue briefly that Frege's methodological choice seems to have been influenced by Kant's philosophy.[2] The same argument could be raised in connection with Husserl's project, although I will not do that here (see Haaparanta 1988, 1990). I will suggest that Peirce's methodological commitments can also be described against this historical background. Since Peirce adopted Boole's logical algebra, he originally relied on Boole's method. This chapter is an attempt to evaluate the hypothesis that a methodological change led Peirce to discover his logical language. I will try to show that Frege's method of discovering the new logical language was the Kantian method of analyzing judgments and that Peirce relied on the very same logic of logical discovery when he discovered the distinction between indices, relative terms, and quantifiers.

Before going into my discussion, I would like to mention a number of limitations of my argument. First, I will not deal with the possible impact of medieval logical theories on Peirce's view. Second, I will not give a detailed treatment of the various changes in Peirce's notation from the 1860s until the 1880s. Instead, I will suggest a new way of looking at

Peirce's development which has turned out to be fruitful in connection with Frege's logic. Third, even though I will speak about the distinction between individuals and concepts or relations in the philosophy of both Frege and Peirce, I will not discuss the two authors' views on the nature of individuals in more detail.

If we chose to investigate this problem, a number of questions would arise. For example, an individual might be construed as a bare substrate, an individuating property, or a bundle of properties which constitute its complete individual concept. Besides asking ontological questions, we might be faced with various epistemological alternatives. For example, if we first argued that we have a direct access to an individual, hence, that there is a direct link between our faculty of knowledge and the individual object in itself, we might then specify our point by saying that the link is constituted by our acquaintance with the object itself, or that we are able to give an adequate description of the essence of the object. If we argued that we have a mediate relation to the object, we would argue that we are able to give a partial description of the object. We would then mean that we can never reach the complete individual concept of the object, or we would maintain that it is only through our characterizations that the object exists for us, and that we have no means of knowing what the object is in itself. In addition to the ontological question concerning what an individual is and the epistemological question concerning how we have access to an individual, we might raise a separate semantic question concerning the relation between the proper names of natural language and the objects of the world. In what follows, I will construe the semantic relations between names and individuals as epistemic relations, but I will not try to find Frege's and Peirce's choices among the numerous alternative answers mentioned above.

On Frege's Method

According to his own testimony, Frege wanted to realize Leibniz's project of an ideal language which would be a mirror of the universe. The preface of Frege's *Begriffsschrift* contains a reference to Adolf Trendelenburg's exposition of Leibniz's idea (Frege 1964b:xi). Trendelenburg

(1867) suggested that philosophers ought to construct a Leibnizian universal language on the basis of Kant's epistemological ideas. In his view, Kant had developed Leibniz's ideal in his transcendental logic by distinguishing between the conceptual (i.e., the formal or the logical) and the empirical (i.e., the material) components of thought and by studying the conceptual component, which consisted of the formal concepts presented in the table of categories. Trendelenburg also mentioned Ludwig Trede, who had written a short study in 1811, which was precisely meant to be a realization of Leibniz's idea on the basis of Kant's epistemological doctrines. He mentioned that Trede's symbolism was merely a codification of formal concepts, because human beings are not able to know all the empirical properties of objects.

Trendelenburg blamed Trede for staying too close to Kant's tables of judgments and categories and thus led the way for those who wanted to develop Leibniz's great idea further. If we place Frege in the chain formed by Leibniz, Kant, Trede, and Trendelenburg, we may conclude that Frege was skeptical about the possibility of realizing Leibniz's original dream and hence agreed with Trendelenburg (Frege 1964b:xii). Nor was Frege true to Kant's tables of judgments and categories, even if he could naturally express Kant's judgments in his new symbolism. However, the distinction between negative and infinite judgments was not made by Frege, and modal judgments could not be expressed in his symbolism, because modal concepts were not taken to be logical concepts in Frege's analysis (Frege 1964b:4–5).

Frege called his symbolism *Begriffsschrift*, which Trendelenburg (1867:4) used as a synonym for *lingua characterica*. However, Frege (1969a:273) also said that the term was misleading, precisely because he did not start from concepts but from judgments. Even if he differed from Aristotle, Leibniz, and Boole at this point, as he himself wrote, he did not differ from Kant. Frege chose the method of analysis just as Kant had done when he proceeded from the table of judgments to the table of categories (A68–69, B93–94).[3]

Frege was thus quite explicit in relying on the method of analyzing judgments. But that was all he told us. However, from his achievement in the *Begriffsschrift*, we may reconstruct the steps of his analysis. Frege started with judgments and first distinguished between judgments and judgeable contents, which were acknowledged to be true in judgments. In his analysis, he then found logical functions such as conditionality, negation, generality, and identity. He also discovered the basic categories of object (or argument) and function, which represented the distinction between individuals and their properties and relations. This distinction was not made in traditional logic, which relied on the concepts of subject and predicate. Frege viewed his distinction as one of his most important logical discoveries and as a distinction which must always be respected (Frege 1934:x, xiv; 1967b:168, 233, 270; 1969b:19–20). This did not mean, however, that he would not have accepted expressions referring to all kinds of curious objects, such as numbers and qualities, for admission into the empty argument places of his judgments. His analysis of judgments finally came up with subjective representations from which empirical concepts can be abstracted. Even if his first-order language was not a mirror of the empirical properties of objects, it contained Greek letters which stood for empirical functions.

Peirce in the Kantian Tradition

There is plenty of evidence that Kant was Peirce's philosophical hero (*CP* 1.560, 563). Peirce confessed his early reliance on Kant's transcendental logic by saying that he "believed more implicitly in the two tables of the Functions of Judgment and the Categories than if they had been brought down from Sinai" (*CP* 4.2). As Murphey (1961:33) remarks, it was the manner in which Kant discovered his categories that interested Peirce most of all.

However, careful consideration revealed problems and insufficiencies in Kant's transcendental analytic. In 1867 Peirce suggested a new list of categories because he had reached the conclusion that Kant's examination was "hasty, superficial, trivial, and even trifling" (*CP* 1.560). One of the problems was that, even if the correspondences between the functions of judgment and the categories were obvious, the preliminary division of judgments was not warranted. Hence, Peirce made a proposal which was meant to provide us with a new list of preconditions for knowing and judging.

The numerous changes in Peirce's theory of categories, and the roles of different types of sign in the logic of relations and quantification theory, have been discussed in great detail in the literature (Freeman 1934; Murphey 1961; Esposito 1980; Hookway 1985; Tursman 1987). Murphey (1961:137–38) has carefully documented the thesis that the discovery of the logic of relations led to certain changes in Peirce's theory of categories after 1867, but that the quantification theory was responsible for even more basic changes which Peirce made in his doctrine of three categories starting from the 1880s. Murphey (1961:4, 153) argues that revisions in philosophy were necessitated by a new logical discovery.

Murphey (1961:153, 298–99) finds a number of crucial changes enforced by the new logic. Contrary to the 1867 theory, the theory of relations showed that monadic, dyadic, and triadic relations were equally abstract and simply the three kinds of predicates in propositions. The same observation is also made by Hookway (1985:87), who argues that Peirce's categories manifested themselves in his new notation as incomplete expressions of valences one, two, and three. Among the changes which Murphey discusses is the revision of the category of secondness, which was caused by the discovery of quantification. Murphey (1961:299–300) claims that, inspired by Mitchell, Peirce introduced quantifiers and indices, which referred directly to individuals, or haecceities, whereas, in 1867 he had no concept of the individual. He concludes that the theory of the referential power of indices brought Peirce back to Kant's position that existence can never be given in concepts but only in intuition (ibid.:309–10).

Even if the Kantian influence has been widely recognized and discussed by scholars, little attention has been paid to the connection between Kant and Peirce from the perspective of the methodology of logic. I believe that, if we adopt this standpoint, we may be able to throw some new light on what happened in late nineteenth-century logic.

O. H. Mitchell and Peirce's Change of Mind

In the 1870s, Peirce discovered his logic of relatives, which was inspired by De Morgan's ideas and Boole's algebra of logic. Peirce's articles, entitled "The Logic of Relatives" (1883) and "On the Algebra of Logic: A Contribution to the Philosophy of Notation" (1885), contain the first presentation of his quantification theory, which he himself called his general algebra of logic and which, as he wrote, he developed on the basis of O. H. Mitchell's ideas (*CP* 3.363, 393). The first important change from Boole's logical algebra was that Peirce added indices to relations (l_{ij}). Indices referred directly to individuals. Second, he introduced the quantifiers "some" (Σ) and "every" (Π). In an article in the *Monist* (1897), he wrote that in his general algebra of logic, every proposition consisted of two parts, namely, its quantifiers and its Boolean part. The Boolean part consisted of relatives united by multiplication and aggregation, and each elementary relative had indices, which denoted haecceities, that is, individuals. According to Peirce, haecceities were fit to receive proper names or to be indicated by "this" or "that," and they were the ones which filled the blanks of relatives (*CP* 3.460, 500).

When he introduced his two improvements of logic, Peirce referred to his student O. H. Mitchell's article "On a New Algebra of Logic" (1883). He expressed his indebtedness to Mitchell regarding both indices and quantifiers. He described Mitchell's way of using indices in the algebra of logic by interpreting Mitchell's formula F_1 as "the proposition F is true of every object in the universe" and formula F_u as "the proposition F is true of some object in the universe." Here, however, Peirce deviated from what Mitchell had said. For Mitchell (1883:74), F was any logical polynomial involving class terms and their negatives, for example, $a+b$, ab, and so forth. He did not take F to be a proposition, but rather called it a predicate or a description of every or some part of the universe (ibid.:75, 96). Moreover, Peirce introduced the concept of the individual, which Mitchell did not use. Otherwise, it is true that Mitchell had both indices and quantifiers, as Peirce declared.

What Peirce himself did, and how he "misinterpreted" Mitchell's text, must be evaluated in more detail. Mitchell (1883:73) supported the view that objects of thought, in which logic is interested, are either class terms or propositions, but that every proposition, in its ultimate analysis, expresses a relation among class terms. Since Mitchell thought that, basically, every proposition expresses a relation among class terms, he relied on the Boolean method, which started from concepts and came up with propositions by combining concepts. It was precisely this way of thinking which Frege attacked. Hence, even if Mitchell did suggest indices and quantifiers, the new logical language cannot be encountered in his treatment.

Peirce's contribution was to take propositions as the starting point of analysis and to generate a distinction between the level of relative terms and the basic level of the names of individuals. In 1885 Peirce did not regard propositions as consisting of relations among class terms, but he gave them a basic structure which was formed by relative terms and indices and enriched by signs of logical aggregation, signs of logical composition, and quantifiers. These observations suggest that between Peirce's "New List" (1867) and his discovery of the new notation (1883, 1885) there was a methodological change, which contributed decisively to Peirce's logical discovery.

As Murphey (1961:152) also stresses, Peirce relied on the subject-predicate form of propositions in 1867 (*CP* 1.548). Even if Kant had taken the same traditional doctrine for granted, his transcendental logic suggested a new way of looking at judgments. This new way was based on the idea that there was a crucially important epistemological distinction between intuitions and concepts (A320, B376–77; Kant 1923:91). Frege made critical comments on Kant's distinction. Nonetheless, he viewed it as a precursor of his own distinction between objects and concepts, which was manifested as the distinction between arguments and functions in the conceptual notation (Frege 1934:37; see Haaparanta 1985:93–96, 102–17). Murphey does not miss the observation that, surprisingly enough, Peirce's logical discovery brought Peirce closer to Kant at this very point. As Murphey (1961:306) reads Peirce (*CP* 1.452), Peirce had come to accept Kant's doctrine that the logical form of knowledge can be known a priori, but its content must come through sensory experience. What Murphey does not suggest is that it was Peirce's change of mind concerning the method of discovery which was, to a great extent, responsible for this realization.

If we want to see the true difference between Peirce's method in 1867 and his method in 1883 and 1885, we must pay close attention to his 1867 concept of substance. His "New List" described a procedure in which the categories of quality, relation, and representation were derived from the manifold of substance. The highest category which he found above the category of quality was the category of being, which was unity and had no content (*CP* 1.548, 550). It was immediately after the "New List" that Peirce adopted and strongly praised De Morgan's theory of relations, by means of which he wanted to enrich Boole's logical algebra (*CP* 3.45). Peirce's new logical language did preserve some of the key elements which he had suggested in his new list. The one-place, two-place, and three-place predicates became proper representatives of the three categories of quality, relation, and representation, even if they had no role as cognitive steps, which characterized the categories in the new list (*CP* 1.562). The term "index" already occurred in Peirce's text in 1867. It meant a sign or a representation which had a relation of factual correspondence to its object (*CP* 1.558). However, Peirce did not give us any clues as to how we ought to understand the concept of an object and the direct relation between an index and an object.

If we try to interpret Peirce's concept of substance in terms of the link between him and Kant, we may find two candidates for its predecessors in Kant's philosophy. One alternative is to connect it with Kant's concept of an object of experience. Objects of experience are present to us via the pure concepts of understanding, shown as the various forms of judgment (A68–69, B93–94; A89, B122; A93, B126; A106). There is another possible interpretation if we combine Peirce's manifold of substance with Kant's concept of intuition. Kant's intuitions are individual representations of objects, that is, objects which are given to us through the medium of sensibility and formed by space and time (A19–22, B33–36). Peirce's wordings, however, contradict both of these interpretations. In 1867 he called substance the universal conception which was nearest to sense and which was the conception of the present in general, or *it* in general (*CP* 1.547). Hence, in the aggregate of a subject and a predicate, which a proposition was for Peirce in those days, the subject represented the substance (*CP* 1.548), but it did not refer to any individual. It was not until the 1880s that individuals in the sense of Kant's intuitions appeared in Peirce's logical notation.

The decisive insight both for Frege and for Peirce was that a judgment is not an aggregate of terms which represent concepts or classes but that its elements have different kinds of roles in their contexts. Two of those basic roles were that of representing relations and that of denoting individuals. My suggestion is that this idea occurred to them because they first confessed the priority of judgments over their constitutive concepts and then started to analyze judgments following Kant's epistemological lessons.

Notes

1. Frege repeats his view in various connections. See, for example, Frege 1964b:xi–xii, 98; 1967b:227, 341; 1969b:9–52.

2. For Frege's Kantian and neo-Kantian background, see, for example, Sluga 1980; Gabriel 1984, 1986; Haaparanta 1985; and Weiner 1986.

3. "A" refers to the first edition and "B" to the second edition of the *Kritik der reinen Vernunft*.

References

Boole, George. 1958. *An Investigation of the Laws of Thought, on Which are Founded the Mathematical Theories of Logic and Probabilities*. 1854. New York: Dover Publications.

———. 1965. *The Mathematical Analysis of Logic, Being an Essay towards a Calculus of Deductive Reasoning*. 1847. Oxford: B. Blackwell.

Esposito, Joseph L. 1980. *Evolutionary Metaphysics: The Development of Peirce's Theory of Categories*. Athens: Ohio University Press.

Freeman, Eugene. 1934. *The Categories of Charles Peirce*. Chicago: Open Court.

Frege, Gottlob. 1903. *Grundgesetze der Arithmetik. begriffsschriftlich abgeleitet*. Vol. 2. Jena: H. Pohle.

———. 1934. *Die Grundlagen der Arithmetik: Eine logisch mathematische Untersuchung über den Begriff der Zahl*. 1884. Breslau: M. and H. Marcus.

———. 1964a. *Begriffsschrift: Eine der arithmetischen nachgebildete Formelsprache des reinen Denkens*. 1879. In Frege 1964b, pp. 1–88.

———. 1964b. *Begriffsschrift und andere Aufsätze*. Hildesheim: Georg Olms.

———. 1964c. Über den Zweck der Begriffsschrift. 1883. In Frege 1964b, pp. 97–106.

———. 1964d. Über die wissenschaftliche Berechtigung einer Begriffsschrift. 1882. In Frege 1964b, pp. 106–14.
———. 1967a. Anmerkungen Frege zu Philip E. B. Jourdain. The Development of the Theories of Mathematical Logic and the Principles of Mathematics, 1912. In Frege 1967b, pp. 334–41.
———. 1967b. *Kleine Schriften.* Darmstadt: Wissenschaftliche Buchgesellschaft; Hildesheim: Georg Olms.
———. 1967c. Rezension von *Philosophie der Arithmetik*, vol. 1, by E. Husserl. 1894. In Frege 1967b, pp. 179–92.
———. 1967d. Über die Begriffsschrift des Herrn Peano und meine eigene. 1896. In Frege 1967b, pp. 220–33.
———. 1967e. Über die Grundlagen der Geometrie II. 1903. In Frege 1967b, pp. 267–72.
———. 1969a. Aufzeichnungen für Ludwig Darmstaedter. 1919. In Frege 1969c, pp. 273–77.
———. 1969b. Booles rechnende Logik und die Begriffsschrift. 1880–81. In Frege 1969c, pp. 9–52.
———. 1969c. *Nachgelassene Schriften.* Hamburg: Felix Meiner.
Gabriel, Gottfried. 1984. *Bedeutung*: Value and Truth-Value. *Philosophical Quarterly* 34:372–76.
———. 1986. Frege als Neukantianer. *Kant-Studien* 77:84–101.
Goldfarb, Warren D. 1979. Logic in the Twenties: The Nature of the Quantifier. *Journal of Symbolic Logic* 44:351–68.
Haaparanta, Leila. 1985. *Frege's Doctrine of Being. Acta Philosophica Fennica* 39.
———. 1988. Analysis as the Method of Logical Discovery: Some Remarks on Frege and Husserl. *Synthese* 77:73–97.
———. 1990. What Was the Method of Modern Logic? In *Language, Knowledge, and Intentionality: Perspectives on the Philosophy of Jaakko Hintikka*, ed. L. Haaparanta, M. Kusch, and I. Niiniluoto, 97–109. *Acta Philosophica Fennica* 49.
van Heijenoort, Jean. 1967. Logic as Calculus and Logic as Language. *Synthese* 17:324–30.
Hintikka, Jaakko. 1979. Frege's Hidden Semantics. *Revue Internationale de Philosophie* 33:716–22.
———. 1981. Semantics: A Revolt against Frege. In *Contemporary Philosophy*, ed. G. Fløistad, 1:57–82. The Hague: Martinus Nijhoff.
Hookway, Christopher. 1985. *Peirce.* London: Routledge and Kegan Paul.

Husserl, Edmund. 1913. *Logische Untersuchungen.* Vol. 1. 1900. Halle: Max Niemeyer.

———. 1929. *Formale und Transzendentale Logik: Versuch einer Kritik der logischen Vernunft.* Halle: Max Niemeyer.

———. 1948. *Erfahrung und Urteil: Untersuchungen zur Genealogie der Logik.* 1939. Hamburg: Claassen and Goverts.

Kant, Immanuel. 1904. *Kritik der reinen Vernunft.* 1781; 2d ed., 1787. In *Kant's gesammelte Schriften*, 3. Berlin: G. Reimer. English trans. by N. Kemp Smith, 1929. London: Macmillan.

———. 1923. *Logik.* In *Kant's gesammelte Schriften*, 9:1–150. Berlin: Walter de Gruyter.

Mitchell, O. H. 1883. On a New Algebra of Logic. In *Studies in Logic by Members of the Johns Hopkins University*, 72–106. Boston: Little, Brown.

Murphey, Murray G. 1961. *The Development of Peirce's Philosophy.* Cambridge: Harvard University Press.

Sluga, Hans D. 1980. *Gottlob Frege.* London: Routledge and Kegan Paul.

———. 1987. Frege against the Booleans. *Notre Dame Journal of Formal Logic* 28:80–98.

Trendelenburg, Adolf. 1867. *Historische Beiträge zur Philosophie.* Vol 3, *Vermischte Abhandlungen.* Berlin: G. Bethge.

Tursman, Richard. 1987. *Peirce's Theory of Scientific Discovery: A System of Logic Conceived as Semiotic.* Bloomington: Indiana University Press.

Weiner, Joan. 1986. Putting Frege in Perspective. In *Frege Synthesized*, ed. L. Haaparanta and J. Hintikka, 9–27. Dordrecht: D. Reidel.

8 Truth, Laudan, and Peirce: A View from the Trenches

Shelby D. Hunt, Texas Tech University

My area is marketing, not philosophy. Marketing is a discipline within schools of business and includes topics such as consumer behavior (why people buy what they do), distributive institutions (such as retailing and wholesaling), and marketing management (how managers should make decisions in the areas of pricing, product, promotion, and distribution). Our discipline is extraordinarily eclectic; we borrow theories and methods from economics, social psychology, sociology, and statistics, as well as (more recently) from philosophy of science. Although no one has formally applied the philosophy of Charles Sanders Peirce to marketing, given the heavily "applied" nature of our discipline, most marketing academicians would sympathetically view the general thrust of pragmatism (here construed in its colloquial form as emphasizing "practical" concerns).

Until recently, marketing academicians generally adopted a rather traditional view of the role of truth in marketing theory and research. For example: "When confronted with any theory, ask the basic question: Is the theory true? Less succinctly, to what extent is the theory isomorphic with reality? Is the real world actually constructed as the theory suggests, or is it not?" (Hunt 1976:130; reprinted in Hunt 1983:244).

Several marketing writers now advocate what they call critical relativism and can find no role for the concept "truth" in marketing inquiry: "I have made it quite clear that 'truth' plays no role in the ontology of critical relativism" (Anderson 1988a:134). Similarly, "The foregoing has demonstrated that 'truth' is an inappropriate objective for science, and that consumer research will do well to abandon such a quixotic idea" (Anderson 1988b:405).

Those marketing scholars advocating the abandonment of truth as a legitimate goal for inquiry have based their case almost exclusively on the works of Larry Laudan. As is well known, a general antipathy toward "truth" underlies much of Laudan's writings. What may be less well recognized is the fact that Laudan's attack on truth seems to have its historical origins in an attack on the "self-correcting thesis" advocated by Charles Peirce. My objective here will be to review briefly several of Laudan's arguments against truth as an appropriate objective for science, starting with his attack on Peirce's self-correcting thesis.

Laudan's Case against Truth

Laudan's case against truth begins, as best I can determine, with his 1973 article "Peirce and the Trivialization of the Self-correcting Thesis." Peirce contended that science had at its disposal certain methods that were self-correcting and that would therefore lead its statements, in the long run, to converge with truth: "Truth is that concordance of an abstract statement with the ideal limit towards which endless investigation would tend to bring scientific belief." Similarly, Peirce held: "The opinion which is fated to be ultimately agreed to by all who investigate is what we mean by truth, and the object represented by this opinion is the real" (*CP* 5.407).

Laudan begins his attack by discussing the history of the self-correcting thesis. By Peirce's time, he points out, the "thesis of instant, certain proof," or TICT, was viewed by many methodologists as untenable. This two-part thesis, that science "infallibly produces true theories" and that progress "could only consist in the accumulation of new truths," could "no longer be justified" (Laudan 1973:277). Therefore, methodologists developed the self-correcting thesis: "In some ways, it was a face-saving ploy, for it permitted the scientist to imagine that his ultimate goal was, as TICT had suggested, the Truth; although the scientist now had to be satisfied with the quest for ever closer approximations rather than the truth itself" (ibid.:278).

Peirce had viewed science as self-correcting because of its adoption of inductive methods: "Induction is that mode of reasoning which adopts a conclusion as approximate, because it results from a method of inference which must generally lead to the truth in the long run" (*CP* 1.67). However, this left for Peirce the problem of justifying inductive methods. Laudan (1973:289) points out that "Peirce distinguished three varieties of induction: crude induction, qualitative induction, and quantitative induction." However, Laudan contends, Peirce was at best able to justify only one of the three methods, quantitative induction, as even potentially self-justifying, thus "trivializing" his self-corrective thesis. Thereafter, "by a combination of bluster and repetition," Peirce argued that he had "established that all forms of induction are self-correcting and, thus, scientific method itself is self-correcting" (ibid.:292). Laudan concludes by contending that quantitative induction—the only form of induction Peirce could justify—has only a modest role in actual scientific practice and that those who believe in "approximate truth" and believe "that science is a self-corrective, progressive enterprise" should "be seeking to show how and why it is so," because "it is the self-correcting nature of science, not the self-corrective nature of a 'puerile' rule, which should be our main concern" (ibid.:298).

Laudan's attack on truth continues in *Progress and Its Problems* (1977). There he attempts to replace the search for truth with the search for maximum problem solving as the overriding goal of science: "It has often been suggested in this essay that the solution of a maximum number of empirical problems, and the generation of a minimum number of conceptual problems and anomalies is the *central aim* of science" (ibid.:111, emphasis added). The history of science and of the philosophy of science reveal, for Laudan, the hopelessness of the search for truth: "Attempts to show that the methods of science guarantee it is true, probable, progressive or highly confirmed knowledge—attempts which have an almost continuous ancestry from Aristotle to our own time—have generally failed, raising a distinct presumption that scientific theories are neither true, nor probable, nor progressive, nor highly confirmed" (ibid.:2). Furthermore, the search for truth is utopian, and utopian goals are undesirable: "We apparently do not have any way of knowing for sure (or even with some confidence) that science is true, or probable, or that it is getting closer to the truth. Such aims are *utopian* in the literal sense that we can never know whether they are being achieved"; thus, "determinations of truth and falsity are irrelevant to the acceptability or the pursuitability of theories and research traditions" (ibid.:127, 120).

In his article "A Confutation of Convergent Realism" (1981), Laudan links modern versions of realism with the search for truth: "If the realist is going to make his case for CER [Convergent Epistemological Realism], it seems that it will have to hinge on approximate truth" (ibid.:29). The article argues that history shows the theory of convergent realism to be false: "The realist's claim that we should expect referring theories to be empirically successful is simply false" (ibid.:24). Furthermore, "S2 [a thesis of convergent realism] is so patently false that it is difficult to imagine that the realists need be committed to it" (ibid.:25). And, "I shall assume that Putnam and Watkins mean that 'most of the time (or perhaps in most of the important cases) successor theories contain predecessor theories as limiting cases.' So construed, the claim is patently false" (ibid.:39). Finally, "the realist's strictures on cumulativity are as ill-advised normatively as they are false historically" (ibid.:43). The article derisively concludes: "Central realist theses . . . are just so much mumbo-jumbo" (ibid.:32).

The case against truth as a goal for science comes to full flower in *Science and Values* (Laudan 1984). Whereas *Progress and Its Problems* argued that maximum problem-solving was the "central aim" of science (Laudan 1977:111), *Science and Values* maintains that there is no overriding goal for science: "There is simply no escape from the fact that determinations of progress must be made relative to a certain set of ends, and that there is no uniquely appropriate set of those ends" (1984:66). However, not all goals are appropriate for scientific communities. First, the goals or "cognitive values" must not be "utopian" (ibid.:52). As in *Progress and Its Problems*, Laudan claims that truth is utopian, and therefore truth as a goal "cannot be rationally propounded" (ibid.:53). Second, goals can be evaluated on the basis of "reconciling theory and practice." This congruency criterion allows us to criticize a scientist or a scientific community that claims to be pursuing one goal, their "explicit" goal, but whose actions and actual choices imply the pursuit of some other "implicit" goal. Scientific rationality, honesty, and the justified fear of the reprobation of one's peers require goal congruency.

> When we find ourselves in a situation where there is a tension between our explicit aims and those implicit in our actions and judgments, we are naturally under significant pressure to change one or the other, or both. On pain of being charged with inconsistency (not to mention hypocrisy, dishonesty, etc.), the rational person, confronted with a conflict between the goals he professes and the goals that appear to inform his actions, will attempt to bring the two into line with each other. (ibid.:55)

The Case against Truth: An Evaluation

Like all of Laudan's writings, his arguments against truth are meticulously crafted, engagingly written, and "come alive" with their many historical examples and anecdotes. Unlike the writings of many contemporary philosophers of science, even those who have not mastered modern logic (which would include most of us in the trenches) can follow his argumentation. Unfortunately, although his case against truth is brilliantly developed, it seems fundamentally flawed.[1]

Consider the attack on truth that comes from Laudan's analysis of convergent realism. Basically, the argument (as described in the preceding section) claims that many of the underlying theses of the theory of convergent realism are either "simply false" or "patently false" or "false historically." Therefore, so the argument goes, the theory of convergent realism is refuted and, with it, its goal of truth. We focus here on the structure of the argument, rather than dispute the truth or falsity of its substantive historical claims. In brief, how can any argument which uses the concept "false" in so many of its premises possibly yield true conclusions such as "determinations of truth and falsity are irrelevant to the acceptability or the pursuitability of theories and research traditions" (1977:127) or "truth is utopian" or the goal of truth or approximate truth "cannot be rationally propounded"? To those of us in the trenches, the claim that an assertion in a theory is *false* (whether historically or otherwise) implies that under other circumstances the assertion might have been *true*. Thus, Laudan attempts to conclude that concepts like "truth," "falsity," and "approximate truth" are inappropriate, utopian, or cannot be rationally propounded, but uses these very same concepts, or their logical implications, in the premises of his own argument. Such attempts fail minimal standards of coherency.[2]

As I understand it, the technical term for this type of argument is "self-refuting." Siegel (1987) has shown, convincingly in my estimation, that all forms of cognitive relativism advanced until now (from Protagoras to Goodman) have been self-refuting. Likewise, it would seem, Laudan's "argument from convergent realism" against truth refutes itself. Ironically, we might add, Laudan's so-called argument from convergent realism actually suggests that *"truth" is such an essential concept that even its most eloquent opponents find its use necessary for meaningful discourse about science.*

A second flaw in Laudan's case against truth is that it fails one of his own criteria for evaluating the goals or cognitive values of science. In particular, it fails the congruency criterion, as the following example will illustrate.

In 1982 the Arkansas "creation science" trial of *McLean vs. Arkansas* came to a close. Federal judge William R. Overton ruled that the Arkansas law requiring that creation science should be taught, along with the theory of evolution, was a violation of the separation of church and state clause in the United States Constitution.[3] Judge Overton's opinion found that creation science was not science, since it did not collectively satisfy five characteristics (suggested to him by numerous witnesses) that can be used to distinguish scientific knowledge from other knowledge claims: (1) guided by natural law, (2) giving explanations that refer to natural law, (3) testable against the empirical world, (4) offering conclusions that are tentative and not necessarily the final word, and (5) falsifiable. Thus, Judge Overton adopted a traditional view in the philosophy of science that there are some significant differences between science and non-science, or science and religion, and on those grounds he argued that what was called "creation science" was actually a religious theory, not science.

In an article entitled "Commentary: Science at the Bar—Causes for Concern" (1982), Laudan vigorously attacked the reasoning of the judge. He did not disagree with the verdict, saying that "the verdict itself is probably to be commended" (Laudan 1982:16). However, he believed it was "reached for all the wrong reasons and by a chain of argument which is hopelessly suspect" (ibid.), and he subsequently questioned the integrity of those who might agree with the court: "The victory in the Arkansas case was hollow, for it was achieved only at the expense of perpetuating and canonizing a false stereotype of what science is and how it works. If it goes unchallenged by the scientific community, it will raise grave doubts about that community's intellectual integrity" (ibid.:19).

Laudan's disagreement over the basis for Judge Overton's decision was congruent with his writings on the "demarcation issue." For example, "It is probably fair to say that there is no demarcation line between science and non-science, or between science and pseudo-science, which would win assent from a majority of philosophers. Nor is there one which *should* win acceptance from philosophers *or anyone else*" (Laudan 1983:112, emphasis added). Therefore, Laudan's congruency criterion barred him from concurring with the court's reasoning.

How should Judge Overton have justified his ruling? Laudan (1982:16) umbrageously argues that "to make the interlinked claims that Creationism is neither falsifiable nor testable is to assert that Creationism makes no empirical assertions whatever. *That is surely false*" (emphasis added). He then discusses many of the empirical assertions that creationism makes and concludes: "In brief, these claims are testable, they have been tested, and *they have failed those tests*" (ibid., emphasis added). He continues, "Indeed if any doctrine in the history of science *has ever been falsified*, it is the set of claims associated with 'creation-science'" (ibid.:17, emphasis in original). Recall that Laudan argues time and again in the philosophical world against science having the "explicit goal" of truth with claims like "determinations of truth and falsity are irrelevant to the acceptability or the pursuitability of theories and research traditions" (1977:127). Being mindful of Laudan's own counsel to scrutinize carefully the "implicit goals" of researchers as evidenced by their "actions" and "choices," we find, most distressingly, that he now excoriates a federal judge working in the world of science and public policy for *not* choosing to base his decision on the determination that the theory of creation-science has been falsified. Thus, Laudan appears to violate his own congruency criterion.[4] Indeed, one could well understand it if those realist philosophers whose work Laudan characterized as "so much mumbo-jumbo" might choose to repay the compliment in kind.

Conclusion

The analysis here is not meant to disparage Professor Laudan's scholarship. His enormously illuminating contributions to the philosophy and history of science are so well known that they need not be detailed here, and in any case, the problems identified in this one short chapter cannot be considered a refutation of his other views about science. Moreover, there could well be subtle argumentation that would restore coherency and congruency to his views on truth. What the preceding does seem to signify is that, although concepts like "truth" and "falsity" may be maligned and pejoratively dismissed in ethereal argumentation, in the real world of science (the "trenches") they appear to be absolutely essential parts of our vernacular. Perhaps this realization is what leads Arthur Fine (1986:149, 170) to conclude: "Truth is the fundamental semantical concept" because "there is no form of life [in Wittgenstein's terms], however stripped down, which does not trade in truth," and this "redundancy property of truth makes truth a part of any discourse that merits the name."[5]

Since it is a thesis of this paper that Laudan's many attacks on the role of truth in scientific inquiry "got off on the wrong foot," historically speaking, with his 1973 paper on Peirce, perhaps a central insight of Peirce would be a fitting way to conclude. As is well known, Peirce firmly espoused "fallibilism" in science; science both as *product* and as *method* is to be considered fallible. There is no guarantee that scientific method (or any other method) will lead to theories that are true, much less to theories that are True (to use the word as antirealists often do). But the fact that truth-with-certainty (Absolute Truth or TRUTH) is beyond the grasp of science does not imply that the concepts "truth" and "falsity" are "utopian" in any pejorative sense of the word, nor does it imply that the concepts are so ambiguous that they should be expunged from the language of science. On the ambiguity issue, the views of Scheffler and Quine seem cogent to many of us in the trenches:

> One source of the trouble is a persistent confusion between truth and estimation of truth, between the import of our statements and the processes by which we choose among them. If, for example, appeal to reality or direct comparison with the facts is defective as a method of *ascertaining* truth, this does not show that the *purport* of a true statement cannot be properly described in ordinary language, as "to describe reality" or "to state the facts." We may have no *certain* intuition of the truth, but this does not mean that our statements do not purport to be true. Now, if the sentence "Snow is white" is true, then snow is (really, or in fact) white, and vice versa, as Tarski insists. As Quine has remarked, "Attribution of truth in particular to 'Snow is white', for example, *is every bit as clear to us* as an attribution of whiteness to snow." (Scheffler 1967:122, emphasis added)

In conclusion, it would be no more ambiguous for Laudan (or anyone else) to claim "it is true that the world did not come into being as the theses of creation-science contend" than it is for anyone to assert "the world did come into being as the theses of creation-science contend." To all those who believe that science is inextricably involved in making these kinds of assertions *when warranted*, truth remains the primary contender for the role of central aim for science. As Judge Overton might put it: the "case against truth" is dismissed on grounds of incoherence and incongruence.

Notes

1. Space limitations dictate that only some of the weaknesses of Laudan's case against truth can be developed here. In particular, his claim that societal institutions, such as science, should avoid utopian goals is highly suspect.

2. Most assuredly, Laudan would not argue for the following position: "Determinations of truth and falsity, though *relevant* to philosophical theories, are *irrelevant* to scientific theories and research traditions."

3. The case is *McLean vs. Arkansas Board of Education*, 529 F. Supp. 1255

(E. D. Arkansas 1982). For a complete discussion of the case, see the numerous articles in *Science, Technology, and Human Values* 7 (40–41).

4. Michael Ruse (1982:20) shows that Laudan's analysis of the *McLean* case is as misguided from a public policy perspective as it is incongruent philosophically: "Unfortunately [for Laudan's position], the U.S. Constitution does not bar the teaching of weak science. What it bars (through the Establishment Clause in the First Amendment) is the teaching of religion."

5. It should be noted that, although acknowledging the fundamental character of truth, Fine is no realist, for he emphasizes (1986:9): "I have no love for realism."

References

Anderson, Paul. 1988a. Relative to What—That Is the Question: A Reply to Siegel. *Journal of Consumer Research*, 15 June, 133–37.

———. 1988b. Relativism Revidivus: In Defense of Critical Relativism. *Journal of Consumer Research*, 15 December, 403–6.

Fine, Arthur. 1986. *The Shaky Game*. Princeton: Princeton University Press.

Hunt, Shelby D. 1976. *Marketing Theory: Conceptual Foundations of Research in Marketing*. Columbus, Ohio: Grid.

———. 1983. *Marketing Theory: The Philosophy of Marketing Science*. Homewood, Ill.: Richard D. Irwin.

Laudan, Larry. 1973. Peirce and the Trivialization of the Self-correcting Thesis. In *Foundations of Scientific Method: The Nineteenth Century*, ed. Ronald N. Giere and Richard S. Westfall, 275–306. Bloomington: Indiana University Press.

———. 1977. *Progress and Its Problems: Towards a Theory of Scientific Growth*. Berkeley: University of California Press.

———. 1981. A Confutation of Convergent Realism. *Philosophy of Science* 48:19–49.

———. 1982. Commentary: Science at the Bar—Causes for Concern. *Science, Technology, and Human Values* 7(41):16–19.

———. 1983. The Demise of the Demarcation Problem. In *Physics, Philosophy, and Psychoanalysis*, ed. R. S. Cohn and L. Laudan, 111–27. Dordrecht: D. Reidel.

———. 1984. *Science and Values*. Berkeley: University of California Press.

Ruse, Michael. 1982. Response to the Commentary: *Pro Judice*. *Science, Technology, and Human Values*, 7 (41):19–23.
Scheffler, Israel. 1967. *Science and Subjectivity*. Indianapolis: Bobbs-Merrill.
Siegel, Harvey. 1987. *Relativism Refuted*. Dordrecht: D. Reidel.

9 Peirce and Statistics

Henry E. Kyburg, Jr., University of Rochester

R. A. Fisher, in his last book, *Statistical Methods and Scientific Inference*, draws a fundamental threefold distinction (1956:35):

> It may be that the data are such as to allow us to apply Bayes' theorem. . . . [And] we may be able validly to apply a test of significance to discredit an hypothesis the expectations from which are widely at variance with ascertained fact. . . . [This] does not authorize us to make any statement about the hypothesis in question in terms of mathematical probability . . . [and] the logical situation we are confronted with may admit of the consideration of a series, or . . . continuum of hypotheses, one of which must be true, and among which a selection may be made, and that selection justified, so far as may be, by statistical reasoning.

We can find passages in Peirce that reflect each of these distinctions, although it is hard to be sure that they were ever held simultaneously, since Peirce's views changed over time.

I

It might be thought, since Peirce has little to say about the use of Bayes's theorem, that the first of these approaches was foreign to his thought. But Fisher himself was no advocate of free-wheeling Bayesianism and advocated the use of Bayes's theorem *only* when the required antecedent probabilities were available. He would agree with Peirce's thesis in "The Fixation of Belief" (*CP* 5.358–87) that all our knowledge is based on observable fact; "that experience alone teaches us anything" (CP 5.360). The antecedent probabilities, for Fisher, were to be based on observed facts, and so the use of Bayes's theorem provided no magical solution to the problem of induction.

If we suppose, however, that the antecedent probabilities required for Bayes's theorem were factual in character, then the application of the theorem is a purely straightforward instance of direct inference, about which Peirce had relatively few misgivings. The misgivings he did have are important and well worth noting. But let us begin with the demonstration that, in fact, some uses of Bayes's theorem can be accounted for in terms of direct inference.

We construe direct inference as inference of the form:

> The proportion p of M's are P's.
> S is an M.
> It follows, with probability p, that S is a P. (see *CP* 2.694–97)

Let us consider a plausible example of Bayesian inference, in which the antecedent probabilities can sensibly be said to be known. Suppose, for simplicity, that the boxes of bolts produced by a certain factory contain either 10 percent defective bolts or 30 percent defective bolts. Reasonably enough, boxes containing 30 percent defective bolts are rare—say that they occur only 5 percent of the time.

As buyers of boxes of bolts, you and I are concerned to avoid the 30 percent boxes. We can employ a method of sampling and Bayesian inference to reduce the likelihood that we will get a 30 percent box below the 5 percent produced by the factory. The procedure is that we draw three bolts at random and examine them for defects. We use the results as a basis for inference.

Of course, what we are ultimately after in this context is a decision rather than an inference—a decision to accept or not to accept that box of bolts. But we can view that decision as being based on, as incorporating, an inference concerning the likelihood that the box is a 30 percent box.

Bayes's theorem allows us to compute the probability that we have a 30 percent box, given the evidence we get by drawing three bolts at random.

Number of defects	Probability of 30 percent box
0	0.024
1	0.087
2	0.316
3	0.500

Let us see how this is just unadorned direct inference. Consider the general reference class to be the following: the set of pairs consisting of the selection of a box and the selection of a sample of three from *that* box, where each box is paired with each distinct sample that could be drawn from it. In the general reference class, 95 percent of the pairs consist of a 10 percent box paired with something, and 5 percent consist of a 30 percent box paired with something. Call this set *G*.

But when we have drawn a sample, the *general* reference class is no longer the appropriate one. Peirce does not appear to make much of this principle of looking for the reference class that embodies as much of our knowledge as possible, but Fisher clearly does. He invokes the principle of "recognizable subsets," to which we will return.

In the present case, if we have drawn a sample of three bolts and found two to be defective, we should look at the marginal distribution of 30 percent boxes. That is, we should compute the proportion of pairs of which the first member is a 30 percent box, not in our general reference class, but in that subset of our general reference class in which the second member of the pair (the sample) contains two defective bolts.

It is important to see that what is going on here, although the computation is exactly in accordance with Bayes's theorem, is simply an instance of direct inference. The general class *G*, prior to the examination of a sample, yields the prior probabilities of 30 percent and 10 percent boxes. This class of pairs may be divided into four parts, according to the number of defective bolts in the sample that constitutes the second member of the pair. Call these G_0, G_1, G_2, and G_3. Before we have examined a sample from our box, we have no knowledge of which subset our box belongs to; *G* is the appropriate reference class. *After* we have examined a sample and found two defective bolts, G_2 is the appropriate reference class. In either case, we are doing direct inference.

Thus, if "the data are such as to allow us to apply Bayes's theorem," it seems quite clear that Peirce would have no more trouble than Fisher in accommodating the probable inference as a direct inference, based on known frequencies or chances.

There is a lot to be said about direct inference, however, and we shall return to Peirce's view of it later.

II

How about the second sort of inference: that in which we discredit a hypothesis by applying a test of significance and by finding results that are at variance with the expectations generated by the hypothesis?

Here again, Peirce is right in line with modern statistical thinking. He says, for example, "One [mode of inference] will carry the truth with it oftener than [another]. . . . We may, therefore, define the probability of a mode of argument [inference of a given kind] as the proportion of cases in which it carries the truth with it" (*CP* 2.647–57). And here is a quotation that could have been taken from Jerry Neyman: "Any individual inference must be either true or false. . . . in reference to a single case considered in itself, probability can have no meaning" (ibid.).

In a chapter entitled "Forms of Quantitative Inference" (cf. Peirce: "The Theory of Probabilities is simply the science of logic quantitatively treated"), Fisher (1956:60) writes of tests of significance, "They do not generally lead to any probability statements about the real world, but to a rational and well-defined measure of reluctance to the acceptance of the hypotheses they test." He refers to "the resistance felt by the normal mind to accepting a story intrinsically too improbable" and says that this improbability is "not capable of finding expression in any calculation of probability a posteriori" (ibid.:44).

Except for the fact that in tests of significance there is a measure, though not a probability, that reflects this reluctance, this sounds much like the attitude that Peirce takes toward plausibility, which he takes as being "entirely different" from probability. Peirce's example is too nice to be paraphrased. I quote it in full:

> Suppose a particularly symmetrical larch tree near the house of a great lover of such trees had been struck by lightning and badly broken, and that as he was looking sorrowfully out of the window at it, he should have happened to say, "I wonder why that particular tree should have been struck, when there are so many about the place that seem more exposed!" Suppose, then, his wife should reply, "Perhaps there may be an eagle's eyrie on some of the hills in the neighborhood, and perhaps the male bird in building it may have used some stick that had a nail in it; and one of the eaglets may have scratched itself against the nail; so that the mother may have reproached the male for using such a dangerous stick; and he, being vexed with her teasing, may have determined to carry the piece to a great distance; it may have been while he was doing this that the explosion of lightning took place, and the electricity may have been deflected by the iron in such a way as to strike this tree. Mind, I do not say that this is what did happen; but if you want to find out why that tree was struck, I think you had better search for an eyrie, and see whether any of the eaglets have been scratched." (*CP* 2.261)

Peirce says, "This is an example of as unplausible a theory as I can think of. We should commonly say it was highly improbable; and I suppose it would be so." There is more wrong with the explanation than its improbability, as Peirce also points out. There is no explanatory virtue, because it is "gratuitous" to suppose that the lightning was deflected at all (*CP* 2.662).

Here is another example mentioned by Peirce in this connection: A woman trickster, many of whose tricks had been explained away by a Mr. Carrington, could nevertheless do some things that Mr. Carrington could not explain. Carrington concluded that these were "supernormal." Peirce comments: "Granted that it is not yet proved that women who deceive for gain receive aid from the spiritual world, I think it more plausible that there are tricks that can deceive Mr. Carrington than that the Palladino woman has received such aid" (ibid.).

Compare Fisher (1956:40–41): "In the studies known as parapsychology enormous odds are often claimed, evidently with a view to raising the resistance felt to accepting what is intrinsically improbable to such a pitch that [an occult explanation], though itself repugnant, shall be accepted in preference."

Although there are no doubt differences between Peirce's "unplausibility" and Fisher's significance tests that could be brought out by more detailed exegesis, it certainly seems clear that both were sensitive to, and appreciative of, the mind's "reluctance" to accept what we might call tall tales, and both were disinclined to waste the quantitative notion of probability on this reluctance.

On the other hand, Peirce clearly stated that fractions between 1 and 0 represented "the degree in which the evidence leans toward one or the other side" (*CP* 2.647–57), and perhaps under some circumstances, small numbers could correspond to the reluctance to believe tall tales.

III

The third case considered by Fisher was that in which we must make a selection among a continuum of hypotheses and justify that selection, so far as possible, by statistical reasoning. Let us first look at Fisher's standard example. Suppose that we know that a certain quantity (function) X is distributed in a certain population according to a normal distribution of unit variance and unknown mean. We select an object a from that population and observe that $X(a) = 10$.

From the fact that X is distributed normally with unit variance, it follows that the quantity $Y = X - m(X)$—the difference between the X-value that an object has and the general mean value of X in that population—is distributed normally with unit variance and a mean of 0.

We have selected the object a. We can say that if a is a random member of the population concerned, the probability that it—that very object a—has a value of Y removed three standard deviations from the mean value of Y (recall that the mean of Y is 0) is .003 (from the table of the normal integral with 0 mean and variance 1). But to say that this is true of the object a, which is known to have an X-value of 10, is exactly to say that the mean of X does not lie between 7 and 13: $m(X) \leq 7$ or

$m(X) \geq 13$. Or, more helpfully, we can say that the probability is .997 that $7 \leq X \leq 13$.

Now Fisher, for some time, wanted to regard this as a special kind of probability: fiducial probability. Its validity depends on the fact that the object a is drawn at random from its population. Fisher provided a characterization of the relevant sense of randomness in various places. Here is one: "The subject of a statement of probability must not only belong to a measurable set, of which a known fraction fulfils a certain condition, but every subset to which it belongs, and which is characterized by a different fraction, must be unrecognizable" (Fisher 1956:57). Fisher originally thought that "fiducial" and other probabilities were different, but he later wrote: "There are in reality no grounds for any such distinction" (ibid.:56).

Now it is not clear that Peirce thought that probabilities could ever become so close to 1 as to amount to practical certainties, but he certainly did believe in practical certainties. He referred to them as "likelihoods."

> The kind of reasoning which creates likelihoods by virtue of observations may render a likelihood *practically* certain—as certain as that a stone let loose from the clutch will, under circumstances not obviously exceptional, fall to the ground—and this conclusion may be that under a certain general condition, easily verified, a certain actuality will be probable, that is to say, will come to pass once in so often in the long run. One such familiar conclusion, for example, is that a die thrown from a dice box will with a probability of one-third, that is, once in three times in the long run, turn up a number . . . that is divisible by three. (*CP* 2.664)

Note that there is no suggestion on Peirce's part that this "practical certainty" was the same concept as implausibility. In fact, Peirce goes to some length to distinguish them. It is also the case that Peirce did not want to identify practical certainty with any sort of probability. This was true, also, for Fisher in 1930. It was only much later that Fisher realized that the constraints he felt necessary to impose in order to derive fiducial probabilities from measures were just the same as the constraints needed in general to derive probabilities from measures.

IV

What can we say about the constraints that Fisher found necessary to bridge the gap between measures and probabilities? And did Peirce consider a similar problem? It seems to me that there is something to be said here. The problem is essentially the problem of "direct inference"—that is, inference from knowledge of a long-run frequency, or a "would-be," to an assertion of probability, properly so-called. Both Peirce and Fisher have expressed views on this matter.

Fisher's views (1956:57) are neatly encapsulated in the quotation already offered: "The subject of a statement of probability must not only belong to a measurable set, of which a known fraction fulfils a certain condition, but every subset to which it belongs, and which is characterized by a different fraction, must be unrecognizable."

This is not quite adequate, since it does not take account of *approximate* knowledge. If we recognize, as is only plausible, that we may know neither that one fraction is "different" from another nor that they are "the same," then we must account for a number of different cases: first, when one bit of knowledge concerning measures is inconsistent with another (there is no possible world in which both can be true); second, when the two items of knowledge are strongly consistent (one known interval falls entirely within the other); and third, when the intervals are known to overlap.

I have proposed that when intervals overlap we regard that as a matter of conflict, and that when all conflicts are dealt with (essentially by means of the principle that subsets are more relevant than their supersets), then we should look at the most precise knowledge that we have. Combined with the principle that statements known to have the same truth value should have the same probability (a principle held in common by any classical probability), this yields a natural and plausible construal of probability.

Peirce's treatment of direct inference leads to the same point. "In probable reasoning . . . account has to be taken of various subjective circumstances—of the manner in which the premisses have been obtained, of there being no countervailing considerations, etc.; in short, good faith and honesty are essential to good logic in probable reasoning"

(*CP* 2.694–97). We have here a clear statement of the principle of total evidence: we are obliged to take account of all of the knowledge we possess. And we also have here a clear foreshadowing of the idea that probable reasoning is nonmonotonic: that taking account of more evidence may undermine a conclusion we have already accepted.

It is not easy to know how to take account of "all our knowledge." It seems fairly clear that even Fisher's proposal to take account of subsets of a potential reference class is inadequate, except in the very special case of complete statistical knowledge. Peirce's exhortation "Be honest!" is too general to offer much guidance in particular cases.

V

What is of interest in all this is the fact that Peirce, at the end of the nineteenth century, and R. A. Fisher, in the early years of the twentieth, segmented uncertain reasoning in much the same way. Peirce's "unplausibility" is driven by the same intuitions as Fisher's "level of significance," and both are sharply distinguished from probability. Peirce's "likelihood" seems to correspond quite closely to Fisher's "fiducial probability," as Fisher construed it in 1930, and is the notion that can lead to conclusions as certain as that the stone will drop. Again, both implausibility and likelihood were to be distinguished from probability, about which, again, our two heroes appear to have been largely in agreement. It is only later that Fisher came to see "fiducial probability" as essentially the same as any other probability. Now Peirce did, and Fisher did not, consider that likelihood, or practical certainty, could give rise to rational acceptance. This is not a doctrine that Fisher rejected, but one that he did not think about. And Peirce did not, as Fisher did, see how to tie likelihood and probability together.

Reference

Fisher, R. A. 1956. *Statistical Methods and Scientific Inference.* New York: Hafner Publishing.

10 Peirce's View of the Vague and the Definite

Joseph Margolis, Temple University

A very good case can be made for the claim that the entire philosophical effort of Charles Sanders Peirce is ultimately focused on, and returns again and again to clarify, the ubiquitous relationship between the vague and the definite and the general and the determinate. Saying so is itself an instance of the kind of executive heuristic Peirce favored in his own search for a conceptual map of the whole of effective human intelligence, of the whole of the intelligent world. He certainly regarded the world as too complex in its structure, too instantly altered by every effort to understand and interpret it, to admit any conceptual mapping as final and true. Every conceptual scheme owes its apparent precision and scope to what, finally, is the link between the vague and the general, on the one hand, and the definite and the determinate, on the other. And yet his own inquiries manifest what many regard as an unresolved paradox, staining the various stages of his work; for he seems to be an advocate at one and the same time of a kind of scientific precision that aims at the definite in all discursive work and yet that regards the vague as insuperable—and that treats the vague and the definite as opposed.

In electing to review the matter and its importance, we are well advised to treat these distinctions congruently, that is, as a provisional heuristic aspiring to the precision of the definite and determinate within the encompassing space of the vague and general, which suggests that the two sets of notions are not really opposed. That seems, in fact, to be Peirce's own best vision. It is also a heuristic that fits extraordinarily well the characteristic thrust of nearly all contemporary philosophical programs, whether they are aware of it or not. In any case, it affords an extremely helpful way of organizing an inclusive picture of the principal puzzles of current philosophy.

The essential themes are quite detachable from Peirce's own account, though there can be no question that his own often characteristic mixture of the florid and the lucid never obscures how remarkably apt his metaphysics, semiotics, and logic are for promoting the lessons of these distinctions. In any event, they include the following:

1. Thinking serves effective life and is a real part of the world in which it functions.
2. In both thought and nature, nothing exists or is intelligible (in effect, nothing is humanly interpretable) that is not relationally structured, hence triadic.
3. The cognizing mind and the cognized world cannot be disjunctively sorted.
4. The discursive elements of thought—words, terms, concepts, propositions, arguments—are linguistically abstracted from, and applied to, the symbiotic continuum of nature and thinking.
5. Linguistic abstraction and application are interpretive or semiotic processes capable of infinite extension or of deliberate finite closure.
6. With regard to the relational continua of thought and nature to which it refers, language is inherently defective where it is most precise—because, in being definite, it remains vague.
7. General distinctions of language express and inform general habits of action, so that habits of mind and action are also defective in a way resembling language.
8. The canons of formal logic (deduction) are correspondingly defective.
9. Both in reasoning and in action, the corrective requires a capacity to apply or reinterpret the fixed, abstracted, detached, and formalized regularities of language to and within the relational order of an evolving world.

These nine themes are not peculiarly Peircean, except in the important sense that there probably is no other philosopher who has analyzed or organized these lessons in as original and prophetic a way as Peirce. They collect the convergent force of Peirce's own theories of signs, of the continuum, of abduction, and of thirdness. But they also set many of the problems of current philosophy. To exhibit the strategic role of the general and the vague, therefore, is to gain a strong sense of the unity of Peirce's own system and something of its permanent contribution to the ongoing work of philosophy.

I

"By all odds," says Peirce, "the most distinctive character of the Critical Common-sensist, in contrast to the old Scotch philosopher [Thomas Reid], lies in his insistence that the acritically indubitable is invariably vague" (*CP* 5.466). Peirce has a great deal to say about vagueness, though his claim to "have worked out the logic of vagueness with something like completeness" seems to have puzzled his editors (*CP* 5.506). The following remarks may be his most succinct and provocative on the topic:

> Perhaps a more scientific pair of definitions would be that anything is general in so far as the principle of excluded middle does not apply to it and is vague in so far as the principle of contradiction does not apply to it.
>
> In those respects in which a sign is not vague, it is said to be definite, and also with a slightly different mode of application, to be precise [in effect, precised]. (*CP* 5.448–49)

Peirce chides the logicians for neglecting "the study of vagueness, not suspecting the important part it plays in mathematical thought. It is the antithetical analogue of generality" (*CP* 5.505). In an unpublished paper, he expands the definitions just given in a most instructive way.

> A sign is objectively general, in so far as, leaving its effective interpretation indeterminate, it surrenders to the interpreter the right of completing the determination for himself. "Man is mortal." "What man?" "Any man you like." A sign is objectively vague, in so far as, leaving its interpretation more or less indeterminate, it reserves for some other possible sign or experience the function of completing the determination. "This month," says the almanac-oracle, "a great event is to happen." "What event?" "Oh, we shall see. The almanac doesn't tell that."

> The general might be defined as that to which the principle of excluded middle does not apply. A triangle in general is not isosceles nor equilateral, nor is a triangle in general scalene. The vague might be defined as that to which the principle of contradiction does not apply. For it is false neither that an animal (in a vague sense) is male, nor that an animal is female. (*CP* 5.505)

These comments would be quite unsatisfactory on a certain canonical reading. It might be argued that, although the "general" is indeterminate as far as definite instantiations of its range of application are concerned, its instantiations are, wherever real, definite—whether we are cognizant of that fact or not; and that, although the "vague" is left indefinite or indeterminate enough to disallow the exclusion of one or the other of contradictory attributes, nothing insofar as it is real can actually possess such paired attributes—again, whether we know or not which member of the contradictory pairs it possesses. On that reading, neither excluded middle nor contradiction is waived or abandoned. Peirce would reject this countermove, not because he would disallow the counterclaim itself taken in the narrowest sense (i.e., on its own supposition), but because he would allow it to acquire the function it has only when placed within an ampler context of encountered reality—or, of our being able to speak of reality—in which it does not yet have application. So Peirce means in a realist sense just what he says: the real world exhibits generality and vagueness; it is not primarily a question of our contingent knowledge. Alternatively put, thinking is itself a part of nature and affects the structure of nature; but we misread the precision of propositions (hence, of formal reasoning as well) if we fail to remember its natural source and natural function.

Part of Peirce's point is that signs are real enough, and they may be (and are) general and vague; part of his point is that the real is always of the nature of a sign—and so what is real may be (and is) general and vague; and part of his point is that the rigorous effort to interpret signs and the sign-infected nature of whatever is real, for the purpose of rendering it definite or determinate (as opposed to vague or general), is a deliberate (volitional) effort made within the encompassing space of the general and vague—an effort that cannot succeed, in real-time terms, in actually eliminating the essentially general or vague. So the paradox grows. At any rate, this way of explicating Peirce's subtle doctrine stands a good chance of making sense of his anti-Cartesian view that "the acritically indubitable is invariably vague," as well as of our suggestion that the vague and the definite are not exclusively opposed in Peirce.

II

Alonzo Church (1956:25 n. 67) saw in Peirce's 1885 paper "On the Algebra of Logic: A Contribution to the Philosophy of Notation" an analysis in which "the explicit use of two truth-values appears for the first time," a paper published six years before Frege's discussion of sentences as the names of truth-values, in *Funktion und Begriff*.[1] Peirce says there: "According to ordinary logic, a proposition is either true or false, and no further distinction is recognized. This is the descriptive conception, as the geometers say; the metric conception would be that every proposition is more or less false, and that the question is one of amount. At present we adopt the former view" (*CP* 3.365).

It is important to bear in mind that Peirce supposes that he is contributing to the precision of logical notation. He therefore introduces the disjunctive truth-values, true and false, in the very context in which definiteness and determinacy are normally provided (paradigmatically, in applications of deductive logic); in the same breath, however, he suggests an ampler context in which that very instrument is first formed and in which larger considerations bearing on its application to the real world are called into play. In logical space, then, according to the "Algebra of Logic" paper, a proposition, a sentence precised for definiteness, is taken to be "limited to that one individual state of things, the Actual"—ideally, an atomic state of affairs lacking, as such, all reference to the conditions under which actual states of affairs are precised so as to serve the usual function of propositions and to preserve (on that condition) the full pertinence of their truth-value assignment. But this is just what the common-sensist's insistence challenges; for, on Peirce's view, the indubitable is "invariably vague," that is, what, in the strong realist reading of the experienced world, yields the firmest knowledge and the most satisfactory true belief. Congruently, the real world is thoroughly vague and general, though it is also hospitable, within limits, to the logician's and scientist's deliberate effort. The world is really precised through science—but on the world's own terms. (It is, incidentally, a well-known fact that Frege never provided a sufficient motivation for a realist reading of the application of his own idealized logic to what Peirce had independently isolated as the generality and vagueness of the experienced world.)

So we may say, without yet having the full argument before us, that Peirce had apparently succeeded in formulating a very large vision of a natural world—which humans evolve in, make inquiry into, and alter by their inquiry and intervention—a formulation suitable for answering, in a nonpsychologistic way (in the sense Frege was to oppose), the relation between the precision and definiteness of the instruments of logic and the conditions for their application to the actual world. Peirce's solution is radically different from Frege's, although it makes room (in an unorthodox way) for Fregean precision.

In this regard, Peirce unmistakably stands at the very beginning of that enormously influential tradition of analytic philosophy that (understandably enough) has been worried about the realism (or the realist-preserving function) of formal logic and mathematics in application to the actual world that is, as in Frege, Carnap, Tarski, Wittgenstein, Quine, and an army of more recent enthusiasts. It was never the tautological character of deductive reasoning that worried or reassured Peirce. He seems to have been remarkably prompt in developing a truth-table analysis of necessary truth (*CP* 3.387; cf. Tarski, 1983:40 n. 2); it was rather the import of applying such precised instruments to, and within, the actual world. None of the figures of the tradition mentioned can claim a command of the metaphysics of thought equal to Peirce's. Put less grandly: none has explored the implications of treating thinking as a natural process blessed and cursed with the ability to generate isolatable sentences.

In fact, in a very short piece written in 1857, in which he considers very tentatively the difference between the meaning of "actual" and that of "real," Peirce pointedly says (the remark opposes the sense given above of "actual" and strongly suggests Peirce's own eventually ramified view) that

> actuality refers to the relation of the object [say, an object perceived or sensed] to our understanding, reality to the existence of the object itself. An examination of their opposites will best explain this. Reality is opposed to shadowiness and non-existence. By shadowiness I mean a sort of half way between existence and non-existence, in the object itself. Actuality stands between Possibility and necessity, and therefore refers you see to the modality of the conception. The same difference exists between asserting and affirming between being and existing ([which is] not strictly observed [in usage]). (*W* 1:18)

Here, at an extremely early date, Peirce in effect contrasts the formal logician's use of "actual" (as determinate and definite enough to support a bivalent logic) and a metaphysician's or, better, a common-sensist's use of "actual"—not explored in the "Algebra of Logic" paper—so as to worry the realist standing of excluded middle and contradiction.

III

A number of interlocking distinctions are needed here, if we are to grasp Peirce's mature view of the issues collected by reflecting on the definite and the vague. First of all, Peirce regarded his pragmatism, in particular his pragmatic semiotics, as not addressed to saying "in what the meanings of all signs consist, but merely to lay down a method of determining the meaning of intellectual concepts . . . those upon which reasoning may turn. Now all reasoning that is not utterly vague, all that ought to figure in a philosophical discussion involves, and turns upon, precise necessary reasoning" (*CP* 5.8).

Peirce does not rule out meaningful signs that cannot be given determinate or definite meaning. For, "the brutes use signs," he says, though "they perhaps rarely think of them as signs. . . . They do not criticize their thought logically. [Doing that marks a] grade of thinking about thought, which my logical analyses have shown to be one of the chief, if not the chief, explanation of the power of mathematical reasoning" (*CP* 5.534). Again, though "all thinking is by signs," and though thinking about thinking "is itself a phenomenon of self-control" (ibid.), there is in all of us a significant reserve of thought and belief that is not under our control (or under our logical or ethical self-control) that is (for us) "the absolute truth" for the interval affected—even if, later, what is now indubitable may be criticized and challenged: "In every stage of your excogitations, there is something of which you can only say, 'I cannot think otherwise,' and your experientially based hypothesis is that the impossibility is [in the nature of things]" (*CP* 5.419).

So Peirce gives experiential priority to the vague over the definite; treats the definite as a critical precising of the vague; construes that precising as a form of self-control, which explains the equation of the logical and the ethical (*CP* 5.419); admits that the objectively vague cannot be eliminated and, correspondingly, that the existence of signs that cannot be brought under voluntary control; treats the vague as actual or real, though it may be precised under favorable conditions; construes the work of logic and mathematics as changing the nature of what is actual by making it more definite; and therefore enlarges the conception of logic to include reflection on the conditions under which the rigor of formal deduction and related instruments are applied to the objectively vague and general.

In the context in which he develops these notions (remarkable enough in their implications), Peirce adds an absolutely extraordinary observation.

> Two things here are all-important to assure oneself of and to remember. The first is that a person is not absolutely an individual. His thoughts are what he is "saying to himself," that is, is saying to that other self that is just coming into life in the flow of time. When one reasons, it is that critical self that one is trying to persuade; and all thought whatsoever is a sign, and is mostly of the nature of language. The second thing to remember is that the man's circle of society (however widely or narrowly this phrase may be understood), is a sort of loosely compacted person, in some respects of higher rank than the person of an individual organism. It is these two things alone that render it possible for you—but only in the

abstract, and in a Pickwickian sense—to distinguish between absolute truth and what you do not doubt. (*CP* 5.421)

For one thing, the continuum of evolving selves helps to explain Peirce's early allusion to the "shadowiness" of things ("between existence and non-existence")—and, consequently, his associating (implicitly) the notion of reality with difficulties regarding the principle of contradiction and the notion of actuality with difficulties regarding the principle of excluded middle. For another, the passage clarifies the sense in which Peirce regards the questions he raises as having a full realist import, not merely a verbal or conceptual or cognitive role. For a third, it helps to explain the sense in which the triadic relation between sign, object, and mind is treated in two quite different ways: as "conventional or arbitrary" (one might add, as discrete, definite, deliberate, reflexive, controlled) when employed in the context of logical reasoning (*CP* 5.360–61) and as vague, general, dependent on habits, relational in a holist sense, inherently and objectively indeterminate—when functional in that sense in which we are confronted with the "acritically indubitable" in the middle of life. This is just the reason Peirce says that "a sign ['every kind of thought'], that is in any respect objectively indeterminate (i.e., whose object is undetermined by the sign itself) is objectively general in so far as it extends to the interpreter the privilege of carrying its determination further" (*CP* 5.447). In carrying "its determination" further, thinking becomes controlled, logical; also, it actually alters the properties of a part of the world.

Thinking does so most dramatically in the case of a person; on Peirce's view, a person is a sign interpreting other signs (including any earlier evolved self of his or her own continuous series of selves), a "thing" that is both eminently real and objectively indeterminate. In fact, on Peirce's rather ironic account, "Every thought . . . is, so far as it is immediately present, a mere sensation, without parts, and therefore, in itself, without similarity to any other, but incomparable with any other and absolutely sui generis." But if it is that, he adds at once, then, since it "is wholly incomparable with anything else[, it] is [also] inexplicable, because explanation consists in bringing things under general laws, or under natural classes. Hence every thought, in so far as it is a feeling of a peculiar sort, is simply an ultimate, inexplicable fact." Furthermore, Peirce says, "we can never think" such a thing: the ephemeral sensation is gone before it can be thought (mere firstness, in effect). For Peirce, then, two important consequences follow: first, thinking is always "mediate," linking what we take to be present, to things that are like it (under some generality); second, the meaning of whatever we think is only virtual, ultimately interminably so. There is no contradiction, Peirce insists, in "making the Mediate immediable"; and "the meaning of a thought is altogether something virtual" (*CP* 5.289; see also 5.504 n. 1), that is, made more determinate in a proximate sense only by some succeeding interpretant.

More profoundly still, Peirce draws out of this reflection the notion—implied in every contemporary version of the realist/idealist symbiosis (i.e., the supposed impossibility of sorting subject and object in some disjunctive way that could ground a simple correspondence theory of truth)—that "everything which is present to us is a phenomenal manifestation of ourselves does not prevent its being a phenomenon of something without us, just as a rainbow is at once a manifestation both of the sun and of the rain" (*CP* 5.283).[2] Since no thought can be instantaneous, "there is no intuition or cognition not determined by previous cognitions," "every thought-sign is translated or interpreted in a subsequent one, unless it be that all thought comes to an abrupt and final end in death," and so "in every case the subsequent thought denotes what was thought in the previous thought" (*CP* 5.283–85).

Correspondingly, in a famous remark, Peirce declares that "the individual man, since his separate existence is manifested only by ignorance and error, so far as he is anything apart from his fellows, and from what he and they are to be, is only a negation" (*CP* 5.327).[3] This catches up Peirce's very early speculation about the use of "being" and "reality" as distinct from the use of "actuality"—in terms of his synechism, which, as he says, "can never abide dualism," neither between being and not-being (against Parmenides), nor between the mental and the physical (against Descartes), nor between individual selves or individual events and phenomena (against nearly all familiar accounts) (*CP* 7.569–73). There are differences, Peirce concedes, or different weightings of given characteristics, but there are no "different categories of substance" (*CP*

7.570). It is, rather, the continuity of a certain regular order of experience and behavior that individuates—without altogether eliminating indeterminacy—one self from another, as, with respect to other features we individuate, one atom from another (*CP* 3.613).[4]

IV

Peirce's theory of signs and their interpretation is, to say the least, rather complicated. But there can be no question that his account depends upon and employs his distinction between the vague and the definite. We must bear in mind that, on Peirce's view, "the entire universe—not merely the universe of existents . . . —all that universe is perfused with signs if it is not composed entirely of signs" (*CP* 5.448 n. 1). This, of course, is Peirce's semiotized version of the realist/idealist symbiosis. By the theory of the triadic nature of every sign situation, by the doctrine of synechism, and by the doctrine that every sign represents a manifestation of ourselves (whatever else it represents), every sign situation is open to infinite interpretation; a fortiori, all self-understanding is potentially infinite (*CP* 5.283; also *CP* 2.228).

And yet Peirce raises the question of bringing interpretation to an end. There are actually two ways in which Peirce considers the end of interpretation: in one, in the circumstances of ongoing inquiry; in the other, at the limit of infinite inquiry. In both, he introduces the joint issue of truth and reality. Regarding the first possibility, he asks us, in order to fix our ideas, to consider "the map-coloring problem" (the so-called four-color problem). The particular puzzle is not essential, except that it requires demonstrative knowledge (deduction), it might or might not lead to a correct general rule but is not known to be solvable by the application of a general rule covering a known class of cases, and it probably is "indefinite" as far as any familiar way of handling problems is concerned (*CP* 5.490). About such a puzzle, construed as a sign-situation—that is, one to be "interpreted," solved in effect—Peirce asks us to consider what we should regard as its final interpretant. His answer is very delicately balanced and very complicated.

He concedes, first of all, that "a concept, proposition, or argument may be a logical interpretant" (of that particular puzzle or, in general, of similar problems). "I only insist," he says, "that it cannot be the final logical interpretant, for the reason that it is itself a sign of that very kind that has itself a logical interpretant" (*CP* 5.491). What Peirce obviously means is that a logical interpretant of that form that is definite only in the way of a proposition introduces an infinite regress of similar interpretants (by way of formal connection). It does so because the proposition in question is not yet interpreted in such a way, with regard to the particular interests, perceptual beliefs, and related constraints that bear on an actual puzzle, that it can be made to yield a definite and determinate solution. It cannot do that because, in the context of action, it remains inappropriately general and vague—because all logical interpretants are (relevantly) general, if not also vague. What is wanted is a thoroughly determinate "solution." Peirce concludes that "the real and living logical conclusion [or interpretant] is [a] habit; the verbal formulation merely expresses it." "The conclusion (if it comes to a definite conclusion)," Peirce says, "is that under given conditions, the interpreter will have formed the habit of acting in a given way whenever he may desire a given kind of result." That is "the final logical interpretant" (*CP* 5.491). It is not, of course, the final interpretant.

The reader may well ask: what sort of solution is that? Peirce's explanation runs as follows:

> The habit alone, which though it may be a sign in some other way, is not a sign in that way in which that sign of which it is the logical interpretant is the sign. The habit conjoined with the motive and the conditions has the action for its energetic interpretant; but action cannot be a logical interpretant, because it lacks generality. The concept which is a logical interpretant is only imperfectly so. It somewhat partakes of the nature of a verbal definition, and is as inferior to the habit, and much in the same way, as a verbal definition is inferior to the real definition. The deliberately formed, self-analyzing habit—self-analyzing because formed by the aid of analysis of the exercises that nourished it—is the living definition, the veritable and final logical interpretant. (*CP* 5.491)

What Peirce means is that the habit, though general—hence, a logical interpretant—is itself interpreted by an individual act. Never itself general, an act is not a logical interpretant. Since the habit that it interprets can (and must) be so interpreted, habits are logical interpretants different from, and superior to, propositions and arguments, which express them "externally." Ultimately, therefore, it is Peirce's pragmatist conception of the function of thought and logic and inquiry that leads him to his particular thesis as well as to the pivotal role of the general and the vague.

This is an extraordinarily refined and tortured argument. But Peirce is drawing attention to the fact that (1) there is no real difference between logic and its application to life (another instance of synechism); (2) propositions and habits are, in different ways, general and vague: in fact, the first expresses the second; and (3) the interpretation required in contexts of application or of action must eliminate both generality and vagueness. Only a particular action can do that, in the sense that it brings the confronted puzzle to a close. (Doubtless, this is close to what Dewey must have had in mind in introducing his own notion of a "problematic situation.")[5] On the argument, only a habit can draw on pertinent experience and interest with a view to narrowing the options of action. The result is rational behavior or activity and entails what may be called a sort of pragmatic consistency—the very point of the "logic" of the ancient practical syllogism. Aristotle's thesis is very close to Peirce's purpose in assigning an executive role to abduction over deduction and induction, though Peirce does not seem to have discussed the practical syllogism with his usual attentiveness.[6]

In any case, this solution per habit and action cannot bring the infinite process of inquiry to a close. It does, however, draw attention to the asymmetry of the structures of propositions and thinking, which decides the fate of logic. Action brings inquiry to a finite end, but inquiry is potentially infinite. The first resolves generality and vagueness—"energetically"; the second is inherently general and vague—in the "logical" issue.

V

The doctrine just sketched is in reasonable accord with the pragmatism of the *Popular Science Monthly* series, titled "Illustrations of the

Logic of Science," except for the fact that Peirce changed a number of key views between 1877–78 and 1905–6, the time the *Monist* series on pragmatism was published. For example, Peirce observes, in "The Fixation of Belief," speaking of "the end of inquiry," that "the irritation of doubt causes a struggle to attain a state of belief." That struggle he terms "inquiry" (*CP* 5.374). In "How to Make Our Ideas Clear," he adds: "The whole function of thought is to produce habits of action" (*CP* 5.400), which leads fairly perspicuously to the thesis we were just rehearsing. The conceptual discrepancies between the earlier and later views really reflect Peirce's growing awareness of the sheer complexity of his own claims: for instance, of the need to resist the idea (in "The Fixation of Belief") that "there are Real things, whose characters are entirely independent of our opinions about them," and that "the method [of inquiry] must be such that the ultimate conclusion of every man shall be the same" (*CP* 5.384). On the issue we are canvassing, the congruity between the early and later views is distinctly stronger. Thus, in explaining "habits of action," Peirce says: "The identity of a habit depends on how it might lead us to act, not merely under such circumstances as are likely to arise, but under such as might possibly occur, no matter how improbable they may be. What the habit is depends on *when* and *how* it causes us to act. As for the *when*, every stimulus to action is derived from perception; as for the *how*, every purpose of action is to produce some sensible result" (*CP* 5.400). There is at least a loose allusion here to the need to give greater determination to the generality ("when") and the vagueness ("how") of any habit of thought.

That is the heart of the matter. But what it means is that habits have two sorts of indeterminacy. First, their generality is a function of their scope of application: change the circumstances in which they apply, and they change. This provides an analogy to Peirce's view of the generality of propositions—hence, of the generality of logic itself. It is for this rather unexpected reason that *the very generality of propositions*, detached from the circumstances of action or doubt (or of habits keyed to action or doubt), *invariably raises the specter of the inapplicability of the principle of excluded middle.* Because generality cannot, by itself, determine its own scope, it cannot determine whether it applies to this case or that. (We must suppose a structural similarity here between propositions and habits as *logical* interpretants of a given sign situation.) The end of inquiry, however, is the production of an opinion (or a habit of action) suited to some particular doubt, which, on another view of Peirce's, already mentioned, signifies the nullity that is an individual self: the very reason Peirce gives for saying that "a person is not absolutely an individual" (*CP* 5.421).

Habits of action, then, are vague: one cannot tell from the general habit alone whether this particular action is or is not a pertinent ("energetic") solution of what, in actual life, is puzzling or doubt provoking. We need to know the when and how, and that depends on particular perceptions and particular purposes. So *the vagueness of a habit* (a fortiori, of a proposition that "expresses" a habit), detached from the particular perception and purpose implicated in a genuine doubt, *invariably raises the specter of the inapplicability of the principle of contradiction.* The indefiniteness of a generalized habit cannot by itself determine whether, when it is pertinently exercised in particular cases, it should affirm or deny the indicated application. Here, the puzzle of vagueness presupposes a relative resolution of the puzzle of generality.

It takes but a small step, now, to see the point of Peirce's grand thesis—that "the question of pragmatism . . . is nothing else than the question of the logic of abduction." On Peirce's view, pragmatism cannot interfere with the "logic" of induction or of deduction. Abduction is, in a sense, the "logic" of life and experience. *It cannot be merely formal*, and it cannot affect the range of abstractly possible experience. It concerns only "the admissibility of hypotheses to rank as hypotheses" (i.e., pertinently for new or puzzling circumstances) (*CP* 5.196). "Abduction," says Peirce, "consists in studying facts and devising a theory to explain them" (*CP* 5.145). A convenient way of putting Peirce's ingenious point is this: deductive reasoning, all propositions and verbal arguments, are confined to merely isolated (abstracted) linguistic encounters; but thinking (in the pragmatist's sense) is addressed to the larger, encompassing activity of interpreting the semiotic possibilities of the actual world—including the application of language to that world from which it is originally generated and misleadingly detached. *Deduction*, therefore, concerns only propositions and related linguistic counters—as if they could guarantee their own realist import; *abduction* concerns, rather, the full semiotic space in which the other is able to function at all. Hence, abduction is not a logic in the formal sense, though it can be formally represented. It involves firstness and secondness in a way deduction cannot.

The obvious problem of abduction is one of the fruitfulness and relevance of infinitely many possible hypotheses under real-time circumstances. But, more than that, since abduction is "the only logical operation which introduces any new idea," it affects both the relative determinacy of the general and the relative definiteness of the vague: "Its only justification is that from its suggestion deduction can draw a prediction which can be tested by induction, and that if we are ever to learn anything or to understand phenomena at all, it must be by abduction that this is to be brought about" (*CP* 5.171).

All this pretty well comes to the plain thesis that logic is an activity rather than a formal canon, and that it depends for its force on how it is interpreted both with regard to its range of application and with regard to what, in that range and relative to some particular purpose and belief, is to be affirmed or denied. Abduction is not capable of being formalized; for that very reason, it is the sole discipline of productive inference (whatever that may be) by which deductive and inductive logic bear at all on whatever confronts us as a puzzle in human experience. For neither deduction nor induction can claim such an ability (*CP* 5.181, 189).[7]

VI

Returning, now, to the first way in which Peirce says that inquiry may be brought to an end: it would not be misleading to maintain that truth and reality are provisionally what we affirm in a finite span of inquiry—that is, in inquiry brought to a close by resolving a genuine doubt. This is why, even in the context of "How to Make Our Ideas Clear," Peirce explains these dual notions in terms of an infinite inquiry, though in a context essentially addressed to inquiries of the finite sort.

> No modification of the point of view taken, no selection to other facts for study, no natural bent of mind even, can enable a man to escape the predestinate opinion. This great hope is embodied in the conception of

truth and reality. The opinion which is fated to be ultimately agreed to by all who investigate, is what we mean by the truth, and the object represented in this opinion is the real. (*CP* 5.407)

Certainly, Peirce's sense here is that truth and reality are virtually determined in finite contexts because they are "predestinate" in infinite inquiry. That must be the sense of Peirce's summary account: "Truth is a character which attaches to an abstract proposition, such as a person might utter. It essentially depends upon that proposition's not professing to be exactly true" (*CP* 5.565).

Peirce's best view of the matter, published in Baldwin's Dictionary (1901), confirms our entire reading:

> Truth is that concordance of an abstract statement with the ideal limit towards which endless investigation would tend to bring scientific belief, which concordance the abstract statement may possess by virtue of the confession of its inaccuracy and one-sidedness, and this confession is an essential ingredient of truth. . . . Reality is that mode of being by virtue of which the real thing is as it is, irrespectively of what any mind or any definite collection of minds may represent it to be. (*CP* 5.565)

His point is the extremely subtle and complex one that (1) what holds for scientific truth holds for normative truth (morality); (2) "the ideal is real" (*CP* 5.566); (3) "the perfect doctrine can neither be stated nor conceived" (either in science or morality or practical life) (ibid.); (4) regarding facts and the summum bonum, we can only hope for an approximative form of progress toward truth (*CP* 5.608);[8] (5) a person's mind and "moral nature" develop by accepting the reality of the ideal (*CP* 5.566–68; cf. 5.213); and (6) perhaps most important for our present purpose, truth—more the product of abduction than of deduction or induction—depends on what (in either of the two senses of closing an inquiry) involves "all the operations by which theories and conceptions are engendered" (*CP* 5.590).

The upshot of all this, to put the point a little abruptly, is that Peirce regards the determination of truth and reality in finite contexts as "virtual" (*CP* 5.289) because they can be determined "perfectly" only at the end of an infinite inquiry. The reason is that *truth is, in the limit, rela-*

tional, internal to the whole of reality. Fortunately, given the difficulties of the doctrine of internal relations, the pragmaticist revision of infinite inquiry makes it quite clear that the limit cannot possibly be reached—and is intended to dramatize the point (optimistically) (cf. *CP* 2.203). The thesis is implicit in the profoundly relational meaning of the doctrine of synechism and of the theory of the infinite (made into a teleology and made endlessly novel) interpretation of signs (*CP* 5.311; cf. 2.203). But that (relational) reading of truth also explains the persistence, in Peirce, of the themes of the general and the vague, which, one may now see, bear directly on the inherent blindness and limited adequacy of any and every finite propositional determination of facts or directives for action—hence, the blindness of deduction itself. Propositions are defective because the meaning of what they affirm depends inherently (in the pragmatist vision) on determinations regarding excluded middle and contradiction that cannot be drawn from those abstract counters themselves.

VII

In a remarkably acute way, Peirce collects his entire philosophy—insofar as it bears on our present issue—in a small distinction between his own pragmaticism and Hegelian absolute idealism, with which it was bound to be confused. Pragmaticism, he says, involves "the recognition that continuity is an indispensable element of reality, and that continuity is simply what generality becomes in the logic of relatives, and thus, like generality, and more than generality, is an affair of thought, and is the essence of thought" (*CP* 5.436).

The point is immensely important, because Peirce means here to oppose the Kantian distinction between the analytic and the synthetic, as well as the entire canonical picture of deductive inference as "a following out of a rigid rule, so that machines have been constructed to draw conclusions. But this conception is not borne out by relative logic. . . . [In the logic of relatives] *matter entirely foreign to the premises may appear in the conclusion*" (*CP* 3.641). There is the connection between the superior function of abduction and the defect of deductive and inductive logic with regard to generality and vagueness.

What Peirce says here is also a very nice anticipation of W. V. Quine's (1953) treatment, for instance, of the empiricist dogma of the analytic/synthetic distinction and suggests the profound possibilities of Peirce's own argument. For example, Peirce claims that "deductive logic can really not be understood without the study of the logic of relatives" (in effect, the logic of the tacit context in which definite, seemingly fully determinate abstractions are treated according to formal rules). One mistake that results from its neglect, he says, is supposing that "demonstrative reasoning is something altogether unlike observation" (*CP* 3.641). By the logic of relatives, Peirce means to consider both a logic of relations and the relativity of fixing any nonrelational element in real contexts, which are always thoroughly relational (*CP* 3.636–41; 3.145–48). In fact, Peirce links the function of contradiction and excluded middle to the logic of relatives (*CP* 3.148). So it is very reasonable to construe his work on relative logic as forming a bridge between nonrelative deductive logic and the logic of abduction.

Proceeding from there, Peirce adds an intriguing observation about what pragmaticism is committed to (as distinct from Hegelianism):

> namely, that the third category—the category of thought, representation, triadic relation, mediation, genuine thirdness, thirdness as such—is an essential ingredient of reality, yet does not by itself constitute reality, since this category (which in [the] cosmology [given] appears as the element of habit) can have no concrete being without action, as a separate object on which to work its government, just as action cannot exist without the immediate being of feeling on which to act. The truth is that pragmaticism is closely allied to the Hegelian absolute idealism, from which, however, it is sundered by its vigorous denial that the third category (which Hegel degrades to a mere stage of thinking) suffices to make the world, or is even so much as self-sufficient. Had Hegel, instead of regarding the first two stages with his smile of contempt, held on to them as independent or distinct elements of this triune Reality, pragmaticism might have looked up to him as the great vindicator of their truth. (*CP* 5.436)

Here, Peirce suggests the analogy between his treatment of generality and vagueness and his treatment of firstness (feeling or perception), secondness (action or reaction), and thirdness (habit or representation) within the triune whole of thought and reality. He means to insist (in accord with scholastic realism) "that general principles are really operative in nature" (*CP* 5.101); to connect generality and thirdness, in that "the general is essentially predicative and therefore of the nature of a representamen" (*CP* 5.212); to link pragmatism to the abductive study of thirdness (as an irreducible element both of thought and reality), strategically important because "the conformity of action to general intentions is as much given in perception as is the element of action itself, which cannot really be mentally torn away from such general purposiveness" (*CP* 5.102); to construe human self-understanding conformably with the notion of a real community of signs, for to say that "no cognition of ours is absolutely determinate" signifies that "man is a sign," in fact man is the thought he uses—a thought or sign that "has only a potential existence, dependent on the future thought of the community" (*CP* 5.314–16).

In pursuing these issues, Peirce distinguishes pragmaticism from Hegelian idealism in the sense that, in the first but not the second, what instantiates firstness and secondness are not mere artifacts of thirdness (*CP* 5.79). Nevertheless, Peirce continues, "not only does Thirdness suppose and involve the idea of Secondness and Firstness, but never will it be possible to find any Secondness or Firstness in the phenomenon that is not accompanied by Thirdness." Contrary to Hegel, however, thirdness is not "the one and sole category" (*CP* 5.90). So Peirce rejects "the idea that the Absolute is One" and replaces it with an evolutionary notion governed by the law of continuity—or habit of mind—which, over time, gathers all firstness and secondness in the formations of thirdness: "The one primary and fundamental law of mental action," says Peirce, "consists in a tendency to generalization. . . . The consciousness of such a habit constitutes a general conception" (*CP* 6.21).

The evolutionary convergences of thirdness, in which alone the distinctions of firstness and secondness obtain at all, signifies the most inclusive sense in which, in all finite intervals of inquiry, generality and vagueness cannot fail to mark the unavoidable defects of discursive reasoning. At the limit of time, a completely adequate bipolar logic may be counted on, but not before: excluded middle and contradiction are forever at risk in the process of natural thought—which is to say, in whatever realism we may assign to thinking.

Notes

1. This matter is expanded somewhat in Turquette 1964. I have benefited from Turquette's survey.

2. See, for example, Putnam 1981:chap. 3 for an account of Putnam's "internalism," which is clearly indebted to Peirce.

3. See Peirce's remarks on children, *CP* 5.227–36.

4. "Everything whose ideality consists in a continuity of reactions will be a single logical individual" (*CP* 3.613).

5. See Dewey 1938, particularly chaps. 1–2 and p. 9 n. 1.

6. See Hare 1952:56–59 for a limited contemporary use of the notion.

7. On Peirce's formal treatment of induction in terms of probabilistic inference, see Cheng 1969.

8. Peirce adds elsewhere: "The logical atom, or term not capable of logical division, must be one of which every predicate may be universally affirmed or denied." That is, it would preclude vagueness altogether. But, says Peirce, "A logical atom, . . . like a point in space, would involve for its precise determination an endless process. We can only say, in a general way, that a term, however determinate, may be made more determinate still, but not that it can be made absolutely determinate" (*CP* 3.93). For any reasonably determinate (but not atomic) term, however, one would need to call on *all* its determinations to ensure the applicability, at the point of *every* predication, of the principle of excluded middle. This is an extraordinarily effective reductio of Wittgenstein's project in the *Tractatus*.

References

Cheng, Chung-Ying. 1969. *Peirce's and Lewis's Theories of Induction*. The Hague: Martinus Nijhoff.

Church, Alonzo. 1956. *Introduction to Mathematical Logic*. Vol. 1. Princeton: Princeton University Press.

Dewey, John. 1938. *Logic: The Theory of Inquiry.* New York: Henry Holt.

Hare, R. M. 1952. *The Language of Morals.* Oxford: Clarendon.

Putnam, Hilary. 1981. *Reason, Truth, and History.* Cambridge: Cambridge University Press.

Quine, Willard Van Orman. 1953. Two Dogmas of Empiricism. In *From a Logical Point of View*, 20–46. Cambridge: Harvard University Press.

Tarski, Alfred. 1983. Investigations into the Sentential Calculus. In *Logic, Semantics, Metamathematics*, 2d ed., trans. J. H. Woodger, ed. John Corcoran, 38–59. Indianapolis: Hackett.

Turquette, Atwell R. 1964. Peirce's Icons for Deductive Logic. In *Studies in the Philosophy of Charles Sanders Peirce: Second Series*, ed. Edward C. Moore and Richard S. Robin, 95–108. Amherst: University of Massachusetts Press.

11 The Test of Experiment: C. S. Peirce and E. S. Pearson

Deborah G. Mayo, Virginia Polytechnic Institute and State University

Peirce's philosophy of experimental testing shares a number of striking features with the Neyman and Pearson theory of statistics developed in the 1930s. For both, statistical methods provide, not means for assigning degrees of evidential support, confirmation, or belief to hypotheses on the basis of data, but procedures for testing (and estimating), whose rationale is their predesignated high frequencies of leading to correct results in some hypothetical long run. Isaac Levi is perhaps the first philosopher to have drawn out these key parallels between Peirce and Neyman-Pearson theory: "Peirce's inductions are inferences according to rules specified in advance of drawing the inferences where the properties of the rules which make the inferences good ones concern the probability of success in using the rules. These are features of the rules which followers of the Neyman-Pearson approach to confidence interval estimation would insist upon" (1980:138). Ian Hacking (1980) also discusses philosophical as well as historical connections between Peirce and Neyman and Pearson. Nor is it just the general view of statistical method that Peirce shares with Neyman and Pearson. One finds specific examples in Peirce that anticipate the mathematics of Neyman-Pearson hypothesis tests and confidence-interval methods. While tracing these mathematical ideas in Peirce is of interest in its own right, my focus in this chapter is to suggest what Peirce's experimental philosophy has to offer to contemporary debates between Neyman-Pearson and its rival statistical philosophies.

To get at the main issues, consider a Neyman-Pearson test of hypothesis *h*. The test is defined by a rule that indicates which of the possible experimental outcomes will be taken to reject *h* in favor of some alternative hypothesis *j*, and which will be taken to accept *h*. The cornerstone of Neyman-Pearson tests is their ability to ensure, *before the trial is made*, that (regardless of the true hypothesis) the rule will erroneously reject *h* and erroneously accept *h* no more than with some specifiably small probabilities. The probability of an erroneous rejection of *h*—the Type I error—and the probability of an erroneous acceptance of *h*—the Type II error—are the test's *error probabilities*. Error probabilities are not probabilities of hypotheses but the probabilities with which certain results would occur in some sequence of applications of such tests.

Despite the widespread use of Neyman-Pearson methods, most philosophers of statistics have questioned their relevance for statistical inference as opposed to the process of making routine decisions. A theory of statistical inference, it is typically thought, should provide measures of the *evidential strength* that data afford hypotheses. Examples of such evidential-strength measures are posterior probabilities and likelihood ratios. Let us call this view the evidential-strength view of statistical inference. Since Neyman-Pearson tests fail to provide measures of evidential strength, but only give error probabilities of procedures, they are inadequate for the evidential-strength view of statistical inference.

Neyman himself held a statistical philosophy that encouraged such criticisms. He continually denied that the tests provided methods for inductive inference, but rather construed tests as rules for *inductive behavior*. For Neyman (1950:258), "The problem of testing a statistical hypothesis occurs when circumstances force us to make a choice between two courses of action: either take step *A* or take step *B*." These are not decisions to accept or believe that what is hypothesized is approximately true or not, Neyman stresses. Rather, "to accept a hypothesis means only to decide to take action *A* rather than action *B*" (ibid.:259). The criticisms of Neyman-Pearson theory are largely directed at Neyman's *inductive-behavior* model of the tests. The critics admit that the Neyman-Pearson test may be sensible in certain decision-theoretic contexts, the paradigm example being acceptance sampling in industrial quality control. Here the choice is whether or not to reject a certain batch of products as containing too many defectives, say, for shipping. The primary interest is ensuring that the long-run risks of such business decisions are no more than can be "afforded," and in such cases, Neyman-Pearson tests can provide the desired guarantees. But the process of testing claims in a scientific context does not seem to have these features.

Although Neyman-Pearson tests are typically associated with this behavioral-decision interpretation, E. S. Pearson (not to be confused with his father, Karl), cofounder of the Neyman-Pearson methods, himself rejected it. In responding to Fisher's (1955) criticism of the term "inductive behavior," Pearson (1955:207) declared, "This is Professor Neyman's field rather than mine." Pearson stressed that, from the start, the Neyman-Pearson theory—with its introduction of the two types of error—reflected his view of statistical *inference*. According to him: "It was not until after the main lines of this theory had taken shape . . . that the fact that there was a remarkable parallelism of ideas in the field of acceptance sampling became apparent. Abraham Wald's contributions to decision theory of ten to fifteen years later were perhaps strongly influenced by acceptance sampling problems, but that is another story" (Pearson 1955:205).

While Pearson continued to speak in his papers of using the tests inferentially, he admitted to never having described a systematic inferential model of Neyman-Pearson tests. The debate as to whether, and if so how, Neyman-Pearson theory is relevant for statistical inference continues. A key question is why information about a test's error probabilities is relevant to particular inferences. I suggest that a plausible answer is provided by the view of experimental tests found in Peirce and that Peirce's answer has much in common with, and even extends, the one Pearson suggests.

What Peirce provides is a philosophy of inductive inference that calls for precisely the sort of tools provided by the Neyman and Pearson theory. In describing his theory of inference, Peirce could be describing that of Neyman-Pearson.

> The theory here proposed does not assign any probability to the inductive or hypothetic conclusion, in the sense of undertaking to say how fre-

> quently that conclusion would be found true. It does not propose to look through all the possible universes, and say in what proportion of them a certain uniformity occurs; such a proceeding, were it possible, would be quite idle. The theory here presented only says how frequently, in this universe, the special form of induction or hypothesis would lead us right. The probability given by this theory is in every way different—in meaning, numerical value, and form—from that of those who would apply to ampliative inference the doctrine of inverse chances. (*CP* 2.748)

The supposition that the probability of the conclusion is needed, Peirce claims, stems from a faulty analogy with deductive inference. Since deductive inference tells us that, if such and such premises are true, then a given conclusion is true, it might be thought that inductive inference should tell us that, if such and such premises are true, then a given conclusion is probable. Peirce denies this, claiming, "In the case of analytic inference we know the probability of our conclusion (if the premisses are true), but in the case of synthetic inferences we only know the degree of trustworthiness of our proceeding" (*CP* 2.693).

Peirce suggests that if probability had always been understood as relative frequency, it would not have been thought that a final probability was wanted. Since the conclusion concerns a hypothesis that either is or is not true about this world, the only probabilities that a frequentist could assign to it are the trivial ones—1 or 0. What we really want to know, according to Peirce, is this:

> Given a certain state of things, required to know what proportion of all synthetic inferences relating to it will be true within a given degree of approximation. Now, there is no difficulty about this problem (except for its mathematical complication); it has been much studied, and the answer is perfectly well known. And is not this, after all, what we want to know much rather than the other? (*CP* 2.686)

The current debate in philosophy of statistics, as in Peirce's day, reflects opposing answers to Peirce's question—Neyman-Pearson theorists agreeing, Bayesians and others disagreeing. This opposition is revealed in what each considers to be relevant information for inference.

A theory based on error probabilities—in Peirce's words, on "how frequently, in this universe, the special form of induction of hypothesis would lead us right" (*CP* 2.748)—is interested not in just the observed result but in other possible outcomes that could result from using the inference rule. That is, it is interested in the entire sample space, which is why the Neyman-Pearson theory is called a sampling theory. For example, the probability that a test commits a Type I error is the probability that the test would reject hypothesis h erroneously. It is a function of all the outcomes that would have led to rejecting h.

This points up the crucial difference between Neyman-Pearson tests and rival approaches which are nonsampling approaches, such as Bayesian or likelihood accounts: The Neyman-Pearson approach views the probability of outcomes that would have happened as crucially important in interpreting what actually did happen. For the nonsampling approaches, the relevant evidence contributed by the data is fully contained in the likelihood ratio actually obtained—a point formally expressed in the *Likelihood Principle*, which follows from Bayes's theorem. According to the Likelihood Principle, to quote from Dennis Lindley (1976:361):

> If we have two pieces of data x_1 and x_2 with the same likelihood function, [i.e., the probability of x_1 given h equals the probability of x_2 given h], the inferences about [h] from the two data sets should be the same. This is not usually true in the orthodox [Neyman-Pearson] theory, and its falsity in that theory is an example of its incoherence. . . . As [Harold] Jeffreys has said, *What has what might have happened, but did not, got to do with inferences from the experiment?* (Emphasis added)

From the Bayesian point of view, the interest in what might have occurred renders Neyman-Pearson tests "incoherent"; but from the Neyman-Pearson point of view, Bayesian (and other nonsampling) methods are unpalatable just because they ignore what the data generation procedure might have produced, once the data are in hand. This reflects the incompatibility between the former's interest in measures of evidential strength and the latter's interest in controlling error probabilities. To defend the inferential use of Neyman-Pearson tests requires a rationale for this interest in "what would have but did not happen"—correspondingly, in error probabilities. Peirce anticipates and even strengthens the rationale to be suggested by Pearson.

The concern with error probabilities for Pearson, like the concern with "the trustworthiness of the proceeding" for Peirce, is directly tied to their view that statistical method should closely link experimental design and data collection with subsequent inferences. Pearson claims: "But looking back I think it is clear why we regarded [the test's error probabilities] as more meaningful than the likelihood ratio [of the given result]. . . . The reason was this. We were regarding the ideal statistical procedure as one in which preliminary planning and subsequent interpretation were closely linked together—formed part of a single whole" (Pearson 1966d:277–78).

Pearson railed against the tendency to see statistical inference as beginning only after "data are thrown at the statistician and he is asked to draw a conclusion" (ibid.:278). Analogously, Peirce criticized opposing inference philosophies, (the so-called conceptualists) for ignoring the manner of collecting data. In criticizing John Stuart Mill, for example, Peirce insists that "an induction, unlike a demonstration, does not rest solely upon the facts observed, but upon the manner in which those facts have been collected" (*CP* 2.766). Peirce introduces the term "quasi-experimentation" to include the whole process of generating and analyzing the data as well as using the data to test a hypothesis. And "this whole proceeding," Peirce declared, "I term Induction" (*CP* 7.115; see editor's note). For Peirce, the "true and worthy" task of logic is to "tell you how to proceed to form a plan of experimentation" (*CP* 7.59).

The reason error probabilities become relevant when one considers the "preliminary experimental planning" is this: by considering ahead of time the probability an experiment has of detecting errors in hypotheses of interest, the statistical theory can then specify what type of and how large a sample would be needed to put hypotheses to an efficient test of experiment. For Peirce "induction begins with a question or theory" (*CP* 2.775).

> The next business in order is to commence deducing from it whatever experiential predictions are extremest and most unlikely, . . . in order to subject them to the test of experiment, and thus either quite refute the

hypothesis or make such corrections of it as may be called for by the experiments; and the hypothesis must ultimately stand or fall by the result of such experiments. (*CP* 7.182, emphasis added)

Introducing the Type II error—an innovation which Pearson declares to be his own improvement on Fisherian tests[1]—gives precision to Peirce's testing conception. Requiring a small Type II error ensures that a result is taken to accept hypothesis *h* only if such a result is unlikely, if in fact *h* was (specifiably) false and a rival hypothesis was true.[2] Analogously, for Peirce:

> When we adopt a certain hypothesis, it is not alone because it will explain the observed facts, but also because the contrary hypothesis would probably lead to results contrary to those observed. So, when we make an induction, it is drawn not only because it explains the distribution of characters in the sample, but also because a different rule *would probably have led to the sample being other than it is*.[3] (*CP* 2.628, emphasis added)

This concern with the probability that the sample *would have been other* than it is in reasoning from the actual sample obtained, then, reflects the same inferential concerns in Peirce and Pearson, namely, the concern that an agreement between data and a hypothesis *h* be taken to accept *h* as approximately true only by being the result of a severe test of *h*—one with a high probability of having rejected the hypothesis, if false. And this, in turn, links to both their concerns with preliminary planning—it is only by appropriate procedures for generating the data and hypothesis that this severity requirement is met. (A fuller discussion of the account of severity I have in mind occurs in Mayo 1991.)

The aim of securing severe or powerful tests underlies the importance that Peirce and Neyman and Pearson place on predesignation of hypotheses. Predesignation requires that the hypothesis be specified before obtaining the data to be used in its test. Again this touches the heart of the contemporary controversy between Neyman-Pearson sampling and nonsampling philosophies, that is, between error-probability principles and the Likelihood Principle. For those who accept the Likelihood Principle, once the data are obtained, it is irrelevant for assessing their evidential import how they were selected, or whether the hypothesis was predesignated. (This is often called "irrelevance of the sampling rule.") For such matters do not alter likelihoods. But they do alter error probabilities. Just as Neyman-Pearson theorists insist, as against their rivals, on the relevance of predesignation, Peirce is highly critical of predesignation being "singularly overlooked by those who have treated of the logic of [induction]" (*CP* 2.738).

In addressing this issue, Peirce and Pearson consider a basic test where they want to argue from a property *P* shared by a proportion of some sample to a hypothesis that the observed regularity is genuine—that is, it may be expected to hold about as well in some population or class *M*. In other words, they want to deny that the observed regularity is merely accidental, or due to chance. As against violating predesignation, Pearson (1966b:127) cautions: "To base the choice of the test of a statistical hypothesis upon an inspection of the observations is a dangerous practice; a study of the configuration of a sample is almost certain to reveal some feature, or features, which are exceptional if the [chance] hypothesis is true." In so doing, Pearson stresses, the question is changed, and therefore also the corresponding error probabilities of the test.

Peirce's points mirror Pearson's. Peirce insists that "the conclusion be confined strictly to the question. If the instances examined are found to be remarkable in any other respect than that for which they were selected, we can draw no inference of the present kind from that" (*CP* 2.784). The reason Peirce gives is the same as Pearson's.

> But suppose we were to draw our inferences without the predesignation of the character *P*; then we might in every case find some recondite character in which those instances would all agree. That, by the exercise of sufficient ingenuity, we should be sure to be able to do this, even if not a single other object of the class *M* possessed that character, is a matter of demonstration. For in geometry a curve may be drawn through any given series of points, without passing through any one of another given series of points. (*CP* 2.737)

That is, we can always find *some property or other* that all members of a sample share—even if no other members of the population do. Hence, if predesignation is violated, one cannot say that the hypothesis that the regularity is genuine has been subjected to a severe test. One cannot say that, were the regularity merely due to chance, the test would probably not have led to the observed regularity. On the contrary, it is highly probable that it would have led to observing some such regularity, even if it were only due to chance. In thus arguing against violating predesignation, Peirce also anticipates proofs by Neyman, and more recently by statisticians Birnbaum and Armitage, that violating predesignation permits tests which are wrong with high probability—in extreme cases, with probability 1! Thus, as Birnbaum (1969:128) concludes, methods based on the Likelihood Principle "cannot be construed so as to allow useful appraisal, and thereby possible control, of probabilities of erroneous interpretations."

Neyman-Pearson tests are often wrongly taken to always require predesignation, when in fact it is called for only in certain types of cases—those in which violating predesignation conflicts with the goal of controlling the error probabilities. I find Peirce to be clearer than Neyman and Pearson as regards the necessity of qualifying the predesignation rule (e.g., *CP* 2.730). As I see it, Peirce's qualification amounts to describing circumstances where violating predesignation "doesn't matter much" to the reliability of the proceeding and thereby does not much hamper the ultimate goal of statistical inference.

This links with the centerpiece of Peirce's philosophy of experimental inference—his argument for the self-correcting rationale of induction. Peirce claims that "in the individual case in hand, if there is any error in the conclusion, that error will get corrected by simply persisting in the employment of the same method" (*CP* 2.781). As Levi (1980:138) stresses, however, "Peirce is not claiming that induction is self correcting in the sense that following an inductive rule will, in the messianic long run, reveal the true value [of the proportion with a given property]."

In the case of hypothesis testing, I suggest that Peirce's idea is that if a particular conclusion is wrong, subsequent severe (or highly powerful) tests will, with high probability, detect it. For example, in a good test, hypothesis h is rejected by outcomes that are extremely improbable if h is true. Then, if we are wrong in rejecting h (h is actually true), we would find we were rarely able to generate such outcomes; in this way, we would discover our original error.

But I want to suggest an even more localized conception of error correction in Peirce—one that is relevant even where no repetitions of an experiment are planned or are even possible. I suggest that Peirce valued considerations about the trustworthiness of a proceeding—or in the more modern terminology we have been using, considerations about a test's error probabilities—because they serve as tools for rooting out specific errors in a case of experimental reasoning. Examples of such errors are mistaking chance correlations for systematic ones, mistakes about the value of a parameter, and mistakes about a causal factor. I limit myself to just one of the many examples I find in Peirce: a test carried out by Dr. Lyon Playfair. It illustrates both a mistake resulting from violating predesignation as well as how, arguing from cases where error probabilities are sustained, the original mistake is corrected.

While considering Dr. Playfair "an accomplished reasoner," Peirce describes how he violates predesignation in testing a hypothesis about a regularity between the specific gravity of a metal and its atomic weight. Looking at the specific gravities of three forms of carbon, Peirce tells us, Playfair seeks out and discovers a formula connecting them: each is a root of the atomic weight of carbon, which is 12. Peirce describes the test Playfair carries out to judge if this regularity can be expected to hold generally for metals, showing that several alleged instances of the formula really involve modifications not specified in advance. If one limits the instances to those for which the formula is predesignate, only half of them satisfy Playfair's formula. Peirce reasons: "Having thus determined the ratio, we proceed to inquire whether an agreement half the time with the formula constitutes any special connection between the specific gravity and the atomic weight of a metalloid" (*CP* 2.738).

Peirce then subjects this to a test of experiment. The falsity of the hypothesis that there is a special connection is the claim that the observed agreement is "due to chance." Peirce asks: How often would such an agreement be found, even if it were due to chance? To answer this, Peirce introduces a hypothetical chance distribution by matching the specific gravity of a set of elements, not with its own atomic weight, but with the atomic weight of some other element, with which it is arbitrarily paired. For example, the specific gravity of carbon is compared to the atomic weight of iodine. Note that Peirce is not running more trials of Playfair's experiment but considering how many agreements with Playfair's formula would occur in a case designed so that such agreements could only be due to chance and, using this information about *what would be*, is arguing about the cause of the agreements actually found. Peirce finds about the same number of cases satisfying Playfair's formula in this chance pairing of elements as Playfair found in comparing specific gravities and atomic weights of a given element. Peirce concludes: "It thus appears that there is no more frequent agreement with Playfair's proposed law than what is due to chance" (*CP* 2.738). So Playfair was mistaken in thinking he had found a special or systematic connection. With procedures that violate predesignation of the formula, in contrast, one can persist in this error.

Pearson, like Peirce, denies that the importance of low error probabilities is a concern with ensuring a low frequency of erroneous judgments in some long run—at least, in typical scientific uses of tests. Rather, Pearson (1966a:172) suggests that "the formulation of the case in terms of hypothetical repetition helps to that clarity of view needed for sound judgment" in understanding the process in a particular case.[4] For example, a consideration of error probabilities is valuable, Pearson explains, because "it helps us to assess the extent of purely chance fluctuations that are possible" (ibid.:176). If purely chance fluctuations can easily account for an observed regularity, as Peirce shows in the case of Dr. Playfair, there is no reason for taking it as a sign of a genuine regularity.

It is important to consider one further way in which the manner Pearson advocates for using error probabilities sheds light on the manner in which induction may be seen as self-correcting according to Peirce. The behavioral-decision model, typically associated with Neyman-Pearson tests, encourages tests to be viewed as automatic methods, or "recipes," for testing claims, often leading to their misuse. However, Pearson specifically denied that the tests are to be used as automatic routines for testing claims, insisting that "no responsible statistician . . . would follow an automatic probability rule" (ibid.:192) in scientific contexts.[5] Likewise, Peirce's argument that induction is self-correcting should not be thought to be claiming that some automatic routine for detecting errors exists.[6] As Nicholas Rescher (1978:14) argues: "It cannot be said too emphatically that Peirce does not hold the ill-advised view that the method of science is 'self-corrective' in providing some sort of automatic, cookbook procedure for devising good theories to put in place of bad ones."

The view Peirce holds, I suggest, is that inductive methods—viewed as methods for severe testing—provide systematic (though not automatic) procedures whose aim is detecting and correcting key errors in experimental reasoning. The modern theory of Neyman-Pearson tests may be seen to provide Peirce's experimental philosophy with mathematically rigorous methods for carrying out this aim. It does so in two ways. First, Neyman-Pearson theory provides standard statistical models in which to couch hypotheses about key types of experimental errors. This includes erroneous assumptions in these models themselves (see note 8). Second, it provides systematic rules for generating tests with specifiably small error probabilities (e.g., so-called Uniformly Most Powerful tests).[7]

To conclude, I have argued that the ability of the Neyman-Pearson test to control low error probabilities may be seen to correspond to Peirce's self-correcting rationale of inductive methods. While, on the one hand, Neyman-Pearson methods increase the mathematical rigor and generality of Peirce's assertions about self-correcting methods, on the other, Peirce goes further in providing an experimental philosophy that can serve to justify Neyman-Pearson statistical methods in science. He does so by providing a clear answer to the question of why error probabilities matter in scientific inference: they are what permit the detection and correction of errors in making inferences from experiments. Just as it is only by planning ahead of time that a test has a good chance of detecting errors, for Peirce "reasoning tends to correct itself, and the more so, the more wisely its plan is laid" (*CP* 5.575).[8]

Notes

This research was conducted during tenure in a National Endowment for the Humanities Fellowship for College Teachers. I gratefully acknowledge that support. I thank Nicholas Rescher for useful comments on an earlier draft.

1. See Pearson's (1955) response to Fisher, particularly p. 204.

2. In "Small Differences of Sensation" (*CP* 7.21–35), Peirce gives an interesting example of an experimental test that he himself conducted which satisfies this Neyman-Pearson requirement.

3. A "rule" for Peirce here is an assertion such as "most *A*'s are *B*'s."

4. Here Pearson is really only *asking* if this is not the real reason we consider hypothetical repetition (in contrast to a concern with a low error rate in the long run), because he wants to avoid being dogmatic. However, it is apparent from many of Pearson's examples that he values this inferential function served by considering hypothetical repetitions.

5. For a fuller argument, see Mayo 1992.

6. Except possibly in the case of observing a sample mean to estimate a population mean—sometimes referred to as quantitative induction by Peirce. The continual taking of samples yields an estimated mean that gets closer and closer to the true population mean.

7. A test with a specified Type I error is Uniformly Most Powerful if at the same time it has the smallest Type II error over the possible alternative hypotheses.

8. Peirce continues: "Nay, it not only corrects its conclusions, it even corrects its premisses." This gives further insight into Peirce's idea of self-correcting, namely, that the methods of testing can be used to check a test's own assumptions, such as that of randomness. Elsewhere Peirce discusses how randomness may be checked, once again anticipating aspects of modern sampling statistics.

References

Birnbaum, A. 1969. Concepts of Statistical Evidence. In *Philosophy, Science, and Method: Essays in Honor of Ernest Nagel*, ed. S. Morgenbesser, P. Suppes, and M. White, 112–43. New York: St. Martin's Press.

Fisher, R. A. 1955. Statistical Methods and Scientific Induction. *Journal of the Royal Statistical Society* (B) 17:69–78.

Hacking, I. 1980. The Theory of Probable Inference: Neyman, Peirce, and Braithwaite. In *Science, Belief, and Behavior: Essays in Honor of R. B. Braithwaite*, ed. D. H. Mellor, 141–60. Cambridge: Cambridge University Press.

Levi, I. 1980. Induction as Self Correcting according to Peirce. In *Science,*

Belief, and Behavior: Essays in Honor of R. B. Braithwaite, ed. D. H. Mellor, 127–48. Cambridge: Cambridge University Press.

Lindley, D. V. 1976. Bayesian Statistics. In *Foundations of Probability Theory, Statistical Inference, and Statistical Theories of Science*, ed. W. L. Harper and C. A. Hooker, 2:353–63. Dordrecht: D. Reidel.

Mayo, D. 1985. Behavioristic, Evidentialist, and Learning Models of Statistical Testing. *Philosophy of Science* 52:493–516.

———. 1991. Novel Evidence and Severe Tests. *Philosophy of Science* 58:523–52.

———. 1992. Did Pearson Reject the Neyman-Pearson Philosophy of Statistical Testing? *Synthese* 90:233–62.

Neyman, J. 1950. *First Course in Probability and Statistics.* New York: Henry Holt.

———. 1971. Foundations of Behavioristic Statistics. In *Foundations of Statistical Inference*, ed. V. P. Godambe and D. A. Sprott, 1–13. Toronto: Holt, Rinehart and Winston of Canada. Comments and Reply, 14–19.

Pearson, E. S. 1955. Statistical Concepts in Their Relation to Reality. *Journal of the Royal Statistical Society* (B) 17:204–7.

———. 1966a. The Choice of Statistical Tests Illustrated on the Interpretation of Data Classed in a 2 × 2 Table. *Biometrika* 34 (1947):139–67. In Pearson 1966c, pp. 169–97.

———. 1966b. The Efficiency of Statistical Tools and a Criterion for the Rejection of Outlying Observations. (Assisted by C. Chandra Sekar.) *Biometrika* 28 (1936):308–20. In Pearson 1966c, pp. 118–30.

———. 1966c. *The Selected Papers of E. S. Pearson.* Berkeley: University of California Press.

———. 1966d. Some Thoughts on Statistical Inference. *Ann. Math. Statistics* 33 (1962):394–403. In Pearson 1966c, pp. 276–83.

Rescher, N. 1978. *Peirce's Philosophy of Science.* Notre Dame, Ind.: University of Notre Dame Press.

12 Pragmatism, Abduction, and Weak Verification

Jeremiah McCarthy, College of Staten Island, City University of New York

A. J. Ayer's first formulation of the empiricist criterion of meaning in *Language, Truth and Logic* (1952:35) resembles Peirce's formulation of pragmatism in the "Lectures on Pragmatism." Peirce's formulation is given in an unpublished notebook that supplies the end of the sixth lecture in the series (*MS* 316). It cannot be gathered from the published material. I will give Ayer's formulation of his weak verificationism and the counterexample that proved it inadequate. Then I will give Peirce's version of pragmatism from the "Lectures" and show how the counterexample and other objections to Ayer apply to it. I will discuss some other objections to Ayer and Peirce and finally consider some applications of the principle.

Weak Verification

Ayer (1952:35) tells us that "a sentence is factually significant to any given person, if, and only if, he knows how to verify the proposition which it purports to express—that is, if he knows what observations would lead him, under certain conditions, to accept the proposition as being true, or reject it as being false." To put it more exactly, "We may say that it is the mark of a genuine factual proposition . . . that some experiential propositions can be deduced from it in conjunction with certain other premises without being deducible from those other premises alone" (ibid.:38–39). An experiential proposition is "a proposition which records an actual or possible observation" (ibid.:38).

This formulation of the criterion of meaning succumbed to the following well-known counterexample. Let *S* be any proposition one pleases and *O* be an experiential proposition. Then *S* together with "If *S*, then *O*" will entail *O*, which does not follow from either of the premises alone. Thus *S* will count as a genuine factual proposition no matter what it says (ibid.:11–12).

Pragmatism

To explain pragmatic meaning Peirce introduces a technical sense of meaning as well as the novel term "ultimate meaning" (*CP* 5.175, 179). "The meaning of a proposition or term is all that proposition or term could contribute to the conclusion of a demonstrative argument" (*CP* 5.179). The pragmatic meaning of a term is to be identified with its ultimate meaning. From the unpublished notebook I previously referred to, we can see that ultimate meaning is whatever cognitive purpose is common to all terms having a cognitive purpose (*MS* 316:44, 57–60). This can be determined by looking at the purpose of assent to a proposition that introduces a new term into reasoning. The only argument that does so is abduction. So the question of pragmatism comes down to the question of the "true Logic of Abduction" (*MS* 316:59). When Peirce talks about the true logic of abduction, he is referring to a minimum standard for the admissibility of hypotheses.

> It is now generally admitted, and it is the result of my own logical analysis, that the true maxim of abduction is that which Auguste Comte endeavored to formulate when he said that any hypothesis might be admissible if and only if it was verifiable. . . . [W]hat must and should be, meant [by "verifiable"] is, that the hypothesis must be capable of verification by induction. Now induction, or experimental inquiry, consists in comparing perceptual predictions deduced from a theory with the facts of perception predicted. (*MS* 316:59–60)

Perceptual predictions are propositions about sensible results to be expected in consequence of actions that bring about or lead us to encounter certain conditions (*CP* 5.168). The cognitive purpose that is common to all propositions having a cognitive purpose is to provide perceptual predictions. That is what verifiability comes down to. So the ultimate meaning of a term or proposition—its pragmatic meaning—is whatever in the way of perceptual predictions the term or proposition can contribute to the conclusion of a demonstrative argument.

This gives the rationale for the principle of pragmatism, and there is nothing like it in Ayer or other positivists. I cannot discuss it at any length here. (For a thorough discussion, see McCarthy 1990.) Whether it works depends on whether conduct has an ultimate purpose, on whether meaning and purpose can be connected, as Peirce tries to do, and on whether meaning in general can be explicated as entailment. I will have something to say about this last point later.

Objections to Weak Verificationism and Pragmatism

Peirce's idea of pragmatic meaning shares an apparently fatal similarity to Ayer's weak verificationism. Peirce talks about "perceptual predictions deduced from a theory" and does not mention additional premises. But it seems fair enough to assume that a hypothesis plus some additional premises would be used. If so, then suppose that *A* is a proposition that specifies an action and *O* a proposition that specifies a sensible result of the action. Let *S* be any proposition whatever. Then, from *S* and "If *S*, then if *A*, then *O*," we can deduce the perceptual prediction "If *A*, then *O*." That makes *S*, whatever it is, pragmatically meaningful, since it conforms to Peirce's verifiability requirement.

Let me stop here to notice an ambiguity present in Ayer's verificationism but not in Peirce's pragmatism. In *Ayer*, John Foster (1985:22) distinguishes two interpretations of Ayer's verificationism. One involves what Foster calls the "evidence-principle," which makes factual significance lie in the possibility of observational evidence for or against a proposition. The other involves the "content-principle," which makes factual significance depend on purely observational content. At the beginning of *Language, Truth and Logic*, Ayer appears to embrace the first type of verificationism, but in the latter part of the book he seems to adhere to the second. This latter involves a commitment to reductionism. I do not find this in Peirce, and his pragmatism otherwise resembles verificationism of the evidence-principle type.

Is pragmatism in better shape than Ayer's verificationism under the evidence-principle interpretation? Here are some reasons for thinking so.

If a term or proposition cannot turn up in an abduction, no question about verifiability arises. Accordingly, the question of the pragmatic significance of a term or proposition will have to begin with the question whether it is abducible. An abducive argument is an argument one premise of which represents an unexpected fact and the other of which represents a hypothesis being arrived at in consequence of the encounter with the unexpected fact. The conclusion, the hypothesis, is drawn in recognition that a hypothesis so arrived at has a significant chance of being true.

It is not clear from anything Peirce says in his "Lectures on Pragmatism" how to apply what might be called the requirement of abducibility. Does a hypothesis need to be abducible under conditions that do or have existed, or is abducibility under some imaginary set of conditions enough? It seems that for a term or proposition actually to be pragmatically meaningful, we need some actual conditions that have led to its abduction.

The requirement of abducibility will rule out the counterexample or at least one problem that it poses. It is clear that S cannot be any proposition at all, since any proposition at all will not be abducible, let alone actually abduced. Obviously, nonsensical propositions like "The Absolute is lazy" will be excluded.

Unfortunately, the counterexample can still cause trouble in a way that Barry Gower (1987:16) has discussed. Ayer's verificationism and, by implication, Peirce's pragmatism seem to "permit anything to be verified by any observation." If pragmatism only requires that from a hypothesis perceptual predictions can be deduced, and we may add the assumption that additional premises are needed, then any perceptual prediction can confirm any abducible hypothesis. For example, from S = "The earth revolves around the sun" and "If *S*, then if *A*, then *O*" = "If the earth revolves around the sun, then if I look down, I'll see my shoes," we can get the perceptual prediction "If I look down, I'll see my shoes," which is obviously useless for verifying *S*.

This is a problem which is not directly solved by putting abducive restrictions on the content of *S*. It can be partly solved by a restriction on other premises that are used to connect the hypothesis with perceptual predictions. The restriction comes from the function of hypotheses as devices for anticipating experience. To determine which additional premises to use in deducing perceptual predictions, it is necessary to think what experiences there is reason to expect, given that the hypothesis is true. There is no magic, infallible, or formal way of making the exclusion, but it is done every day and should not cause trouble. "The mode of suggestion by which in induction the hypothesis suggests the facts is by contiguity, familiar knowledge that the conditions of the hypothesis can be realized in certain experimental ways" (*CP* 7.219). Then if there are experiences which the hypothesis, if true, would give us reason to expect, the hypothesis counts as verifiable.

For a proposition actually to be pragmatically meaningful, it must be (1) actually abduced and (2) verifiable. Any successful counterexample to Peirce's pragmatism will have to include an abduced proposition and an additional premise that makes explicit some sensible result that the proposition gives a reason to expect. Given that these conditions are met, it should be the case that the proposition in question is evidently meaningless or at least without empirical significance. It appears to me that these requirements cannot be met.

An additional problem with Ayer's formulations of verificationism that may apply to Peirce is that propositions having meaningless components can be classified as verifiable (Foster 1985:16–17; Wright 1987:264–65). If so, then sometimes something meaningless can be deduced from meaningful premises. In Peirce's case the abducibility requirement prevents components from being added to premises ad libitum. Hence we cannot go from *S* which is pragmatically meaningful to "*S* and *N*," where *N* is anything and where, with additional premises, "*S* and *N*" gives the same perceptual predictions as *S* alone.

There may be a related problem, however, because it is not possible to guarantee that abduction will introduce only hypotheses with no pragmatically meaningless components. If a hypothesis with a meaningless component is verifiable, it is admissible. Peirce seems to have thought that the admissibility of such hypotheses was a point in favor of pragmatism. Pragmatism "allows any flight of imagination, provided this imagination ultimately alights upon a possible practical effect; and thus many hypotheses may seem at first glance to be excluded by the pragmatical maxim that are not really so excluded" (*CP* 5.196). The only difficulty here is that deduction will not always preserve pragmatic meaningfulness. There might be a problem if abduced content that is useless for generating perceptual predictions could not be identified, but there is no reason to think that.

Meaning and Observational Content

Foster (1985:29) raises an objection different from these about allowing meaningless content. If the evidence-principle version of verificationism is adopted and factual content exceeds observational content, Foster says that meaning will not be confined to the possibilities of experience. He believes that this will keep us from excluding as meaningless on semantic grounds factual statements that are not open at all to observational evidence.

For example, suppose that physical-object statements are understood realistically as referring to some mind-independent reality. Then "given our realistic construal of physical statements, we could hardly regard their openness to this kind of observational evidence [i.e., observational data that they can explain] as a prerequisite of their having the significance we attach to them" (Foster 1985:29). Why? Apparently because pressing a skeptical view that these statements are not open to evidence in the way that we suppose would not be to challenge their meaning. The skeptic would not force us to deny that the statements are factually significant and concerned with a mind-independent reality. He would only raise an epistemological issue to be decided independently of the semantics of the statements.

Foster's point, I take it, is that once we admit that there is something or some class of things independent of mind, there is no way to avoid the possibility that mind is independent of them, that is, that they never affect human thought at all or at least never in such a way as to produce a true representation of themselves. Hence, it is not part of the meaning of any physical-object statement that the physical object has observational effects. The content-principle version of verificationism does not have this problem because it makes physical objects logical constructions from sense-data.

Peirce's pragmatism resembles the evidence-principle version of verificationism. Nevertheless, the objection is not a problem for Peirce for two reasons, the first of which is his understanding of meaning. Meaning in general (of which pragmatic meaning is a species) is anything that a term or proposition can contribute to the conclusion of a demonstrative argument (*CP* 5.175–76, 179). So part of the meaning of a physical object statement is that certain visual phenomena will occur. The physical object statement together with other premises will imply that proposition.

This case signals a difference of opinion about the meaning of "meaning" with regard to fact-stating discourse. Peirce starts from ordinary usage, in which one meaning of "meaning" is what something entails. For example, the atheist says, "All the arguments for the existence of God are invalid or have false premises." The theist replies, "But that doesn't mean that there is no God." Also, informal logic texts often give as an indicator of the conclusion of an argument, the phrase "which means that."[4] And the word "means" can be a sign of a conditional, which connects meaning and implication. Peirce's notion of meaning in general is anchored in ordinary usage, and he cannot be accused of making a novel contribution to lexicography. Given Peirce's concept of meaning in general, it follows that observational conclusions would be part of the meaning of physical-object statements.

The second reason why the objection is not a problem is that Peirce would disallow the skeptical challenge that it relies on. If the skeptic urges his objection by using a device like the Evil Demon Hypothesis or urges the hypothesis that there are unknowable physical objects, these will be ruled out. The hypotheses do no explanatory work, and they have not been, and cannot be, legitimately introduced by abduction.

Status of the Principle

Another kind of objection made to verificationism has to do with the status of the principle. Ayer allows for only two categories of meaningful discourse. A proposition needs to be either analytic or verifiable. But the principle itself does not easily fit into either category. Ayer (1952:15–16) said that the principle was a definition but had to admit that it explicates one sense of meaning among others. Whatever Ayer's problems, they are not Peirce's. Pragmatism turns out to be pragmatically meaningful. The thesis about meaning or the cognitive use of terms and propositions is based on a hypothesis about the nature of belief, namely, that the purpose of belief is to guide action. This is abducible and appears to have been abduced as an explanation of the difference between the inquiry that is involved in scientific method, which has produced genuine knowledge, and that involved in less successful cognitive endeavors. The resolution of doubt by the establishment of a belief, the goal of inquiry, should give something expressible in terms of perceptual predictions. Hence, the proposition is verifiable that content about perceptual prediction is the common meaning component of whatever has cognitive meaning. If the purpose of belief were something else—for example, to produce pleasing fantasies—then the pragmatic principle would be false and could be found so.

Pragmatism as a Critical Principle

The next-to-last paragraph of the "Lectures on Pragmatism" gives a metaphoric statement of pragmatism as a critical principle. "The elements of every concept enter into logical thought at the gate of perception and make their exit at the gate of purposive action; and whatever cannot show its passports at both those two gates is to be arrested as unauthorized by reason" (*CP* 5.212). Peirce means to include abduction when he mentions perception. The abductive conclusion is like a perceptual judgment except that it is criticizable (*CP* 5.181, 185–86).

A notable case in which Peirce takes some trouble to conform to the principle is the discussion in "A Neglected Argument for the Reality of

God" (*CP* 6.452–91). The usual way with the question of the existence of God is to examine arguments that have been or might be given without any worry about whether the concept of a supreme being is admissible as an explanation. Worries about logical problems with the concept center on its consistency. But Peirce is evidently concerned about whether the idea of God can be introduced into reasoning by abduction. That is the purpose of musement, the play of ideas that will suggest to the muser the reality of God as an explanation for the three universes and their interrelations (*CP* 6.465).

I have already touched on another application of the principle, in the case of skepticism. Descartes introduces the Dream Hypothesis and the Evil Demon Hypothesis as devices to produce doubt. On pragmatic grounds the hypotheses are unmotivated. They do no explanatory work, and there is no abductive occasion to introduce them. They are inadmissible, and there is no reason to argue against a hypothesis that it was a logical mistake to entertain in the first place.

Theological ethics presents us with another sort of case in which the principle might be applied. According to Catholic moralists, sex has an inseparable twofold purpose: to express love and to reproduce within marriage. This is a hypothesis that was introduced in order to harmonize older church teachings and more modern ideas about sex. It can be found in *Humanae Vitae*, the 1968 encyclical of Pope Paul IV. The intellectual developments leading up to it are discussed in Noonan 1966:491–504. The hypothesis was not intended to explain anything, but it might be abducible nevertheless. However, facts about sexual behavior, where we frequently find the parts of the purpose dissociated or entirely absent, could hardly suggest it. Hence, it gets ruled out as pragmatically meaningless.

Goodman's (1983:73–75) problem gives another case in which the principle is violated. The problem is that whenever we have data to confirm the hypothesis that all emeralds are green, we will also have data to confirm the hypothesis that all emeralds examined before future time *t* are green, but otherwise are blue. The complex predicate "green and examined before *t* or else blue" may be expressed by "grue." Peirce would consider this a problem about abduction rather than induction. The hypothesis, however expressed, has to be introduced somehow, and if the data to go on are observations of green emeralds, it is clear that nobody would entertain the grue hypothesis. There is nothing to suggest that emeralds examined after *t* will be blue. Hence, the grue hypothesis is ruled out as inadmissible or pragmatically meaningless.

A final objection to pragmatism might be based on the fact that it classifies some hypotheses as pragmatically meaningless which can generate perceptual predictions. The hypothesis about the purpose of sex and the grue hypothesis are two cases in point. It may seem that on its own grounds pragmatism ought to classify these hypotheses as meaningful. The idea behind pragmatism is that a proposition counts as meaningful if it can guide conduct. Any proposition that gives perceptual predictions can guide conduct because it tells one what to expect in consequence of certain conduct. The hypothesis about sex tells us that we should expect human sexual behavior for the most part to occur if it expresses love within marriage and has a significant chance of leading to conception. But when conditions are not right for it to achieve its supposed end, then we should find that it generally does not occur. If a time is specified for *t* in the grue hypothesis, then we should expect emeralds first encountered after *t* to be blue. These hypotheses, the objection would run, may be false, but they certainly are not meaningless.

The question at issue in the objection is, What is it for a proposition to be capable of guiding conduct? The objection assumes that providing perceptual predictions is the answer. However, a proposition that nobody will take seriously—the grue hypothesis is an example—is not capable of guiding conduct, no matter whether it gives predictions. Furthermore, adding a normative element is justifiable. Peirce never said that a proposition subserves the common purpose of all propositions when it has been mistakenly introduced into reasoning. So a proposition must be capable of *rationally* guiding conduct. Whether a proposition can do this depends on whether it results from a valid abduction, not merely on whether it yields predictions.

References

Ayer, Alfred Jules. 1952. *Language, Truth and Logic*. 1946. New York: Dover Publications.

Foster, John. 1985. *Ayer*. London: Routledge and Kegan Paul.

Goodman, Nelson. 1983. *Fact, Fiction, and Forecast*. 4th ed. Cambridge: Harvard University Press.

Gower, Barry. 1987. Introduction: The Criterion of Significance. In *Logical Positivism in Perspective: Essays on Language, Truth and Logic*, ed. Barry Gower, 1–22. Totowa, N.J.: Barnes and Noble.

McCarthy, Jeremiah. 1990. An Account of Peirce's Proof of Pragmatism. *Transactions of the Charles S. Peirce Society* 26:63–113.

Noonan, John T., Jr. 1966. *Contraception*. Cambridge: Harvard University Press, Belknap Press.

Wright, Crispin. 1987. Scientific Realism, Observation, and the Verification Principle. In *Fact, Science and Morality: Essays on A. J. Ayer's Language, Truth and Logic*, ed. Graham Macdonald and Crispin Wright, 247–274. Oxford: B. Blackwell.

13 Peirce's Theory of Statistical Explanation

Ilkka Niiniluoto, University of Helsinki

Explanation in the Mid-twentieth Century: Hempel, Salmon, and Fetzer

Carl G. Hempel's seminal papers "The Function of General Laws in History" (1942) and "Studies in the Logic of Explanation" (1948, with Paul Oppenheim) started a new epoch in the philosophy of science. In the spirit of logical empiricism, Hempel employed exact conceptual tools from logic to make precise, or to explicate, in Rudolf Carnap's sense, the notion of scientific explanation.

According to Hempel's *deductive-nomological* (D-N) model, scientific explanations are arguments which answer why-questions about particular events or general regularities by "subsuming" them under general laws and particular antecedent conditions. This "covering-law model," or "subsumption theory" of explanation, can be summarized as follows:

	$L_1, L_2, \ldots, L_r$	general laws	explanans
	$C_1, C_2, \ldots, C_k$	statements of antecedent conditions	
logical deduction	E	description of the event or regularity to be explained	explanandum

A D-N explanation is *true* if its explanans is true. Without this requirement, an explanation is said to be *potential*. Thus, a potential D-N explanation contains "nomic," or "lawlike," premises which may be false. If the logical form of such lawlike statements is expressed by universal generalizations,[1] paradigmatic simple examples of D-N explanations[2] are represented by

(1) $(x)(Fx \rightarrow Gx)$
Fa

$$\frac{\qquad}{Ga}$$

(2) $(x)(Fx \rightarrow Gx)$
$(x)(Hx \rightarrow Fx)$

$$\frac{\qquad}{(x)(Hx \rightarrow Gx)}$$

Already in 1942 Hempel hinted that some explanatory arguments may replace universal, or "deterministic," laws with "probability hypotheses," which (together with the antecedent conditions) make the explanandum event "highly probable" (Hempel 1965:237). In 1948 he pointed out that the subsumption under "statistical laws" has "a peculiar logical structure" which "involves difficult special problems" (ibid.:251). Later Hempel (1962:138) has reported that he invented in 1945 the problem of the "ambiguity of statistical syllogisms."[3] If $P(G/F)$ is the statistical probability of outcome G in the class F, then two arguments of the forms (3) and (4) may both have true premises, even if their conclusions are inconsistent with each other.

(3) $P(G/F)$ is nearly 1.
Fa.
So, it is almost certain that a is G.

(4) $P(\sim G/H)$ is nearly 1.
Ha.
So, it is almost certain that a is not G.

Rudolf Carnap, who started to develop his logical theory of induction in the 1940s, assured Hempel that this ambiguity is "but one of several apparent paradoxes of inductive logic which result from violations of the requirement of total evidence." In "Inductive Inconsistencies" (1958), Hempel offered the solution that, in a statistical syllogism, probability should not be understood as a *modal qualifier* of the conclusion but rather as a *relation* between the premises and the conclusion. Thus, instead of (3), we should say that *Ga* is highly probable relative to the statements $P(G/F) \approx 1$ and *Fa* (Hempel 1965:60). Schema (3) can now be written in the following form:

(5) $P(G/F)$ is nearly 1
Fa
═════ [makes almost certain]
Ga

The double line here indicates that (5) is an inductive rather than a deductive argument.

With this reasoning, Hempel was finally ready to formulate his model of *inductive-statistical* (I-S) explanations in the paper "Deductive-Nomological vs. Statistical Explanation" (1962). This model can be expressed by the following schema:

(6) $P(G/F) = r$
Fa
═════ [r]
Ga

Here r in the lawlike premise $P(G/F) = r$ is a *statistical probability*, and r in the brackets indicates the *inductive probability* of the explanandum *Ga*, given the explanans. Furthermore, Hempel required that r should be close to 1,[4] and the relevant knowledge situation K should satisfy the *Requirement of Maximal Specificity*:

(RMS) If K contains statements F_1a and $(x)(F_1x \rightarrow Fx)$, then K also contains a law $P(G/F_1) = r_1$, where $r_1 = r$, unless that law is a theorem of probability theory.

RMS is a counterpart for explanation of Hans Reichenbach's (1938) advice in the context of prediction: use the statistical probability statement with the narrowest available reference class for determining the "weights" of single cases. When RMS is satisfied, the probability r expresses the *nomic expectability* of Ga on the basis of the explanans (see Hempel 1965, 1968).

Later in 1977 Hempel reformulated RMS by dropping the condition F_1a from its antecedent. In this form, it requires that the reference class F is *epistemically homogeneous* for G in Wesley Salmon's sense.

If schema (6) is a statistical counterpart to D-N inference (1), Hempel's concept of *deductive-statistical* (D-S) explanation corresponds to (2). A D-S explanation is an argument where a statistical probability statement is derived from other such statements by means of the theory of probability (Hempel 1965:381).

It seems that no one before 1962 took the idea of probabilistic explanation seriously. Hempel's hint at statistical explanation in 1942 did not stimulate any interest among philosophers for two decades. In the debates following the appearance of Hempel's *Aspects of Scientific Explanation* (1965), some philosophers still denied altogether the idea that probabilistic arguments could serve as explanations. In particular, G. H. von Wright (1971) and W. Stegmüller (1973) argued that Hempel's I-S model is valid only for inductive predictions and other reason-giving arguments. But many other philosophers accepted Hempel's model with modifications or started to develop its alternatives.

Since 1970 an important rival to Hempel's model has been Wesley Salmon's *statistical relevance* (S-R) model of explanation (Salmon 1965, 1984; Salmon et al. 1971). With Richard Jeffrey (1969), Salmon rejects Hempel's high probability requirement. He also replaces the epistemic principle RMS with a requirement that the reference class F should be *objectively homogeneous* for attribute G, that is, no property H (independent of G) divides F in a "statistically relevant" way to a subclass $F \cap H$

such that $P(G/F) \neq P(G/F \cap H)$. According to Salmon's S-R model, an answer to the question Why does this member a of B have the property G? is obtained by partitioning B into subclasses $B \cap F_i$, $i = 1, \ldots, n$, such that each $B \cap F_i$ is objectively homogeneous for G, and $P(G/B \cap F_i) \neq P(G/B \cap F_j)$ if $i \neq j$. The answer $a \in B \cap F_k$ thus locates the place of individual a in this classification and assigns it to a *maximal* homogeneous reference class. Here it may happen that $P(G/B \cap F_k)$ is higher, smaller, or equal to the initial probability $P(G/B)$ (positive relevance, negative relevance, irrelevance, respectively).[5]

While Hempel and Salmon have relied mainly on a long-run *frequency interpretation* of statistical probability,[6] James Fetzer (1974) formulated a *propensity model* of probabilistic explanation. According to the single-case propensity interpretation, a lawlike probability statement $(x)[Hx \rightarrow P_x(G/F) = r]$ asserts that every chance setup x of type H has a dispositional tendency of strength r to produce a result of type G on each single trial of kind F. If $0 < r < 1$, such a law implies that setups of type H are *indeterministic* systems. A typical explanation with such single-case propensity laws is then as follows:

$$(7) \qquad \begin{array}{l} (x)[Hx \rightarrow P_x(G/F) = r] \\ Ha \;\&\; Fa \\ \hline\hline Ga \end{array} \; [r]$$

The r in brackets is again the degree of nomic expectability of outcome G on the relevant trial with chance setup a. In this case, a separate RMS condition is unnecessary, since already the law in (7) presupposes that F is objectively homogeneous for G.[7]

Explanation in the Nineteenth Century: Comte, Mill, and Peirce

When Morton White in 1943 attributed the deductive-nomological pattern of historical explanation to Hempel's 1942 paper, Karl Popper quickly complained that Hempel had only reproduced *his* theory of causal explanation, originally presented in *Logik der Forschung* (1935) (see Popper 1945:chap. 25, n. 7; 1957:144). With his characteristic modesty, Hempel (1965:231) pointed out that his account of D-N explanation is "by no means novel" but "merely summarizes and states explicitly some fundamental points which have been recognized by many scientists and methodologicists." Hempel goes on to quote definitions of explanation as subsumption under laws from the 1858 edition of John Stuart Mill's *System of Logic* (1st ed., 1843), Stanley Jevons's *Principles of Science* (1st ed., 1873), and from the books of Ducasse (1925), Cohen and Nagel (1934), and Popper (1935) (see Hempel 1965:251).

Von Wright (1971:175) makes the sarcastic remark (directed at Popper) that "in point of fact the 'Popper-Hempel' theory of explanation had been something of a philosophic commonplace ever since the days of Mill and Jevons." Indeed, in an often-quoted passage, Mill (1906:305) says:

> An individual fact is said to be explained by pointing out its cause, that is, by stating the law or laws of causation of which its production is an instance. Thus a conflagration is explained when it is proved to have arisen from a spark falling into the midst of a heap of combustibles; and in a similar manner, a law of uniformity in nature is said to be explained when another law or laws are pointed out, of which that law itself is but a case, and from which it could be deduced.[8]

This is a restatement of Auguste Comte's assertion in 1830 that, instead of searching "after the origin and hidden causes of the universe and a knowledge of the final causes of phenomena," science endeavors to discover the "actual laws of phenomena," or "their invariable relations of succession and likeness," so that "the explanation of facts . . . consists henceforth only in the connection established between different particular phenomena and some general facts" (Comte 1970:2).

Many later positivists and instrumentalists excluded explanations from science, since they thought that science should drop why-questions in favor of descriptive how-questions, or they were afraid that the aim of explanation would "subordinate" science to metaphysics (Duhem 1954:10). Comte and Mill, instead, defended the subsumption theory of explanation in a form where the laws express verifiable general connections between phenomena.

More liberal versions of the subsumption theory were supported by the champions of the *hypothetico-deductive method*, such as René Descartes, Robert Boyle, and Isaac Newton in the seventeenth century or John Herschel and William Whewell in the nineteenth century, who demanded that a good hypothesis or theory should "explicate," "explain," "prove," "demonstrate," or "account for" known facts.[9] In this view, a theoretical hypothesis, even if it is not directly verifiable, receives inductive support from those observed facts that it successfully explains. In this sense, induction can be understood as "converse deduction."

But the beginning of the theory of deductive explanation goes even further back in history—to the Aristotelian ideal of demonstrative science.[10] In distinguishing in *Physica* the four types of "causes" or "explanatory factors" (Greek *aitia*), Aristotle argued that inquiry proceeds from knowing *that* (Greek *oti*, Latin *quia*) to knowing *why* (Greek *dioti*, Latin *propter quid*): "Men do not think they know a thing till they have grasped the 'why' of it (which is to grasp its primary cause)" (*Physics* 194b 19). The answer to a why-question is provided by a *scientific syllogism* which demonstrates a fact as an effect of its cause. For example, "Let *C* be the planets, *B* nearness, *A* not twinkling. Then *B* is true of *C*, and *A* (not twinkling) of *B*. Consequently *A* is true of *C*, and the syllogism gives us the reason, since its middle term (*B*) is the proximate cause" (*Posterior Analytics* 78b1–4). The paradigmatic examples (1) and (2) of Hempelian D-N explanations are arguments which can be formulated in the mode of Aristotle's *Barbara* syllogism.

Aristotle's *Posterior Analytics* is thus the first systematic attempt to disclose the structure of scientific explanation, and his account was defended and developed by ancient, medieval, and Renaissance philosophers in their theories of *demonstratio propter quid*, synthesis, or composition.

The recognition of Aristotle as the founder of the logical theory of deductive explanation does not diminish our appreciation of Hempel's achievement. What Hempel did was to reopen the study of the logical form of explanatory arguments without relying on Aristotelian essentialism and without restriction to the simple forms of Aristotle's syllogisms.

Hempel's *Aspects of Scientific Explanation* (1965) does not mention at all the name of Charles S. Peirce. Also, others who have listed nineteenth-century advocates of D-N explanations (e.g., Comte, Whewell, Mill, and Jevons) have failed to refer to Peirce. This is surprising because Peirce's paper "Deduction, Induction, and Hypothesis" (1878), which gives very clear examples of deductive explanations, was available in Morris Cohen's edition *Chance, Love, and Logic* (1923). Furthermore, Peirce's *Collected Papers* (1931–35) and J. Buchler's edition of *Selected Writings* (1940) contain a reprint of the article "A Theory of Probable Inference" (1883), which formulates a detailed theory of probabilistic explanation, with deductive explanation as a special case.

Peirce's writings on probability and induction were well known to many philosophers before 1950 (Keynes, Nagel, von Wright, Carnap, Williams, etc.). It was recognized that he was interested both in inductive inference from a sample to a population and in "probable deduction" from a population to a sample or to single cases. But it was almost always thought that the latter type of inference, variously called "the use of *a priori* probabilities for the prediction of statistical frequency" (Keynes 1921), "the problem of the single case" (Reichenbach), "statistical syllogism" (Williams 1947), or "direct inference" (Carnap), was concerned with prediction rather than explanation.[11] However, Peirce himself never made such a restriction.

In a recent review article, I pointed out that "strangely enough, it seems that the modern literature on statistical explanation does not contain even a single reference to Peirce's theory of explanatory statistical syllogisms" (Niiniluoto 1982:160). In another article, I ventured to suggest that "Peirce should be regarded as the true founder of the theory of inductive-probabilistic explanation" (Niiniluoto 1981:444).

Salmon, who takes 1962 as "the year in which the philosophical theory of scientific [statistical] explanation first entered the twentieth century" (Salmon 1983:179), disagrees with my judgment, since "one isolated and unelaborated statement" about explanatory statistical syllogisms "can hardly be considered even the beginnings of any genuine theory" (Salmon 1984:24).

In the following, my aim is to argue, against Salmon, that Peirce had a serious and systematic concern for scientific explanation ever since his

Harvard lectures in 1865 and his Lowell lectures in 1866 "On the Logic of Science," and that his 1883 account of "probable" and "statistical deduction" gives a rich and detailed model for the structure of statistical explanation. Indeed, it seems to me that the relation of Peirce's work to the I-S model parallels the relation of Aristotle to the D-N model.

Peirce on Probabilistic Explanation

Peirce's interest and insight in statistical explanation arose from three different sources: (1) Aristotle's logic, which provided the grounds for distinguishing between deduction, induction, and hypothesis; (2) probability theory, which helped to make precise the idea of probable inference; and (3) science in the mid-nineteenth century, which gave examples of scientific theories with a statistical character.

Deduction, Induction, and Hypothesis

In his Harvard lectures during the spring of 1865, the young Peirce set himself the task of discovering "the elementary processes which lie at the bottom of all scientific reasoning" (*W* 1:175). Starting from Aristotle's doctrine of induction as the inference of the major premise of a syllogism (of the first figure), Peirce observed that there is "a large class of reasonings" that are neither deductive nor inductive: reasoning a posteriori to a physical hypothesis, or inference of a cause from its effect (*W* 1:180).[12] This reasoning, which Peirce called hypothesis (and later abduction), can be represented as the inference of the minor premise of a syllogism.

The classification of inferences into deduction, induction, and hypothesis was elaborated in Peirce's Lowell lectures in the fall of 1866 and was published in the article "On the Natural Classification of Arguments" in the *Proceedings of the American Academy of Arts and Sciences* (1867). In Peirce's series for the *Popular Scientific Monthly* ten years later, this distinction was presented in the article "Deduction, Induction, and Hypothesis" (1878).

Peirce illustrated the two ways of inverting deductive arguments by the following example with a *Barbara* syllogism. *Deduction* is an inference of a result from a rule and a case:

(8) *Rule* All the beans from this bag are white.
Case These beans are from this bag.
∴ *Result* These beans are white.

Induction is the inference of the rule from the case and result:

These beans are from this bag.
These beans are white.
∴ All the beans from this bag are white.

Hypothesis is the inference of the case from the rule and result:

All the beans from this bag are white.
These beans are white.
∴ These beans are from this bag. (*W* 3:325; *CP* 2.623)

Already in the Harvard lectures of 1865, Peirce made it perfectly clear that hypothesis is an inference *to an explanation*: "We find that light gives certain peculiar fringes. Required, an explanation of the fact. We reflect that ether waves would give the same fringes. We have therefore only to suppose that light is ether waves and the marvel is explained" (*W* 1:267).

In the Lowell lectures of 1866, Peirce says that it is hypothesis which "alone affords us any knowledge of causes and forces, and enables us to see the *why* of things," adding that hypothesis is the inversion of the corresponding *explaining syllogism*:

Ether waves are polarizable.
Light is ether waves.
∴ Light is polarizable.

In general, "to explain a fact is to bring forward another from which it follows syllogistically"; that is, "We say that a fact is *explained* when a proposition—possibly true—is brought forward, from which that fact follows syllogistically" (*W* 1:428, 425, 440, 452).

In stating his definition of deductive explanation, Peirce does not refer to Comte or Mill—whom he knew well but did not appreciate too much. The inspiration came rather directly from Aristotle and from Kant, "the greatest philosopher of modern times," whose definition of hypothesis is quoted from *Logic* (1800) in his sixth Lowell lecture: "A hypothesis is a holding-to-be-true of a judgment of the truth of a ground, for the sake of its sufficiency for consequences" (*W* 1:451; see Kant 1974:92).

Probable Inference

In his Harvard lectures Peirce still suspected that "the question what is the *probability* of an induction or hypothesis is senseless" or "absurd" (*W* 1:289, 293). But already in his 1867 review of John Venn's *Logic of Chance* (1866), the first systematic treatment of the frequency interpretation of probability, Peirce discussed inferences of the form:

(9) *A* is taken at random from among the *B*'s.
$\frac{2}{3}$ of the *B*'s are *C*.
∴ *A* is *C*.

The justification for schema (9) is in terms of *truth frequencies*: in the long run an argument of form (9) would yield a true conclusion from true premises two-thirds of the time (*CP* 8.2).

Peirce formulated induction, hypothesis, and analogy as probable arguments in 1867. He employed the concept of truth frequency, or the proportion of cases in which an argument "carries truth with it," to analyze probable inference in "The Doctrine of Chances" (1878) and "The Probability of Induction" (1878). And in "Deduction, Induction, and Hypothesis" (1878), he formulated probabilistic versions of the *Barbara* syllogism (8) and its inversions by replacing the universal rule "All the beans from this bag are white" with a statistical generalization:

(10) Most of the beans in this bag are white.
This handful of beans are from this bag.
Probably, most of this handful of beans are white. (*CP* 2.508–16, 627)

The article "A Theory of Probable Inference," published in *Studies in Logic* (1883), which Peirce edited at the Johns Hopkins University, gives several models of probable "deduction" from a statistical premise. *Simple Probable Deduction* is a statistical version of a singular syllogism in *Barbara* (cf. [1]):

(11) The proportion *r* of the *F*'s are *G*'s.
a is an *F*.
It follows, with probability *r*, that *a* is a *G*.

As Peirce noted, the conclusion here is "*a* is a *G*," and probability indicates "the modality with which this conclusion is drawn and held to be true" (cf. the patterns [3]–[5]). Furthermore, it is required that *a* "should be an instance drawn *at random* from among the *F*'s."

> The volition of the reasoner (using what machinery it may) has to choose *a* so that it shall be an *F*; but he ought to restrain himself from all further preference, and not allow his will to act in any way that might tend to settle what particular *F* is taken, but should leave that to the operation of chance.
>
> . . . the act of choice should be such that if it were repeated many enough times with the same intention, the result would be that among the totality of selections the different sorts of *F*'s would occur with the same relative frequencies as in experiences in which volition does not intermeddle at all. In cases in which it is found difficult thus to restrain the will by a direct effort, the apparatus of games of chance—a lottery-wheel, a roulette, cards, or dice—may be called to our aid. (*CP* 2.696)[13]

This condition guarantees that the result *G* is obtained with the long-run frequency *r* within the unlimited population of possible drawings from the class of *F*'s (*CP* 2.731). We can restate the inference schema (11) as in (12).

(12) $\text{fr}(G/F) = r.$
a is a random member of F.
$\therefore a$ is a G.

In turn (12), with a relative frequency statement as a premise, can be formulated as a probabilistic argument with a lawlike statistical premise:

(13)
$$\frac{\begin{array}{l} P(G/F) = r \\ Fa \end{array}}{Ga} \; [r]$$

Here r in the brackets is a long-run truth frequency.[14]

The schema (13) corresponds to Hempel's model (6) of I-S explanation with the difference that $[r]$ indicates an objective probability rather than an epistemic or inductive probability. But even this difference with Hempel is not very great, since Peirce adds—by appealing to Fechner's law—that the objective odds excite in us a sensation or feeling of confidence, and "it is plainly useful that we should have a stronger feeling of confidence about a sort of inference which will oftener lead us to the truth than about an inference that will less often prove right—and such a sensation we do have" (*CP* 2.697).

If probability is understood as a propensity, as Peirce suggested in 1910 (*CP* 2.664), pattern (13) exhibits the propensity model (7) of probabilistic explanation.

Besides Simple Probable Deduction (11), Peirce formulated a schema for *Statistical Deduction*:

(14) The proportion r of the F's are G's.
a', a'', a''', etc. are a *numerous set*, taken at random from among the F's.
Hence, *probably* and *approximately* the proportion r of the a's are G's.

This inference, which has received surprisingly little attention in the contemporary literature on statistical explanation, can be formalized in the following way:

(15) $\text{fr}(G/F) = r.$
$\{a', a'', a''', \ldots\}$ is a random sample of F's.
$\therefore \text{fr}(G/\{a', a'', a''', \ldots\}) \approx r.$

Again the condition for randomness allows us to reformulate (15) as follows:

(16) $P(G/F) = r$
$Fa' \ \& \ Fa'' \ \& \ Fa''' \ \& \ \ldots$

$$\overline{\overline{\qquad\qquad\qquad\qquad}}$$

$\text{fr}(G/\{a', a'', a''', \ldots\}) \approx r$

As Peirce showed by the binomial formula, r is the most probable value of the relative frequency in the conclusion. Furthermore, by Bernoulli's theorem, the probability of the conclusion given the premises approaches 1 when the size of the sample $a', a'', a''', \ldots$ increases without limit.[15]

It is interesting to note that Mill's *System of Logic* already contained a brief discussion of the application of an approximate generalization (like "Most A are B" or "Nine out of ten A are B") to its individual instances. Mill required that we should know nothing about such instances "except that they fall within the class A" (Mill 1906:391; Niiniluoto 1981:444). This guarantees that Hempel's RMS is satisfied, without implying the stronger condition that the class A itself is objectively or epistemically homogeneous. Similarly, Venn's *Logic of Chance* formulated the rule that statistical inferences about an individual case should refer it to the narrowest series or class which still secures "the requisite degree of stability and uniformity" (Venn 1888:220; Reichenbach 1938).

However, Mill and Venn never say that such a statistical inference could be applicable for the purpose of *explanation*. It is clear that, for Mill, approximate generalizations were important primarily "in the practice of life," but in science they are valuable as "steps towards universal truths." Moreover, at least Venn—as many of his followers in the twentieth century (such as Reichenbach and von Wright)—explicitly restricted his attention to attempts "to make real inferences about things as yet unknown," that is, to *prediction* (Venn 1888:213).

Peirce is equally explicit in reasserting his earlier view that "inductions and Hypotheses are inferences from the conclusion and one premiss of a statistical syllogism to the other premiss. In the case of hypothesis, this syllogism is called the *explanation*." Indeed, Peirce repeats, "We commonly say that the hypothesis is adopted *for the sake of* the explanation." A statistical syllogism "may be conveniently termed the explanatory syllogism" (*CP* 2.716–17).

Peirce's formulation of Hypothesis in the 1883 article is not, in fact, an inversion of (11) or (14) but rather of *Statistical Deduction in Depth*:

(17) Every F has, for example, the numerous marks G', G'', G''', etc.
a has an r-likeness to the F's.
Hence, probably and approximately, a has the proportion r of the marks G', G'', G''' , etc.

The special case of (17) is *Simple Probable Deduction in Depth*:

(18) Every F has the simple mark G.
The S's have an r-likeness to the F's.
Hence, the probability is r that every S is G.

Here a has an *r-likeness* to b if the proportion of the shared properties of a and b is r. (If $r = 1$, [18] reduces to *Barbara*.) Thus, (17) and (18) are probabilistic formulations of inference by *analogy*.[16]

It is nevertheless clear that Peirce's remarks about the connection between hypothesis and explanation were intended to cover probable and statistical deduction of forms (11) and (14). For example, in 1878 he formulated an example of hypothesis which is an inversion of the statistical deduction (10).

It might be objected that the schema (12) is not applicable to explanation, since, in explaining why a is a G, we already know the individual a—and therefore cannot draw it randomly from the class F. Here is Peirce's reply:

> Usually, however, in making a simple probable deduction, we take that instance in which we happen at the time to be interested. In such a case, it is our interest that fulfills the function of an apparatus for random selec-

tion; and no better need be desired, so long as we have reason to deem the premiss "the proportion *r* of the *F*'s are *G*'s" to be equally true in regard to that part of the *F*'s which are alone likely ever to excite our interest. (*CP* 2.696)

This is not a very clear statement, but the intuition evidently seems to be that the schema (12) can be used for explanation as long as the reference class *F* is epistemically homogeneous.

It is interesting to add that Peirce was aware also of problems of *relevance*, later discussed by Salmon, Kyburg, and others. In his eleventh Harvard lecture (1865), Peirce observed, "That contrary inductions and contrary hypotheses are possible is notorious"; "the predicate of a hypothesis may contain superfluous comprehension" which is "limitative of the extension" and "which being taken away, the fact is still explained." For example, suppose that the fact "*D* is *C*" can be explained by two (true) hypotheses:

All *I* is *C*.	All *J* is *C*.
D is *I*.	*D* is *J*.
∴ *D* is *C*.	∴ *D* is *C*.

Then, Peirce said, "All *I* and *J* are *C*," and thereby a better explanation with a *more extensive* predicate *K* is possible:

All *K* is *C*.
D is *K*.
∴ *D* is *C*. (*W* 1:293)

This idea can be expressed by saying that the reference class to which *D* is attributed should not be too narrow. In this example, this is achieved by defining the predicate Kx as $Ix \vee Jx$.[17]

Tychism

In his later work Peirce did not revise his 1883 account of the structure of probabilistic inference. But his views of the nature of chance and of the role of chance in nature went through changes which deepened his insight of the indispensability of statistical explanation.[18]

In "The Order of Nature" (1878), Peirce argued that we do not live in "a thoroughly chance-world," since there are "uniformities" in nature. The idea of law, regularity, and generality, expressed in the category of thirdness, came to play a central role in Peirce's antinominalistic "scholastic realism" (Fisch 1967). At the same time, evidence from science suggested that some of the best theories have a statistical character: in "The Fixation of Belief" (1877), Peirce referred to the 1851 theory of gases by Clausius and Maxwell and the 1859 theory of evolution by Darwin.

> Mr. Darwin proposed to apply the statistical method to biology. The same thing has been done in a widely different branch of science, the theory of gases. . . . In like manner, Darwin, while unable to say what the operation of variation and natural selection in any individual case will be, demonstrates that in the long run they will, or would, adapt animals to their circumstances. (*CP* 5.364; see also *CP* 1.104; 6.47, 613)

In his 1892 papers for the *Monist*, especially in "The Architecture of Theories" and "The Doctrine of Necessity Examined," Peirce formulated his evolutionary metaphysics, with its principles of absolute chance (*tychism*) and continuity (*synechism*).

> Now the only possible way of accounting for the laws of nature and for uniformity in general is to suppose them results of evolution. This presupposes them not to be absolute, not to be obeyed precisely. It makes an element of indeterminacy, spontaneity, or absolute chance in nature. Just as, when we attempt to verify any physical law, we find our observations cannot be precisely satisfied by it, and rightly attribute the discrepancy to errors of observation, so we must suppose far more minute discrepancies to exist owing to the imperfect cogency of the law itself, to a certain swerving of the facts from any definite formula. (*CP* 6.13)

In the first years of the twentieth century, Peirce radicalized his criticism of Hume's "Ockhamist" views of the laws of nature (see Wiener 1966:chap. 18), adopted a realist view of dispositional and modal conceptions, and proposed a propensity interpretation of probability as a "would-be" of a physical chance setup (cf. Niiniluoto 1988a, 1988b). Peirce's tychism, as a via media between necessitarianism and nominalism, thus led him to a philosophical position where the world is governed by evolving probabilistic laws and where all explanation of natural phenomena is probabilistic.

Conclusion

Salmon is certainly wrong in claiming that Peirce's statements about statistical explanation are "unelaborated": the structural richness of Peirce's models supersedes that of many contemporary accounts. It is also unfair to suggest that Peirce's statements are "isolated," since they are systematically connected to the most central tenets of his work on probability, scientific method, and evolutionary metaphysics.

Still, Salmon may be right that the epithet "the true founder of the theory of statistical explanation" (if we wish to use anything like that) should be reserved for Hempel: unlike his great successor in 1962, Peirce failed to convince his contemporaries of the existence of statistical explanations.

Notes

1. According to the Humean view, laws of nature are simply extensional "constant conjunctions" between properties or events. The statistical counterparts of such generalizations are statements about the relative frequency $fr(G/F)$ of an attribute G in a class F. Hempel (1965:340) supplemented this view with the requirement that a lawlike generalization $(x)(Fx \rightarrow Gx)$ has to be "essentially generalized": the predicates F and G should be "nomic," that is, not restricted to a finite number of instances on purely logical grounds. This requirement is valid also for lawlike probabilistic statements of the form $P(G/F) = r$ (Hempel 1968). As lawlike statements sustain counterfactual and subjunctive conditionals, non-Humean intensional analyses equate lawlikeness with "physical necessity" (i.e., truth in all physically possible worlds). Probability in physical laws

$P(G/F) = r$ (where $r < 1$) is then a modal operator which is weaker than physical necessity. See Fetzer 1981 and Niiniluoto 1976, 1978a, 1981.

2. The task of giving sufficient and necessary conditions for adequate scientific explanations has turned out to be surprisingly difficult. For a review of the literature, see Tuomela 1977. In my view, it is not correct to say that this difficulty arises from the fact that Hempel's syntactic-semantic approach excludes pragmatic aspects of explanation. Hempel's treatment of explanation is not antipragmatic, but rather it presupposes as given a particular pragmatic context, namely, inquiry within the scientific community.

3. This problem was not recognized in the extended treatment of statistical syllogisms by Williams (1947) but was discussed in detail by Barker (1957).

4. This condition is rejected in Hempel 1977.

5. For summaries of critical discussions of Salmon's S-R model, see Niiniluoto 1981, 1982 and Tuomela 1977.

6. Cf., however, Hempel's (1965:378) reference to Popper's long-run propensity interpretation, and Salmon's (1988) acceptance of propensities as causes of frequencies.

7. For the propensity view of explanation, see Fetzer 1974, 1981; Mellor 1976; Niiniluoto 1976, 1981; and Tuomela 1977. Railton's (1978) model derives from a propensity law the singular consequence "*a* has probability *r* to be *G* at time t_0." I find this misleading, since the relevant explanandum is *Ga* rather than $P(Ga)=r$.

8. Popper (1957:125) claims that (unlike himself) Mill fails to distinguish here universal laws and initial conditions. But this is not supported by Mill's own examples—and, as Ryan (1970:5) observes, "Mill would hardly forget that a syllogism requires a minor premise."

9. For the history of the method of hypothesis, see Laudan 1981. See also Niiniluoto 1978b.

10. This is acknowledged by Jeffrey (1969).

11. Williams (1947:57), who relies heavily on Peirce, mentions that a statistical syllogism could be used for explanation but does not develop this idea further.

12. Many scholastic and Renaissance Aristotelians believed that inference from effects to causes—analysis, resolution, or *demonstratio quia*—can proceed through deductive syllogisms. See Randall 1961. For Peirce, hypothesis is typically not deductively valid.

13. I have changed Peirce's notation here.

14. Among contemporary philosophers, Henry E. Kyburg, Jr., uses relative

frequency statements, and Isaac Levi probability statements, as premises of direct inference. Cf. Bogdan 1982.

15. See *CP* 2.698–700 and Niiniluoto 1976:358; 1981, 1982. Note that (16) is not a deductive-statistical explanation in Hempel's sense, since its conclusion is not a probability statement.

16. See Niiniluoto 1988a. For anticipation of (18), see Mill 1906:365–67.

17. Cf. Salmon's (1971) requirement of *maximal* homogeneous reference classes and the criticism in Niiniluoto 1982.

18. Peirce's later discussions of abduction—defined as the "operation of adopting an explanatory hypothesis" (*CP* 5.189)—demonstrate his continued interest in the notion of explanation.

References

Aristotle. 1930. *Physics*. In W. D. Ross, ed., *The Works of Aristotle*, vol. 2. Oxford: Oxford University Press.

———. 1964. *Prior and Posterior Analytics*. London: Dent.

Barker, S. 1957. *Science and Hypothesis*. Ithaca: Cornell University Press.

Bogdan, R., ed. 1982. *Henry E. Kyburg, Jr., and Isaac Levi*. Profiles, vol. 3. Dordrecht: D. Reidel.

Comte, A. 1970. *Introduction to Positive Philosophy*. Indianapolis: Bobbs-Merrill.

Duhem, P. 1954. *The Aim and Structure of Physical Theory*. Princeton: Princeton University Press.

Fetzer, J. H. 1974. A Single Case Propensity Theory of Explanation. *Synthese* 28:171–98.

———. 1981. *Scientific Knowledge*. Dordrecht: D. Reidel.

———, ed. 1988. *Probability and Causality*. Dordrecht: D. Reidel.

Fisch, M. 1967. Peirce's Progress from Nominalism toward Realism. *Monist* 51:159–78.

Hempel, C. G. 1962. Deductive-Nomological vs. Statistical Explanation. In *Minnesota Studies in the Philosophy of Science* 3, ed. H. Feigl and G. Maxwell, 98–169. Minneapolis: University of Minnesota Press.

———. 1965. *Aspects of Scientific Explanation and Other Essays in the Philosophy of Science*. New York: Free Press.

———. 1968. Maximal Specificity and Lawlikeness in Probabilistic Explanation. *Philosophy of Science* 35:116–33.

———. 1977. *Aspekte wissenschaftlicher Erklärung*. Berlin: Walter de Gruyter.
Jeffrey, R. C. 1969. Statistical Explanation vs. Statistical Inference. In *Essays in Honor of Carl G. Hempel*, ed. N. Rescher, 104–13. Dordrecht: D. Reidel. Reprinted in Salmon et al. 1971, pp. 19–28.
Kant, I. 1974. *Logic*. Indianapolis: Bobbs-Merrill.
Keynes, J. M. 1921. *A Treatise on Probability*. London: Macmillan.
Laudan, L. 1981. *Science and Hypothesis*. Dordrecht: D. Reidel.
Levi, I. 1977. Subjunctives, Dispositions, and Chances. *Synthese* 34:423–55.
Mellor, D. H. 1976. Probable Explanation. *Australasian Journal of Philosophy* 54:231–41.
Mill, J. S. 1906. *A System of Logic*. London: Longmans, Green.
Niiniluoto, I. 1976. Inductive Explanation, Propensity, and Action. In *Essays on Explanation and Understanding*, ed. J. Manninen and R. Tuomela, 335–68. Dordrecht: D. Reidel.
———. 1978a. Dretske on Laws of Nature. *Philosophy of Science* 45:431–39.
———. 1978b. Notes on Popper as Follower of Whewell and Peirce. *Ajatus* 37:272–327. Reprinted in *Is Science Progressive?* 18–60. Dordrecht: D. Reidel, 1984.
———. 1981. Statistical Explanation Reconsidered. *Synthese* 48:437–72.
———. 1982. Statistical Explanation. In *Contemporary Philosophy: A New Survey*, ed. G. Fløistad, 2:157–87. The Hague: M. Nijhoff.
———. 1988a. Analogy and Similarity in Scientific Reasoning. In *Analogical Reasoning*, ed. D. H. Helman, 271–98. Dordrecht: Kluwer.
———. 1988b. From Possibility to Probability: British Discussions on Modality in the Nineteenth Century. In *Modern Modalities*, ed S. Knuuttila, 275–309. Dordrecht: Kluwer.
———. 1988c. Probability, Possibility, and Plenitude. In Fetzer 1988, pp. 91–108.
Popper, K. R. 1945. *The Open Society and Its Enemies*. London: Routledge and Kegan Paul.
———. 1957. *The Poverty of Historicism*. London: Routledge and Kegan Paul.
Railton, P. 1978. A Deductive-Nomological Model of Probabilistic Explanation. *Philosophy of Science* 45:206–26.
Randall, J. H. 1961. *The School of Padua and the Emergence of Modern Science*. Padua: University of Padua.
Reichenbach, H. 1938. *Experience and Prediction*. Chicago: University of Chicago Press.
Ryan, A. 1970. *John Stuart Mill*. New York: Random House, Pantheon Books.

Salmon, W. C. 1965. The Status of Prior Probabilities in Statistical Explanation. *Philosophy of Science* 32:137–46.

———. 1983. Probabilistic Explanation: Introduction. In *PSA 1982*, ed. P. D. Asquith and T. Nickles, 2:179–80. East Lansing, Mich.: Philosophy of Science Association.

———. 1984. *Scientific Explanation and the Causal Structure of the World*. Princeton: Princeton University Press.

———. 1988. Dynamic Rationality: Propensity, Probability, and Credence. In Fetzer 1988, pp. 3–40.

Salmon, W. C., et al. 1971. *Statistical Explanation and Statistical Relevance*. Pittsburgh: University of Pittsburgh Press.

Stegmüller, W. 1973. *Personelle und statistische Wahrscheinlichkeit*. Berlin: Springer-Verlag.

Tuomela, R. 1977. *Human Action and Its Explanation*. Dordrecht: D. Reidel.

Venn, J. 1888. *The Logic of Chance*. 3d ed. London: Macmillan.

Wiener, P. P., ed. 1966. *Charles S. Peirce: Selected Writings*. New York: Dover Publications.

Williams, D. C. 1947. *The Ground of Induction*. Cambridge: Harvard University Press.

von Wright, G. H. 1971. *Explanation and Understanding*. Ithaca: Cornell University Press.

14 Peirce on Problem Solving

Peter Robinson, The University of Texas at El Paso

Nearly everything that Charles Peirce wrote could be construed as being on methods of problem solving. Philosophy is method. The method is pragmaticism. Pragmaticism is problem solving in its widest sense. Peirce's interest in problem solving is an interest not only in its use in the development of science and philosophy but in the development of the human problem solver. There are two poles to problem-solving activity. At one end is the world, which is changed through the problem being solved. At the other end is the problem solver, who is also changed through the discovery of the solution. The dialectical interaction of the individual, acting as part of a community of thinkers, with the external world, comes to Peirce from German idealism. The dyadic relationship of the two is mediated in Peirce by the description of a world which is changed in the process of problem solving. Its internalization by the knower changes the problem solver, and the changed description approaches the truth about the world. The world is more lawful as it evolves into habitual patterns, and it gains the added complexity of the problem solvers and their solutions.

Peirce, as a source of problem-solving techniques, is quite rich because he was involved in promoting almost every human intellectual endeavor. He had something interesting to say about analysis, method, and experiment in every area of science and philosophy. Peirce's pointers on problem solving are important because they are formulated out of, and directed toward, particular cases. Thus he can be cited, for example, for his foundational work on statistical inference, randomization, and double-blind experiments from his critiques of the Weber-Fechner law and his research on mental telepathy (Hacking 1988:427–51). His interests in logic and mathematics may be found in differing ways in his problem-solving research in the sciences and philosophy. He might use them in his analyses or draw a logical or mathematical analysis from an investigation, or both.

Peirce's three modes of logical reasoning—deduction, abduction, and induction—are the general methods of problem solving in their formal presentation. Abduction is peculiarly Peircean and has been the locus for the analyses of Peirce's methodology of problem solving. An attempt to understand his notion of abduction as problem solving leads away from his formal description of abduction as the fitting of minor premises and then being able to infer to a result. An analysis of abduction as problem solving leads toward the sources of the hypotheses and their major premises, whereas in the formal analysis the abduction is completed and the problem solved.

In comparing Peirce's description of abductive method with the problem-solving practices of fictional detectives, the contributors to *Dupin, Holmes, Peirce: The Sign of Three* (Eco and Sebeok 1983) are faced with a similar metaproblem about abduction. They are led away from the formalities only to be faced with the same form of investigative dead end. Holmes and Dupin, being fictional detectives, are somehow, mysteriously, always right. The completed work that their authors published is all there is to go on. We cannot grab a crumpled sheet out of the air after they exclaimed, "Oh, hell, this won't work," before it hits the wastebasket. Abductions, to be abductive, must fit the formal description of abduction, and therefore if the major premise is also true and the syllogism is valid, the abduction is always right.

When philosophical analysts of Peirce's methods of problem solving turn from formal analyses of the place of hypotheses within arguments and how they lead to the solution of problems, criminal detections, or scientific discoveries, they find there are two kinds of guesses to be distinguished, but there is an odd Peircean continuum between them. The first kind of guess is the formulation of a minor premise that fits with the major to infer the result validly. This is pure abduction as formalized.

The second kind of guess is one which could be formalized as a minor premise, but if it is, the major premise is guessed at simultaneously or else selected from an indefinitely large number of possibilities simultaneously with the minor premise. This is creative guessing de novo and not really abduction, in which the major premise and the conclusion are logically prior.

The continuum of degrees in between corresponds to the number of candidates for major premises. If there are a small number of candidates for a major premise and evidence or some reasons are given for the selection of a particular major premise, then the whole scheme could be presented as properly abductive, as being a formalization of a general problem-solving procedure. Thus from "All the beans in this bag are white" and "These beans are white" arises the explanatory hypothesis "These beans are from this bag." Or, "All Great Northern beans are white" and "These beans are white" give rise to the hypothesis "These beans are Great Northerns." Which abduction one performs depends on what is bothering one about beans. In other words, the abduction and the selection of a major premise depend on the problem to be solved, for problems do not exist without goal-oriented organisms. Problems, as well as things to be detected or discovered, exist only for such creatures. The major premises in each problem-solving case must be known, believed true, and relevant. The problem for problem-solving as well as for abductive inference is laying one's hands on the major premise. Abductions in which the major premise comes out of the blue or is not regarded as relevant beforehand are suspicious, unless an account of how the major premise is known and can be relevant is given as part of the abductive process.

Massimo Bonafantini and Giampaolo Proni distinguish three principle types of abduction with three ascending degrees of originality and creativity in their chapter "To Guess or Not to Guess" in *The Sign of Three.* Of their type 1 abduction they say, "The mediation law to use in inferring the case from the result is given in an obligant [unusual] and automatic or semi-automatic way" (Eco and Sebeok 1983:133–34). According to my way of describing abduction, this occurs when the relevant major premise has been settled on and there is only a minor premise to be fitted in to complete the enthymeme. They describe their type 2 abduction by saying, "The mediation law to use for inferring the case from the result is found by selection in the available encyclopedia" (ibid.:129). They cite Peirce's own example of Kepler's discovery of elliptical orbits as a case of the second type and join their distinction of type 1 and type 2 abductions to Kuhn's distinction between normal and revolutionary science (ibid.). But these are on the continuum. It is a matter of degree, which turns on not merely the number of possible major premises to be selected from but how far they are removed from the collection of major premises that are under the purview of the community of scientists. An abduction made from a major premise in the available encyclopedia may or may not constitute a revolutionary solution to a problem in science, although it is certainly more than just filling in a blank.

Of their type 3 abduction they say, "The mediation law to use for inferring the case from the result is developed de novo, *invented.*" They go on to say, "It is in this last type of abduction that the real guesswork comes in" (ibid.:134). They question the legitimacy of this procedure and Peirce's justification for it, claiming that Peirce supports type 3 abductions "with his theory of natural bent, biologically rooted and accumulated in man in the course of evolution: *lume naturale*, ever-increasingly modeled by the influence of the laws of nature and so more and more spontaneously likely, by secret affinity, to reflect the patterns of reality" (ibid.). They criticize this justification because "it implies the biological inheritance of culturally acquired cultural characters," and they say that Peirce needs a *lume culturale*, which they say, "being steeped in bad metaphysics, is too generic in that it explains everything and nothing" (ibid.). I believe that they are correct in distinguishing an abduction type 3 and that they have identified Peirce's theoretical support for it, but I wish to save it from their critique or, if I cannot do that entirely, at least to elaborate on its sources and its character.

Their abduction type 3 appears to me to correspond to my abduction in which the major premise appears to be selected arbitrarily or by guess from an indefinitely large number of alternatives, so that it is as good as invented along with the abductive hypothesis, which together may be used to deduce the result. The first remarkable point of difference between Bonfantini and Proni's typology and some other contributors to *The Sign of Three* is that they describe their abduction type 1 as abduction derived from direct perceptual judgment (ibid.:129–30), whereas, for Thomas Sebeok, for example, the flashes of insight, the hunches or the pure guesses, arise from direct perception (ibid.:17–18). It seems to me that we are dealing here with two different aspects of perception, the first aspect being the immediate appearance of the abduction as falling under the most relevant major premise and therefore appearing direct and commonsensical, so that the abduction is logically mechanical as Bonfantini and Proni describe it. The second aspect is perception as inferential in itself, continuous in its nature with logically conceived abductions but also providing a source for the consciously conceived logical abductions. Here, Peirce's seemingly conflicting accounts of the source of abduction in perception, as well as his various theoretical supports for abductive inferences, need to be sorted out.

Bonfantini and Proni conclude that the source of support for the guesses of type 3 abduction are philosophical assumptions embedded in the culture, thus they need a *lume culturale* as well as a *lume naturale*. They say:

> When men have to guess, they find themselves guided by systematic and complex visions of reality, philosophical conceptions, of which they are more or less distinctly aware but which anyway shape their cast of mind, their deep habits which determine the bearings of judgment. These philosophies synthesize and organize, by processes of generalization, analogy, and hierarchical ordering, the knowledge and cultural acquisitions deposited in the course of the centuries and derived from extensive social practices. (ibid.:134)

In their description of the sources for type 3 abductions they appear to agree with Richard Tursman (1987:19), who says, "From the point of view of originalian logic, such an inference, a remarkable inference to a new scientific idea or hypothesis, takes place, typically, in a neighborhood conducive to its emergence. The building of such a neighborhood involves historical, institutional, biographical, and educational factors, but most importantly it must include the presence of appropriate ideas." Tursman is specifically referring here to Peirce's example of Dalton's hypothesis of the chemical atom, which Peirce considered to be more remarkable than Kepler's discovery of the elliptical orbits.

However, in some high-level hypothesizing as well as low-level automatic abductions, the guess appears to be directly perceived, both cases arising from the given in perception. There are also the interesting intermediate cases, where the hypotheses are clothed in the perception of dream, such as Descartes's dreams and Kekule's benzene ring. Thomas Sebeok, in *The Sign of Three*, is concerned with the case of Peirce's own hypothetical guesswork in his recovery of his watch and his overcoat which were stolen by one of the waiters while he was traveling on the old Fall River Line. The identity of the thief is low level as a theory, but the dyadic relationship between the waiter and Peirce's watch is nothing directly perceivable. Sebeok cites Peirce's own descriptions of the direct perceptual appearance of the abductive inference, such as "an act of insight" and "like a flash," as well as the passage in which Peirce says that perceptual judgments shade into abductive inferences (Eco and Sebeok 1983:18). If the abductive hypothesis appears as directly perceived and if the knowledge on which it is based is culturally ingrained, then the knowledge and the logical processing both show up as part of the perceptual process itself. In other words, observation is theory laden.

Although Peirce says in that same passage (*CP* 5.181) that perceptual judgments are "absolutely beyond criticism," the immediate dyadic relationships are not present to consciousness infallibly, even if they are automatically. Peirce's phaneroscopy, his psychological analysis of the phaneron (*MS* 1334), goes far toward radical empiricism and Husserlian pure perception, but perception is always judgmental, and the guesses that arise unself-consciously in perceptions are only judgments unnoticed initially by consciousness. They are neither something entirely new, newly created, nor do they carry with them epistemological certainty. Even perceptual judgments about the percept itself (Peirce's percipuum) are not intuitedly infallible but only pragmatically warranted (Bernstein 1964:176).

Perceptual judgments pick out relationships from an infinite number of their relationships in which they are embedded. The selections are not unerring or "given," as they might be regarded by empiricists. The selection is guided by experience that may not be conscious but may be biologically or culturally ingrained. The question remains, What makes a particular guess, arising in perception from an infinite panoply of other relations present with it, the serendipitous one? This is a different question from one that could be answered by invoking the pure tychism which might be required in characterizing an abductive inference as arriving at a completely new hypothetical idea in which the major premise must come into being along with the minor one. A new hypothetical idea of this kind would be part of the larger creative process that is ongoing, adding novelty to the universe. Following a reference to Dalton's theory, Peirce characterized that kind of creative process as being given hints that come straight from a divine Creator and that we ought to labor to cultivate this privilege (*NEM* 3:206). In other words, we ought to be working with this Creator in producing a world that is in progressive development.

The aspects of critical thinking and problem solving that Peirce tries to capture in his notion of abduction, but which escape from his formalization of abductive thinking, are contained in the sources for the hypothetical guess. It is a guess that appears to arise from unconscious cues in perception and appears to be as unquestionable as perception itself. It is the appearance of a pattern or interpretation not to be found in the elements of the empirically given themselves but arising from them anyway. This is the most interesting aspect of Peirce's studies of problem solving. The insightful understanding that such things occur, and that their explanation must follow from correct scientific and philosophical assumptions, makes Peirce the first modern psychologist of problem solving. His answers are the ones that chart today's inquiries after an intervening period dominated by an empirical atomism and a syntactic logic that has now partially failed. Peirce went further in his philosophical assumptions than the science of the time could justify. By not separating the knower from the known or from the community of knowers, Peirce allowed for an explanation of mental development that did not separate, as orthodox Darwinism did, the genetic contributions from the environmental ones, so that he perceived that the explanation of mental development, including problem-solving abilities, must also include developing interaction with the environment and among a community of thinkers. Both Mead's and Dewey's accounts as well as contemporary neurophysiological accounts of mental development are prefigured in Peirce's view.

Peirce's belief in the universal application of continua was affirmed in Darwin's theory of evolution, wherein human characteristics were continuous with those of other animals. What makes one truly human, and how humanness can continue to develop, must be understood in terms of the structures from which it arises. The two older contemporaries of Peirce who held positions similar to his on the details of biological continuity were Chauncey Wright and Wilhelm Wundt. Chauncey Wright's article "The Evolution of Self Consciousness" makes clear his position that the degrees of abstraction and the extended trains of inference in human thought differ only in degree of complexity from animal thought. He says:

> The line of distinction between such results of reflection, or between scientific knowledge and the common-sense form of knowledge, is not simply the dividing line between the minds of men and those of other animals; but is that which divides the knowledge produced by outward attention from that which is further produced by reflective attention. The former, throughout a considerable range of the higher intelligent animals involves veritable judgments of a complex sort. It involves combinations of minor premisses in the enthymematic reasonings, commonly employed in inferences from signs and likelihoods, as in prognostications of the weather, or in orientations with many animals. This knowledge belongs both to men and to the animals next to men in intelligence, though in unequal degrees. (Wright 1877:207)

The enthymematic reasonings involving minor premises appear to be the precursors to Peircean abductions.

Similarly, Wilhelm Wundt, in his *Lectures on Human and Animal Psychology*, bases his studies on the continuity of human and animal thought (Wundt 1977:xxi). But although human intellectual processes developed from animal ones, Wundt takes the object of his own investigations of animal psychology "only so far as they throw light on the evolution of consciousness in man" (ibid.:340). He considers that there is now a gulf between human mentality and animal mentality (ibid.:365). However, in conclusion to his lecture 24 in section 5, "Man and the Animals," he says, "When once the mind has reached the level of logical thought and the constructive imagination, it has before it that prospect of unlimited advance which must inevitably at some point in time be realized in civilization and history" (ibid.:365–66). This projection into the future of the development of human logical thought he shares with Peirce, although for Peirce, if it is to occur, it will not be through Spencerian inevitability but through individual and community effort. That mental processes permeated all of nature was enthusiastically embraced in the nineteenth century. So, the notion that there was a continuity of intellectual processes running from lower forms of animal life through higher, and from simpler mental processes to more complex, was a commonly accessible one, to be found in the more popular philosophers such as Eduard Von Hartmann, Herbert Spencer, and George Lewes.

The continuity of animal and human thought was also affirmed by Hermann von Helmholtz when he said, "Animals make the same kind of inductive inferences, and they make more incorrect ones than we do" (Helmholz 1971:508). In the same article he refers to his *Handbook of Physiological Optics* of 1866, in which he introduced the famous phrase "unconscious inference," referring to inductive inferences operating in perception (ibid.:371). In a 1878 address entitled "The Facts in Perception," Helmholtz refers to the phrase "unconscious inferences" in his earlier works as inferences that are unconscious insofar as "their major premise is not necessarily expressed in the form of a proposition; it is formed from a series of experiences whose individual members have entered consciousness only in the form of sense impressions which have long since disappeared from memory. Some fresh sense impression forms the minor premise, to which the rule impressed upon us by previous observations is applied" (ibid.:381).

Having a major and minor premise must constitute a deductive inference, so that, for Helmholtz, both inductive and deductive unconscious inferences operate in perceptual judgments. Peirce needed only to augment this account with abductive inferences operating unconsciously and appearing as perceptual judgments to find in perception itself a source for hypothetical guesses. Peirce's view that the mentality that constituted human life was continuous, although of a higher degree, with the rest of the living world, was well supported by the best physiological and psychological thought of the time. Only the additional Lamarkian assumption—that one could, by an act of will, participate in the evolutionary advance of one's species—needed to be added in order for problem solving to be crucial for furthering human cosmic progress. That assumption of a vitalistic evolutionism, accepted by Peirce, was a common enough one, and although he saw it as one that separated him from Spencer, it also separated him from Chauncey Wright (Wiener 1949:77–96).

Peirce's analysis of the relationship between perceptual judgments and conceptual ones operates with two different notions of complexity. One notion might be called intrinsic and the other extrinsic. While problem-solving activity discovers and adds complexity to the external world, the thought processes and the solutions to the problems are still simple, internal elements drawn out from the complexity that makes up the world. In making inferences, an infinite number of inferences occur before the ones that we consciously make. Perception involves inferences that are infinitely more complex than conceptions. Peirce says:

> In short, what I wish to say, is that although the act of perception cannot be represented as whole, by a series of cognitions determining one another, since it involves the necessity of an infinite series, yet there is no perception so near to the object that it is not determined by another that precedes it for when we reach a point which no determining cognition precedes we find the degree of consciousness there to be just zero, and in short we have reached the external object itself, not a representation of it. (*W* 2:178–79)

In arguing against simple contemplation or intuition, Peirce uses the example of the blind spot in the middle of the retina to support the claim that in looking with one eye, the sensation of a continuous and unbroken space nearly circular is an inference (*W* 2:166). The blind spot was discovered by Marriote in 1668 but was of recent scientific interest in Peirce's day because of the further physiological researches of Jan Purkinje and Helmholtz in the 1850s. Peirce uses Helmholtz's notion of unconscious inference there, but he expands and reverses the order of complexity, claiming that the "simpler" psychological processes of Helmholtz are not only continuous with conceptual processes but are more—in fact, infinitely—complex. Peirce says, "It is plain that the process we have found to compose any step of perception, a process of the determination of one judgment by another, is one of inference in the strict sense," and he further says, "But if we were to discuss with equal fullness the act of attention which determines the subject, we should find that this also is determined by previous acts of attention and that there is no more a first in this case than in any other" (*W* 2:180). Our direct experience which constitutes our known world is infinitely more complex, in the same terms, as is our conceptualization of our methodology which we use to analyze and describe it, yet the ability to analyze and describe is a higher-order evolutionary advance. The cognitions are not only less complex but, following Hume, less lively than the sensations and perceptions which they are about (*W* 2:177).

While we are continuous with the world biologically and conceptually, and are its most evolutionarily advanced development, the world is infinitely more complex than our understanding of it. Individuals exist only by participating in the world, and they exist as conscious intellects only by tracing out some of the individual strands of inferences already found in perceptual judgments. Peirce concludes in his Lowell lectures of 1866, "We have already seen that every state of consciousness is an inference: so that life is but a sequence of inferences or a train of thought" (*W* 1:494).

Some of Peirce's insights have become commonplaces, such as that observation is theory laden, so that both perceiving something as a problem and the problem one perceives it as are judgments that are shaped by theoretical assumptions. The patterns that appear to be the perception itself are judgments that may show forth a clue to solving a problem, or they may also obscure clues hidden in an alternative interpretation. The key to problem solving lies in pattern making. The seeing of relationships is not merely a first step in a continuum that ends in formal, conceptual analyses, but the formal analyses are a later development out of perceptual patterns. Formal logic consists in ignoring most of the pattern. Rather than going beyond the significant context for knowledge, it merely breaks with it for the sake of a kind of clarity and elegance (Wells 1964:304–22). But problem solving does not consist in following out those thin strands of pattern that are described in formal accounts of reasoning, much less actually consisting of those accounts; rather, it is having in mind the needed relationships, codable into major premises of arguments, at least if one is to articulate one's thought and to cast it into the form of inferences from which some abductive inferences can be drawn in the form of hypotheses. One is not aware of most of one's stored knowledge or the inferences one draws from it (Margolis 1987).

How major premises are sifted through, and how abductive guesses just come in a flash of insight or come as directly perceived, may seem automatic or beyond human control, but one can place oneself in the proper context to have them. If one does not know the proper context, it only pushes this metaproblem back a step to say that it is accomplished through a kind of hunch or blind-sight, but at least if one knows the kind of problem, then one has a better chance at finding the context. Peirce, like Sherlock Holmes, immersed himself in the details of his various problem situations. Their styles come from a form of life designed to give an advantage in a mode of problem solving. It is all anyone can do. The ability to extract complex explanatory patterns out of experience is a product of ongoing interaction with the whole environment. It does not occur in isolation or out of its own internal dynamic. The world grows. The problem solver grows. The world description grows. In a realist view of scientific law, each one is something in itself, a first. Each forms a real relationship with the other two, a secondness, and each serves as an interpretant for the others, a third. Analyses of what constitutes problem solving is by its nature static. The cases are all of problems successfully solved. Elaboration of method does not bring with it much in the way of understanding of how to solve problems. But problem solvers, depending on their problems, may have different kinds of lives. Their lives might be described as various attempts to place themselves in a fertile context for finding solutions to their problems. If one is to aspire to be a master problem solver, one must have the agility and flexibility to move into the proper context. One might appear to be "mercurial," an adjective used to describe Peirce.

References

Bernstein, Richard J. 1964. Peirce's Theory of Perception. In *Studies in the Philosophy of Charles Sanders Peirce: Second Series*, ed. Edward C. Moore and Richard S. Robin, 165–89. Amherst: University of Massachusetts Press.

Cummins, Robert. 1983. *The Nature of Psychological Explanation.* Cambridge: MIT Press.

Eco, Umberto, and Sebeok, Thomas A., eds. 1983. *Dupin, Holmes, Peirce: The Sign of Three.* Bloomington: Indiana University Press.

Hacking, Ian. 1988. Telepathy: Origins of Randomization in Experimental Design. *Isis* 79 (298):427–51.

Helmholz, Hermann von. 1971. The Origin and Correct Interpretation of Our Sense Impressions. 1894. In *Selected Writings of Hermann Von Helmholz*, ed. Russell Kahn. Middletown, Conn: Wesleyan University Press.

Margolis, Howard. 1987. *Patterns, Thinking, and Cognition: A Theory of Judgement.* Chicago: University of Chicago Press.

Rosenthal, Sandra B. 1969. Peirce's Theory of Perceptual Judgement: An Ambiguity. *Journal of the History of Philosophy* 7 (3):303–14.

Tursman, Richard. 1987. *Peirce's Theory of Scientific Discovery: A System of Logic Conceived as Semiotic.* Bloomington: Indiana University Press.

Wells, Rulon. 1964. The True Nature of Peirce's Evolutionism. In *Studies in the Philosophy of Charles Sanders Peirce: Second Series*, ed. Edward C. Moore and Richard S. Robin, 304–22. Amherst: University of Massachusetts Press.

Wiener, Philip P. 1949. *Evolution and the Founders of Pragmatism.* Cambridge: Harvard University Press.

Wright, Chauncey. 1877. *Philosophical Discussions of Chauncey Wright*, with a biographical sketch of the author by Charles Eliot Norton. New York: Henry Holt.

Wundt, Wilhelm. 1977. *Lectures on Human and Animal Psychology.* First published in German in the 1860s; first English translation in 1894. In the series Contributions to the History of Psychology, pt. D (Comparative Psychology), vol. 1. (W. M. Wundt). Ed. Daniel N. Robinson. Frederick, Md.: University Publications of America.

Part 2

The Physical Sciences

15 Peirce as Participant in the Bohr-Einstein Discussion

Peder Voetmann Christiansen, Institute of Mathematics and Physics, Roskilde University Center, Roskilde, Denmark

The development of the triadic system of categories in 1885 served for Charles Peirce as a bridge back from general ideas of epistemology and semiotic to specific questions in physics. In the years 1891–93 he presented to the public these new ways of his thinking in a series of five papers to the new philosophical journal the *Monist*. Today, after almost one hundred years of radical changes in physics, these papers still present a surprising wealth of relevant ideas.

In the last decade of the nineteenth century the main questions in physics concerned the theories of heat and energy and the existence of atoms and molecules. The central person in this debate was the Austrian physicist L. Boltzmann, who had two main groups of opponents. One group was the *energeticists*, who favored the new theory of heat thermodynamics but did not believe in the reality of atoms. Another group, the *mechanicists*, were more willing to believe in atoms but could not accept thermodynamics because the thermodynamic concept of irreversibility, as expressed by Clausius in the law of increasing entropy, contradicted the principles of Newton's classical mechanics. At the same time, in the United States, arguments were developed that might have helped Boltzmann in his struggle against these two groups, had they been appreciated in time: J. W. Gibbs created the new discipline of statistical mechanics, and Peirce introduced the viewpoints of *tychism* and *synechism* in his *Monist* papers. Gibbs's contribution was to become one of the most important founding blocks of twentieth-century physics, whereas Peirce's remained unknown. Today we are in a better position to understand the importance of Peirce's ideas for several reasons: first, because his criticism against classical mechanics and his prophetic vision of a new cosmology, where laws are developed out of chaos, are in accordance with modern physics; second, because some fundamental problems in the interpretation of quantum mechanics remain a puzzling source of disagreement, in continuation of a long debate between Bohr and Einstein. Peirce appears as a highly relevant third party in this discussion, whose ideas are in many ways similar to Bohr's, but in other ways closer to Einstein's.

In the debate between Boltzmann and his two groups of opponents, Peirce did not hesitate to express his opinion. Although he does not mention Boltzmann in the five *Monist* papers, he clearly shares his views. Atoms exist, as a long chain of experimental evidence shows. The simplest and most convincing proof, according to Peirce, is a small apparatus known as Crookes radiometer, which can still be found on the shelves of physics classrooms. It consists of a closed glass bulb filled with air at a rather low pressure. Inside the bulb is a small mill wheel whose three or four wings are reflective on one side and black on the other side. When light is shone on the mill, it rotates with the black side of the wings to the rear in the direction of rotation. If one does not believe in molecules, one would be tempted to say that the mill is driven by "light pressure," but this cannot be true because the momentum change of the light particles is greater on the reflective side of the wings, which therefore ought to be behind. The fact that the black side is behind is easily explained by the molecular hypothesis, because the black side gets heated by absorbing the light, and the molecules are therefore reflected with greater momentum here than on the light reflective side.

So much for the energeticists. The other group of Boltzmann's opponents, the mechanical reductionists, who did not believe in thermodynamics because it contradicts the doctrine that mechanical laws explain everything, got a longer and much more carefully elaborated refutation in the second *Monist* paper "The Doctrine of Necessity Examined." This paper is without doubt one of Peirce's most remarkable. Not only does it present philosophically strong arguments against the universal validity of classical mechanics and its metaphysical follower, necessitarianism, it also presents its adverse doctrine, *tychism*—the viewpoint that chance is a genuine factor in the physical universe and precedes order and laws of nature in its evolution. Apart from being able to bring the concept of evolution to harmonize with physics, tychism also presents some thoughts about the nature of physical observables that are very pertinent to the early phase of the Bohr-Einstein discussion.

Does God Play Dice?

In 1927 the first mathematical formulation of quantum mechanics was being developed by Heisenberg, who worked at Copenhagen in close collaboration with Bohr. At the same time, Bohr struggled to shape philosophical ideas he had cultivated since his early youth into an epistemology of the new atomic physics. The results were Heisenberg's matrix formulation of quantum mechanics, which included the uncertainty relations, and Bohr's theory of *complementarity*.

One of the main points of discussion between Bohr and Heisenberg was the ontological status of the uncertainties in Heisenberg's relations. Heisenberg tried to maintain a classical view, that the quantities in question—the position and momentum of an atomic particle—had some *true values* unknown to the observer. When a measurement tried to determine these values, they would be disturbed from their true values because of the interaction with the measuring apparatus, and this disturbance could not be arbitrarily small because of the finite value of the quantum of action. Bohr, however, objected strongly to this theory of disturbance, because he objected to the metaphysical idea of quantities having true values when these values in principle never could be known. By using all his famous powers of persuasion, Bohr was finally able to convince Heisenberg that the complementarity viewpoint was correct and the disturbance theory wrong, but other physicists outside the Copenhagen circle, such as Schrödinger and Einstein, were not convinced.

Bohr regarded the act of measurement as a two-stage process: first, the apparatus would tell the system what kind of property was being measured; second, the system, having accepted the assumption of this property, would choose a numerical value for it. Therefore, the properties of an undisturbed system are only potential properties until the proper setting occurs, and then the property suddenly appears by an abrupt and random choice called the *collapse of the wave function*. The situation is like a human choice, as described by the existentialist Danish philosopher S. Kierkegaard, for example, in his book *The Concept of Dread* (1844), which Bohr admired greatly and had studied in his youth. As a simple example one may take the case when person A asks person B, "What number are you thinking of now?" The question will force B first to think of a number, which B presumably did not do before the question, and then to choose a value before answering (e.g., "17"). Einstein strongly objected to the element of chance, in connection with a choice, having a legitimate place in fundamental physics. The legend tells us that Einstein exclaimed, "God doesn't play dice!" to which Bohr replied, "Stop telling God what to do!" (Einstein indeed wrote such a remark in a letter to Max Born in 1926, but it is more difficult to document Bohr's reply.)

Peirce's 1892 paper "The Doctrine of Necessity Examined" might have given Bohr some other more elaborate possible replies, if he had known it. There are no signs that Bohr knew any of Peirce's writings in 1927, but his teacher of philosophy H. Höffding had been visiting two of Peirce's close friends, William James and Lady Victoria Welby, the English semiotician, in the autumn of 1904, and Bohr had read James's *Principles of Psychology* in 1905. There are strong similarities between the Copenhagen school of quantum mechanics and American pragmatism, and it seems a reasonable conjecture that Bohr would have accepted Peirce's arguments in favor of tychism.

Peirce, after having refuted several necessitarian arguments, considered the idea that continuous observables had exact values prior to measurement. As an experienced experimentalist, he could easily reject the claim that this idea was supported by experiments: "An error indefinitely small is indefinitely improbable; so that any statement to the effect that a certain continuous quantity has a certain exact value, if well founded at all, must be founded on something other than observation" (*CP* 6.44). Thus, Peirce did not exclude the possibility that some continuous quantities could have exact values independent of measurement;

in fact he gave several examples where such a claim would seem reasonable, but he did not accept exactitude a priori.

To Einstein, the independent existence of exact values was a necessary axiom because it seemed to be equivalent to the notion of an independently existing physical reality, and realism was not to be given up lightly. Bohr was not eager to give up realism either, but his realism became suspicious in the eyes of many physicists because it was not supported by the doctrine of exact values of physical observables *an sich*.

It became increasingly difficult for Bohr to maintain realism. In 1935, in the answer to Einstein, Podolski, and Rosen, Bohr seemed to have abandoned it completely in favor of a special kind of nominalism with the quantum mechanical formalism in the role of a supreme judge of what might correctly be regarded as reality. We shall return to a discussion of this viewpoint in the next section.

Peirce could maintain his realism without the notion of exact numerical values of potential observables because he had developed semiotic to a point where it was able to treat classes of signs existing independent of the human consciousness. The index category of signs seemed to have acted as a "secret weapon" (it was not mentioned directly in the *Monist* papers) that gave him the strength to withstand the nominalistic temptation. In general, a sign is conceived as a genuine triadic relation between (1) a sign vehicle, (2) an object, and (3) an interpretant. However, the interpretant may be latent, and in its absence the sign is of a "degenerate" type, called an index, expressing a dyadic relation between sign vehicle and object. Thus, a physicist is concerned with translating the indexical signs of nature into the symbolic signs of physics, and this process involves the setting of an interpretant. The quantum mechanical measurement process fits well to this general description with the measuring apparatus as the embodiment of the interpretant.

The final mathematization of quantum mechanics was made by Dirac, who, without knowing of Peirce's semiotic, invented a system of notation that contains just enough structure to distinguish between the indexical signs of physical systems by themselves and the symbolic values of measured observable quantities. The system, by itself, is described by a state vector without a coordinate representation (the "kets" and "bras"), and the measurement apparatus serves to define a coordinate system in the vector space of the system. Finally, the symbolic values pertaining to an observable are represented as eigenvalues of the operator representing the process of measurement. The state vectors of Dirac are pure indices, which makes it possible to refer to the system *an sich* without the metaphysical assumption of symbolic "hidden variables" belonging to it.

Dirac's notation is now applied by most physicists. It facilitates an unphilosophical workday realism, but on festive occasions, when verbal declarations are needed, Bohr's somewhat nominalistic statements are often quoted literally.

Synechism and Local Realism

In the third paper of the *Monist* series, "The Law of Mind," Peirce began to expose a theme that was to become very important for his thinking in his last years, the philosophy of continuity, called *synechism.* This development was partly inspired by Cantor's invention of the mathematical theory of sets, particularly the notion of cardinal numbers of different degrees of infinity. Peirce expressed great admiration for Cantor, but in later papers and letters he gradually turned away from the notion of the continuum as a collection of points and came closer to what is now known as nonstandard analysis (Dauben 1984). This is already apparent in "The Law of Mind," where he describes the continuum as made up of infinitesimal intervals.

The synechistic philosophy of Peirce is built on a realistic conception of signs as living entities existing inside and outside the human mind. The signs live in a continuum where they propagate by a wavelike or diffusive motion, and when they meet other signs, a merging and fusion occurs whereby new meaning arises. The essence of continuity is that no jumps occur in the propagation of signs, and no action-at-a-distance is possible. Thus, synechism at first sight is unfavorable to phenomena like telepathy, but as Peirce cautiously remarks near the end of the article, there may be continua other than space and time wherein such things may propagate.

The existence of a maximum velocity for signal propagation, the velocity of light, is a consequence of Einstein's special theory of relativity from 1905 and was not known to Peirce in 1892, but it would have been a strong argument for synechism which in some ways is very close to Einstein's concept of *local realism.*

In 1935 Einstein, Podolsky, and Rosen (EPR) wrote a paper with the title "Can Quantum Mechanical Description of Physical Reality Be Considered Complete?" This paper is Einstein's comeback in his discussions with Bohr after his defeat at the Solvay meetings in Brussels in 1927 and 1930. The idea of EPR is to show that quantum mechanics is incomplete because, for some special situations, it violates the principles of local realism (Einstein locality) and seems to invoke action-at-a-distance in connection with the collapse of the wave function. The situation described in the EPR paper is one where two particles are emitted from an atom such that their total momentum is 0. By measuring the momentum of particle 1, an experimenter is at the same time forcing particle 2 into a state of well-defined momentum, even though the particle is far away and did not possess this property before the remote measurement.

Immediately after the reception of the EPR article, Bohr produced an answer with the same title. Surprisingly, Bohr did not deny that quantum mechanics would have such strange implications. As for the principle of local realism, he did not question the locality principle but only Einstein's conception of realism. According to Bohr, the measurement on particle 1 is not able to influence particle 2 by physical means but only by something called "the very conditions which define the possible types of predictions regarding the future behavior of the system." These mysterious "very conditions" belong exclusively to the formalism and have no counterpart in physical reality. By introducing them into his philosophy, Bohr finally abandoned realism and adopted a peculiar sort of nominalism, in which the quantum mechanical formalism decides how one may speak correctly about reality. Strangely enough, Bohr did not consider the possibility that one could make an independent measurement on particle 2 in order to see if it would act in accordance with "the very conditions." Between the lines the article states that if such a measurement did not agree with the prediction, the measurement must be in error.

In 1964 the Irish physicist John S. Bell, working at CERN in Geneva, published a paper which finally made such experiments possible. Instead of momentum, Bell proposed to measure the components of spin, or polarization perpendicular to the direction of motion, of the two particles. Bell made a mathematical formulation of Einstein's principle of local realism by assuming that the state of the particles could be described locally by a set of hidden variables. When a measurement was performed on one particle, the outcome would depend on the values of the hidden variables and the setting of the local measuring apparatus, but it would be independent of the setting of the remote apparatus measuring the other particle of the pair. This formulation of the locality principle led to a set of inequalities that might be tested experimentally.

Bell's inequalities were subsequently tested in a long series of experiments measuring correlations of the polarizations in pairs of photons emitted by an atomic cascade process. The first experiments gave conflicting results—some of them seemed to confirm Bell's inequalities, others agreed with quantum mechanics. Finally, the experiments conducted by Aspect and his coworkers at Orsay (finished in 1982) have convinced most physicists that quantum mechanics is correct and Bell is wrong.

The concept of local realism, formulated with local hidden variables having exact but unknown values, thus seems to be refuted, although some weak loopholes in the analysis of the experiments still exist. Many physicists and popular essayists have then jumped to the conclusion that nature permits some sort of action-at-a-distance, even though EPR-like experiments do not make superluminal communication possible. A. Shimony has accordingly proposed the name "passion-at-a-distance" for this strange "effect" that apparently has no practical consequences except to violate Bell's inequalities under certain special circumstances.

Many extravagant ontological interpretations of the quantum mechanical formalism have gained credibility because of the Aspect experiment, like splitting universes and/or backward causality (Wheeler), nonlocal hidden variables (Bohm), or just a vague holistic notion that everything depends on everything else (Capra and many others). This is unfortunate because the experiments just confirm the formalism which is suited to describe simple experiments with atomic particles and not to be extrapolated to a "theory of everything," including telepathy and synchronicity. In the opinion of the author, the situation calls for a rethinking of the philosophical principle of local realism in connection with the formalism, and for this purpose Peirce's concept of synechism may be used.

As emphasized earlier, Peirce's realism does not need the notion of hidden variables having exact values independent of measurement. Dirac's state vector is sufficient because it has the character of an index; this category, according to Peirce, has just the right sort of signs to expect in reality prior to the setting of an interpretant (i.e., a measurement). The symbolic values that are the outcomes of measurements are always contextual; that is, they depend on the whole context of measurement. If one can point to a common context for the two measurements on the particles of the pair in the EPR experiments, one may expect another outcome than if the two measurements are independent.

By pursuing this kind of contextual logic, one can show that synechism, Peirce's concept of local realism, does not lead to Bell's inequalities. It is possible to have a situation where no communication is possible between the two "twins" but where Bell's inequalities are violated. If, for example, two human twins are asked about their political persuasion by *the same person*, we have a situation where the inequalities may be violated. This is because the answers may depend on unconscious "signals" emitted by the interviewer. For two statistically independent interviewers, however, the inequalities will always be satisfied, provided no possibility of communication exists between the twins.

In all of the EPR-like experiments that have been performed, it is easy to point to a common context for the two single-particle measurements. All registrations take place in a centrally placed coincidence counter. It is impossible to detect one of the particles without simultaneously detecting the other. However, in most popular accounts of the experiments, the situation is described as if the measurements are independent, such that in principle the distance involved could be "cosmological." It is very difficult to draw wires from one planet to another, so an experiment over cosmological distances will have to wait until it becomes possible to perform the experiment as it is described in the popular accounts, that is, without the use of coincidence counters. If such an experiment was performed (it does not need to be from one planet to another, from one end of a laboratory to the other would be sufficient) without a clearly visible common context for the single-particle detections, and if it was found that Bell's inequalities are still violated, then one could say that local realism is in trouble.

Until then, we can safely believe in synechism as the sort of local realism that has not been violated by any experiments, and we can skip the extravagant ontologies in order to understand quantum mechanics as a logically consistent theory within its limited field of application. The future may show that neither Bohr nor Einstein but Peirce had the correct interpretation of a theory he did not even know.

References

Bell, John S. 1964. On the Einstein-Podolsky-Rosen Paradox. *Physics* 1:195–200.

Bohr, N. 1935. Can Quantum Mechanical Description of Physical Reality Be Considered Complete? *Physical Review* 48:696–702.

Dauben, J. W. 1984. C. S. Peirce's Philosophy of Infinite Sets. In *Mathematics: People, Problems, Results*, ed. D. Campbell and J. Higgins, 218–22. Belmont, Calif.: Wadsworth Publishing.

Einstein, A.; Podolsky, B.; and Rosen, N. 1935. Can Quantum Mechanical Description of Physical Reality Be Considered Complete? *Physical Review* 47:777–80.

Peirce, C. S. 1891. The Architecture of Theories. *Monist* 1:161–76. (*CP* 6.7–34)

———. 1892a. The Doctrine of Necessity Examined. *Monist* 2:321–77. (*CP* 6.35–65)

———. 1892b. The Law of Mind. *Monist* 2:533–59. (*CP* 6.102–63)

———. 1892c. Man's Glassy Essence. *Monist* 3:1–22. (*CP* 6.238–71)

———. 1893. Evolutionary Love. *Monist* 3:176–200. (*CP* 6.287–317)

16 From Peirce to Bohr: Theorematic Reasoning and Idealization in Physics

Eliseo Fernández, Linda Hall Library and
University of Missouri at Kansas City

Introduction: The Philosophical Lesson of Quantum Physics

If one were given the impossible task of encapsulating Peirce's philosophy of science into a single term, perhaps the least ridiculous choice would be the word "synechism," his label for the doctrine of the primacy of continuity. By properly unfolding what he sought to pack into that term, one would be led to articulate the multiple dimensions of Peirce's outlook from their common unifying root: his view of the nature of continuity. Pedagogically speaking, however, this would be an unhappy choice. Perhaps no other Peircean conception has encountered more difficulties and opposition than synechism. No less a Peircean scholar and sympathetic commentator than Charles Hartshorne (1973:196) saw in Peirce's "unreasonable" attachment to this doctrine the cause of his "failure" to predict an essential feature of quantum physics:

> I find something pathetic in Peirce's failure to anticipate both basic aspects of quantum physics, instead of only one of them [indeterminism], for he had all the conceptual tools needed for the second anticipation [discontinuity]. Moreover, indicating the right direction for the development of physics was one of his professed ambitions and, he thought, a test for the soundness of his philosophy.

I do not cite this statement in order to take issue with Hartshorne's criticism of Peirce on the obscure and complicated subject of the metaphysical import of synechism (although developments in nonstandard analysis and category theory invite a reconsideration of these issues). My reason for using it is that it illustrates some fundamental and pervasive misconceptions about the far-reaching conceptual changes introduced by quantum physics and their consequences for our views on scientific knowledge in general. These consequences were often referred to by Niels Bohr as "the philosophical lesson of quantum physics." I would like to use Hartshorne's quotation as a point of departure for sketching a path along which, aided by some of Peirce's most original ideas, we may finally bring one part of this lesson home.

There is a paradoxical irony in Hartshorne's statement in that it fails to grasp the radical *discontinuity* between quantum mechanics and what we have learned to call "classical physics." Peirce did not, and could not, predict the peculiar empirical findings that forced this transition, but as I hope to show, he was able to foresee the discontinuity itself and also the general character of the principles by which it was ultimately overcome.

Current research seems to indicate that the most distinctive features of quantum physics are neither indeterminism nor discontinuity. Indeterminism is necessarily present in classical physics, as Peirce, along with Maxwell, Poincaré, and others, concluded on philosophical grounds. Recent developments in nonlinear dynamics and chaos have shown that, in its weaker form of practical unpredictability, it is an inherent feature of Newtonian dynamics. I will have more to say on discontinuity later.

The original rift between classical physics and quantum physics consists precisely in the fact that quantum physics cannot be conceptually anticipated from within classical physics. This kind of rift is not at all a requisite for revolutionary theories. The special theory of relativity, for instance, was able to resolve deep-seated tensions between classical mechanics and electrodynamics by a single revolutionary revision and generalization of some basic concepts of mechanics and was thereby able to account for anomalous empirical findings (such as the Michelson-Morley experiment) without sacrificing the basic explanatory schemes of classical physics.

Quantum mechanics, in contrast, is a radically empirical theory that, in the face of recalcitrant facts, was developed piecemeal through a painful, protracted revision not only of basic concepts but also of some of the most cherished explanatory ideals and claims of "classical" physics.

I intend to outline here an interpretation of the general explanatory scheme of classical physics and to trace the changes it had to undergo in order to assimilate the empirical revelations of early atomic physics. With the help of Peirce's notions of theorematic reasoning and abduction, both classical and quantum explanations will then appear as based on a balanced interaction between the processes of idealization and experimentation. I will finally return to the issue of Peirce's anticipations to assess them through the insights gained by this analysis.

Theorematic Reasoning and Idealization

Classical physics originated in the wedding of mathematical speculation and experimental praxis, a process that was well underway by the early seventeenth century. A new ideal of scientific knowledge was forged in which mathematical and experimental representations of natural phenomena simultaneously constrained and stimulated each other. The experimental praxis of instrumental measurement provided mathematically manipulable raw material to the theory builders, while also restricting their output through the experimental testing of their precise predictions. Conversely, mathematical theory dictated to experimenters the targets for their observations, while at the same time supplying them with the questions to be settled through experiments.

This fertile partnership between formerly divorced and mutually exclusive fields of endeavor was achieved through the mediatory offices of a third essential element: idealization. Scientific idealizations are diagrammatic representations of aspects of natural phenomena which appear relevant to a theorizer in view of a certain explanatory project. In the Galilean explanation of the fall of an apple, for example, not only air friction but also color, flavor, and so forth are ignored. The actual apple is replaced by a sphere endowed with mass and a few other mathematically describable attributes: its space and time coordinates and some of their relationships. Idealizations can act as mediators between mathematical theory and the world disclosed by experimental observation because they partake of features essential to each.

On the mathematical side, their construction is guided by simplicity: simple, unambiguously defined elements that interact according to simple, unambiguously defined rules. The simplicity of idealizations resides in the fact that their components have *no properties* other than those stipulated and those which follow as logical consequences of those stipulations. On the empirical side, their construction is guided by analogy: idealizations are caricatures, iconic representations of those features of the phenomena which appear relevant to the explanatory goals of a theory.

Peirce's concept of theorematic reasoning offers a key to the understanding of the mediatory function of idealizations in classical science and their altered role in quantum physics. It also serves to explain the crucial but seldom acknowledged function of thought experiments in science. For the purposes at hand, Peirce's own condensed explication of "theorematic reasoning" must suffice:

> I found out . . . that all mathematical reasoning is diagrammatic and that all necessary reasoning is mathematical reasoning, no matter how simple it may be. By diagrammatic reasoning, I mean reasoning which constructs a diagram according to a precept expressed in general terms, performs experiments upon this diagram, notes their results, assures itself that similar experiments performed upon any diagram constructed according to the same precept would have the same results, and expresses this in general terms. (*NEM* 4:47–48)

This reasoning, based on active experimentation on diagrams in the imagination, is what Peirce generally called "theorematic" and what he contrasted to "corollarial" reasoning, which is based on the straightforward application of deductive rules. Theorematic reasoning characterizes creative mathematical proofs (theorems) and essentially involves the experimental introduction of new elements, such as the auxiliary lines of geometric demonstrations, *not previously given* in the premises. In mathematical demonstrations these auxiliary elements act as catalysts in a chemical reaction: they elicit the process but do not appear in the final result. As Ketner (1985:411) has noted, their introduction marks the employment of abduction, Peirce's name for the type of reasoning that creates new hypotheses for future tests.

When theorematic reasoning is applied to the explanation of physical processes via the mediatory function of idealizations, these abductive steps no longer introduce ephemeral auxiliary elements. On the contrary, the new elements are retained as essential building blocks of a new, idealized, diagrammatic representation of those processes, now generally called a *model.*

Models remain associated with explanatory theories and grow and suffer modifications in partnership. They often become so closely linked that scientists use the terms "model" and "theory" interchangeably. The hypothetical elements introduced by abduction take on a life of their own through the explanatory success of their associated theories. Molecules, genes, neutron stars, the earth's mantle, and so forth, once mere abductive fictions, have later acquired ontological status along with chairs and trees and have become, in due time, "observable" (McMullin 1985).

These considerations lead to a new interpretation of the rise of classical physics as a mathematico-experimental science of nature. Experimentation had been present all along in mathematics, not only in the heuristic function performed by the exploration of paths leading to possible theorems, but also in the demonstrative role played by theorematic reasoning in proving those theorems. Similarly, exploratory manipulation of physical objects and processes had always occupied tinkers and artisans in their quest for new results and new tools. The mediatory function of idealization made possible the creation, within empirical praxis, of a counterpart to the demonstrative role played by theorematic reasoning in mathematics. This is what is now most frequently meant by "experimentation": the active, imaginative search for the enactment of those physical circumstances which would test the predictions of theories.

Thought experiments, which have played a crucial role in the development of physical theory from before the time of Galileo, may be understood as empirical exercises of theorematic reasoning. They are experiments performed not upon physical phenomena but upon their diagrammatic representations. Imaginative, exploratory tinkering with these representations may lead to demonstrations of new results or to structural changes in the theories themselves (Fernández 1988:51–53).

From this perspective, mathematics may be seen as the experimental science par excellence, and the simultaneous rise of mathematization and experimentation in the "scientific revolution" may be understood as a single process based on the exportation of theorematic reasoning into empirical praxis through the mediatory offices of idealization.

Limits of Classical Idealizations

Classical physics uses some basic, elementary idealizations as building blocks in the construction of models, such as particles and trajectories, fields and waves. The rift between classical and quantum physics is marked by the emergence of unforeseen limitations on the use of the wave and particle "pictures" for the representation of quantum phenomena. While in classical physics particles are identified with individual things (apples, planets, etc.) and waves with some well-defined processes of energy transfer (sound, light, etc.), quantum physics is forced to associate first one and then the other of these idealizations with one and the same situation, although they exclude each other as simultaneous idealizations of the same phenomenon. Particles are individual bearers of physical properties (momentum, charge, etc.) confined to a sharply defined spatial location at any given time. Waves, however, lack well-defined spatial boundaries and sharp individuation (any wave may be analyzed as a superposition of arbitrarily many individual waves). The alternative application of both representations and their associated concepts is demanded by the extraordinary empirical revelation that, while quantum objects display both wave and particle characteristics, the experimental arrangements set up to manifest and measure their particle or wave attributes are also incompatible: any setting capable of revealing the particle features necessarily precludes the manifestation of wave characteristics, and vice versa. The need to employ both of these mutually exclusive and partial representations led Bohr (1949:210) to the concept of *complementarity*: "Evidence obtained under different experimental conditions cannot be comprehended within a single picture, but must be regarded as *complementary* in the sense that only the totality of the phenomena exhausts the possible information about the objects."

Since the concepts of classical physics can be applied to the phenomena disclosed by experience only through the mediatory agency of idealizations, complementarity dictates that any given application of some of the concepts of classical physics in the quantum domain necessarily precludes the simultaneous application of other classical concepts, which nevertheless are required for an exhaustive explanation of the phenomena.

In classical physics, explanations of events often proceed by the empirical determination of the instantaneous state of a system of objects causally isolated from their environment. This is followed by the application of laws which generate predictions of future states of the system. A tacit assumption behind this scheme is that the measurements of the quantities characterizing a given state can be carried out in such a way that the effect of the measuring procedures on the measured state can be completely ignored or compensated for. The passage from classical to quantum physics coincides with the breakdown of this scheme.

Bohr's Correspondence Principle

In the historical development of classical mechanics, one can discern two successive stages: (1) the establishment of empirical generalizations in mathematical language (such as Kepler's and Galileo's laws), and (2) the discovery of general unifying principles (Newton's laws) which, in conjunction with suitable idealizations, explain those empirically discovered regularities as consequences of these principles. A similar pattern may be seen in the development of quantum mechanics during the first three decades of the twentieth century. Bohr's conception of complementarity served as a guiding rule in the second stage, the formulation of the quantum formalism, through the work of Heisenberg, Schrödinger, Pauli, Born, and others. Bohr's correspondence principle played an even more essential role during the first stage, in the translation of disconcerting experimental results into intelligible patterns of interaction between subatomic particles.

Bohr's correspondence principle was a guide for the abduction of hypothetical rules of particle and radiation interaction. It stipulates that these hypotheses must lead to predictions in asymptotic agreement with those of classical physics in the limit of very large quantum numbers. This constraint brought extraordinary heuristic power to the search for quantum regularities in the "invisible realm" of the atom.

On the heuristic value of the correspondence principle, few would disagree with Jammer's (1966:118) remark: "There was rarely in the history of physics a comprehensive theory which owed so much to one principle as quantum mechanics owed to Bohr's correspondence principle."

But the correspondence principle is something more than an instrument of discovery. It is also a principle of conceptual and theoretical unification for the whole of physics through the articulation and internalization, within quantum theory, of a metatheoretical insight. One comes to this realization upon reflecting that the historical evolution of scientific theories tends toward ever-increasing levels of generality. Superseded theories appear as rough approximations to superseding theories which, although logically incompatible with them, generate closely similar predictions for some limiting range of values. A paradigm example of this is the special theory of relativity, which "reduces" to Newtonian physics for low velocities; indeed the entire history of modern physics can be presented in the light of a "generalized correspondence principle" (Fadner 1985).

Laws of Laws and Simplicity

In the first of a series of five articles which appeared in the *Monist* in 1891, Peirce set out to explain his views on the architectonic of theories by pursuing to its bitter end the analogy between the construction of a theory and the building of a house. Something like the Kuhnian distinction between "normal" and "revolutionary" change appears in the guise of the difference between minor repairs and substantial remodeling, when "the repairs of the dilapidations have not been sufficiently thoroughgoing, and . . . not sufficient pains have been taken to bring the additions into deep harmony with the really sound parts of the old structure" (*CP* 6.8).

A few paragraphs later, Peirce considered the inauguration of modern physics. "A modern physicist in examining Galileo's work is surprised to find how little experiment had to do with the foundations of mechanics. His principal appeal is to common sense and *il lume naturale*. He always assumes that the true story will be found to be a simple and natural one" (*CP* 6.10).

Peirce's evolutionary epistemology explains why the assumption of simplicity led to success in classical dynamics: "Our minds having been formed under the influence of phenomena governed by the laws of mechanics, certain conceptions entering into those laws become implanted in our minds, so that we readily guess at what the laws are" (ibid.). The assumption of simplicity would thus become less and less warranted, the more we stray away from that domain of experience: "The further physical studies depart from phenomena which have directly influenced the growth of the mind, the less we can expect to find the laws which govern them 'simple', that is, composed of a few conceptions natural to our minds" (ibid.).

Peirce foresaw that science would enter alien territory precisely in the realm of atomic phenomena: "When we come to atoms, the presumption in favor of a simple law seems very slender. There is room for serious doubt whether the fundamental laws of mechanics hold good for single atoms, and it seems quite likely that they are capable of motion in more than three dimensions" (*CP* 6.11). Peirce's prophetic insight does not rest with the adumbration of this formidable obstacle. His cognitive optimism prompted him to anticipate clearly the nature of the principles which were to overcome it: "To find out much more about molecules and atoms we must search out a natural history of laws of nature which may fulfill that function which the presumption in favor of simple laws fulfilled in the early days of dynamics, by showing us what kind of laws we have to expect" (*CP* 6.12). Commenting on these same quotations, Max Fisch (1969:220) observes that "what Peirce proposes, then, is a cosmogonic or second-order hypothesis of the evolution of the laws of nature, by which we may be guided in forming first-order hypotheses for experimental testing in the several branches of physics."

Bohr's correspondence principle, I believe, was the first one of these laws of laws to be discovered. Twentieth-century physics has continued advancing down the invisible, and unvisualizable, realm of microphysics toward the discovery of new substructures, the unification of the fundamental forces of nature, and the foundation of a scientific, experimental, and evolutionary cosmology. This progress was made possible by the discovery of new, powerful principles which may also be regarded as second-order, metatheoretical laws: the symmetry principles, such as the principle of gauge invariance.

Conclusion: Peirce and Contemporary Physics

Quantum physics has introduced into natural philosophy several controversial ontological novelties that some physicists have accepted only reluctantly as a price to be paid in exchange for the incredible power of prediction and explanation bestowed by the quantum formalism. These peculiarities include objective chance, objective indefiniteness, real possibility (Heisenberg's potentia), nonlocality, and "entanglement" (Shimoni 1989). I cannot discuss these concepts here but will note only the remarkable fact that Peirce advocated the first three, on philosophical grounds, decades before the introduction of the quantum formalism. If this prophetic gift is an indication of the soundness and depth of his ideas, we would be justified in the conviction that they shall also help us in dealing with the other two.

Although I clearly disagree with Hartshorne on the matter of Peirce's failure to anticipate the novelties of quantum physics, I believe he was right in focusing on the issue of discontinuity. This issue has received little attention or been ignored in recent discussions of the concepts I have just listed; it seems to me, however, that it always lurks behind them. It was certainly important to the creators of quantum physics, especially Einstein, Bohr, and Schrödinger. It appears that Bohr, just like Peirce, may have been motivated to save continuity, under the influence of Höffding's ideas (Witthansen 1980:52–57). The principle of complementarity could perhaps be construed as an instrument to neutralize the continuity/discontinuity antithesis in quantum physics.

Sixty years after the celebrated debates that culminated in the EPR papers, we are as far as ever from reaching consensus on a consistent philosophical interpretation of quantum physics. Recent progress in this area, associated with the names of Bell, Bohm, Aspect, and many others, has served to open new horizons but has also confronted us with new perplexities. I think the task may prove impossible unless we succeed in overcoming the present-day fragmentation and parochialism by building a comprehensive philosophy of nature, by integrating the new quantum conceptions into the whole of science, and by establishing their links with both the philosophical tradition and the life of experimental praxis. This monumental task may benefit greatly not only from Peirce's many original and revealing ideas but even more extensively from the spirit of universal concern and incorruptible reasonableness that animated his life and work.

Perhaps the best way to bring Peirce's ideas to bear in this endeavor is to confront them with those of other original thinkers and with the newly acquired results of scientific research. The modest aim of this paper is to take a few tentative steps in that direction.

Note

I would like to express my gratitude to Siegfried Ruschin of Linda Hall Library for carefully reading the manuscript and suggesting several improvements of form and content, and also for his kind encouragement throughout the years.

References

Bohm, David. 1985. On Bohr's Views concerning Quantum Theory. In French and Kennedy 1985, pp. 153 59.

Bohr, Niels. 1949. Discussion with Einstein on Epistemological Problems in Atomic Physics. In Schilpp 1949, pp. 200–241. As reprinted in Wheeler and Zurek 1983, pp. 9–49.

———. 1963. *Essays, 1958–1962, on Atomic Physics and Human Knowledge.* New York: Interscience.

Chauviré, Christiane. 1987. Schématisme et analyticité chez C. S. Peirce. *Archives de Philosophie* 50:413–37.

Davies, Paul, ed. 1989. *The New Physics.* Cambridge: Cambridge University Press.

d'Espagnat, Bernard. 1985. *Une incertaine réalité: Le monde quantique, la connaissance, et la durée.* Paris: Gauthier-Villars.

Drieschner, M.; Görnitz, T.; and von Weizsäcker, C. F. 1988. Reconstruction of Abstract Quantum Theory. *International Journal of Theoretical Physics* 27:289–306.

Fadner, W. L. 1985. Theoretical Support for the Generalized Correspondence Principle. *American Journal of Physics* 53:829–38.

Falkenburg, Brigitte. 1988. The Unifying Role of Symmetry Principles in Particle Physics. *Ratio*, n. s., 1:113–34.

Fernández, Eliseo. 1988. Theoria y praxis en los orígenes de la física clásica. *La Ciencia y el Hombre* 1:37–59.

Fisch, Max H. 1969. Vico and Pragmatism. In Ketner and Kloesel 1986, pp. 201–48.

Folse, Henry J. 1985. *The Philosophy of Niels Bohr: The Framework of Complementarity.* Amsterdam: North Holland.

French, A. P., and Kennedy, P. J., eds. 1985. *Niels Bohr: A Centenary Volume.* Cambridge: Harvard University Press.

Hartshorne, Charles. 1973. Charles Peirce and Quantum Mechanics. *Transactions of the Charles S. Peirce Society* 9:191–201.

Hintikka, Jaakko. 1980. C. S. Peirce's "First Real Discovery" and Its Contemporary Relevance. *Monist* 63:304–15.

Honner, John. 1987. *The Description of Nature: Niels Bohr and the Philosophy of Quantum Physics.* Oxford: Clarendon Press.

Jammer, Max. 1966. *The Conceptual Development of Quantum Mechanics.* New York: McGraw-Hill.

Jepsen, Torsten. 1987. Epistemological Problems in Quantum Physics. *Danish Yearbook of Philosophy* 24:19–42.

Joos, E. 1986. Quantum Theory and the Appearance of a Classical World. *Annals of the New York Academy of Sciences* 480:6–13.

Ketner, Kenneth Laine. 1985. How Hintikka Misunderstood Peirce's Account of Theorematic Reasoning. *Transactions of the Charles S. Peirce Society* 21:407–18.

Ketner, Kenneth Laine, and Kloesel, Christian J. W., eds. 1986. *Peirce, Semeiotic, and Pragmatism: Essays by Max H. Fisch.* Bloomington: Indiana University Press.

MacKinnon, Edward. 1985. Bohr on the Foundations of Quantum Theory. In French and Kennedy 1985, pp. 101–20.

McMullin, Ernan. 1985. Galilean Idealization. *Studies in History and Philosophy of Science* 16:247–73.

Murdoch, D. 1987. *Niels Bohr's Philosophy of Physics*. Cambridge: Cambridge University Press.

Peres, A., and Zurek, W. H. 1982. Is Quantum Theory Universally Valid? *American Journal of Physics* 50:807–16.

Petersen, Aage. 1966–68. On the Philosophical Significance of the Correspondence Argument. *Boston Studies in the Philosophy of Science* 5:242–52.

Radzabov, U. A. 1980. The Correspondence Principle: History and Present State. *Danish Yearbook of Philosophy* 17:59–82.

Schilpp, P. A. 1949. *Albert Einstein: Philosopher-Scientist*. Evanston, Ill.: Library of Living Philosophers.

Shimony, Abner. 1989. Conceptual Foundations of Quantum Mechanics. In Davies 1989, pp. 373–95.

Stapp, Henry Peirce. 1972. The Copenhagen Interpretation. *American Journal of Physics* 40:1098–1116.

Teller, Paul. 1980. The Projection Postulate and Bohr's Interpretation of Quantum Mechanics. In *PSA 1980*, ed. Peter D. Asquith and R. N. Giere, 201–23. East Lansing, Mich.: Philosophy of Science Association.

Tiles, J. E. 1988. Iconic Thought and the Scientific Imagination. *Transactions of the Charles S. Peirce Society* 24:161–78.

Wheeler, John A., and Zurek, W. H. 1984. *Quantum Theory and Measurement*. Princeton: Princeton University Press.

Witthansen, Johs. 1980. Leibniz, Høffding, and the "Ekliptika" Circle. *Danish Yearbook of Philosophy* 17:31–58.

17 The Role of Potentiality in Peirce's Tychism and in Contemporary Discussions in Quantum Mechanics and Microphysics

Demetra Sfendoni-Mentzou, Aristotle University of Thessaloniki, Greece

"The hypothesis of chance-spontaneity," Charles Peirce claimed in 1892, "is one whose inevitable consequences are capable of being traced out with mathematical precision into considerable detail." He continued as follows: "I mention it . . . to indicate to future mathematical speculators a veritable gold mine, should time and circumstances and the abridger of all joys, prevent my opening it to the world" (*CP* 6.62). In fact, the gold mine was opened a few years later. The dramatic development of quantum mechanics brought to the forefront the concept of indeterminacy and its claim that the scientific world-picture, based on a perfectly deterministic scheme, could no longer be maintained. Furthermore, the subsequent development of high-energy physics during the last decades has revealed the qualitative richness of the inner structure of matter with a considerable number of transmutations taking place in the realm of microphysics.

To appreciate Peirce's evaluation of his own scientific speculation, I wish to follow a path quite different from those that have been followed in the past.[1] I shall move along two lines: My primary concern will be to explore the inner structure of the various aspects through which Peirce gave expression to his doctrine of chance, in an effort to discover one fundamental idea that can serve as an explanatory basis for a unified account of his tychism.[2] This, I will argue, is the idea of "potentiality" in the sense of the Aristotelian concept of *dynamis*. Then I hope to use this very same idea to establish a conceptual relationship of Peirce's theory with quantum mechanics and microphysics as well. My general approach will be that not of a scientist but of a philosopher, whose main concern is to pass the conceptual string through the intellectual pearls so as to establish a common underlying explanatory pattern to the theories of quantum mechanics and microphysics.

The working assumption throughout this chapter is that Peirce's theory of chance cannot be understood apart from his Aristotelian-Scholastic realism.[3] In this respect, my concern is to elaborate those features of Peirce's ontological scheme which have important implications for a deeper understanding of his tychism. Specifically, I wish to show in what particular way *firstness*, as a manifestation of spontaneity-indeterminacy-chance, and *thirdness*, as a manifestation of generality-probability-habit, function through the key concept of potentiality.[4]

I

To begin with, firstness, the category of chance (*CP* 6.33), is described by Peirce as a mere possibility, expressed by the characters of freedom and spontaneity. Furthermore, firstness is the category of the indeterminate, having in this respect a character of generality as regards that which will be actualized in the future (CP 1.422). Our attention, however, must be confined to the fact that in Peirce's scheme there are two types of generality, corresponding to two types of potentiality: (1) the generality of thirdness and (2) that of firstness (see Sfendoni-Mentzou 1991). The latter is interrelated with indeterminacy and hence with potentiality: "The word 'potential,'" claims Peirce, "means *indeterminate, yet capable of determination in any special case*" (*CP* 6.185). He thus defines potentiality in a way analogous to that of Aristotle (*Metaphysics* 9.7.1048b38–1049alff.).[5] According to this definition, which will be my guide in the following analysis, potential is the "as yet undeveloped, since not presenting itself in actually objectified form, but capable of doing so at some future time, when all the conditions of its realization occur." It thus "connotes some inherent tendency to actuality which, if not thwarted, leads to final completeness of being" (*CP* 6.364).

The crucial thing to note here is that this inherent tendency needs a mediating link in order to be carried out, and this is offered by thirdness and more specifically by the triadic relation *generality-continuity-law*. All three elements, grounded not only in logical but also in ontological possibility (i.e., potentiality), contribute to the establishment of the third character of law, the *law of habit* (see *CP* 5.102, 103; 6.170, 172–73). Generality, through the concept of potentiality, acquires the character of a "generalized tendency" which is irreducible to any number of individual instances. The same feature is attributed to law, which is "essentially a character of an indefinite future" (*CP* 8.192),[6] because as a general it "has an admixture of potentiality in it, so that no congeries of actions here and now can ever make a general fact" (*CP* 1.420).[7] Hence, law as a general is a potentiality and has an "*esse in futuro*" (*CP* 2.148). In other words, law is itself an expression of a *tendency toward generalization*, which is carried through a process of habit taking.

It is important to note here that the habit-taking tendency is intimately related, in Peirce's thought, to the concepts of probability and "would-be,"[8] which appear as another name for potentiality.

> If, then, I say that the probability that, if a certain die be thrown in the usual way, it will turn up a number divisible by 3 . . . is $\frac{1}{3}$, what do I mean? I mean, of course, to state that that die has a certain habit or disposition of behaviour in its present state of wear. It is a *would be* and does not consist in actualities or single events in any multitude finite or infinite. Nevertheless a habit does consist in what *would* happen under certain circumstances if it should remain unchanged throughout an endless series of actual occurrences. (*CP* 8.225)

It must also be added that the statement of a would-be is characterized as "essentially an assertion of a *general* nature" which "establishes a *habit* in the mind" (*CP* 8.380).[9] Thus "Real habits" (or would-be's) are defined by Peirce as that "which Really *would* produce effects, under circumstances that may not happen to get actualized and are thus Real generals" (*CP* 6.485).

In virtue of the above accounts, the law of habit, as opposed to mechanical law, has two distinctive characters: The first is its *finality*, which leaves room for variations in the intermediate stages, which are followed for the accomplishment of the final end. It thus represents final causality of exactly the Aristotelian type, as Peirce himself makes clear (*CP* 1.211, 212).[10]

This leads to a flexibility or "plasticity of habit" (*CP* 6.86, cf. 6.614) quite similar to Popper's world of clouds:[11] law is endowed with a liberty to operate not as a blind mechanical process but as a dynamic law. The doctrine of thirdness is thus an assertion that freedom of the mind has its roots in the habit-taking character of the universe itself, which "would at once cease to operate if it were rigidly obeyed" (*CP* 6.613). "The hypothesis suggested by the present writer is that all laws are results of evolution; that underlying all other laws is the only tendency which can grow by its own virtue, the tendency of all things to take habits" (*CP* 6.101).

Consequently, growth and evolution represent a "positive violation of law" (*CP* 6.613) which is made possible through the factor of chance. This is the second essential characteristic of habit: it can function only through *chance*. The reason Peirce uses chance, he says, is "to make room for a principle of generalization, or tendency to form habits" (*CP* 6.63). So habit is "the only bridge that can span the chasm between the chance medley of chaos and the cosmos of order and law" (*CP* 6.262).

The crucial thing to note here is the fact that habit is not only an expression of freedom and flexibility, springing from firstness and chance, but that it also has its laws. Peirce's tychism therefore could not possibly be identified with lawlessness, as has sometimes been claimed.[12] Peirce, time and again, explicitly states his position: "It is a mistake to suppose that there would be no laws in a world of chance. . . . So that *chance* is not the abrogation of all laws" (*CP* 7.136). Certainly he attributes all variety and diversification of the world to chance—but "to chance in the form of spontaneity which is to some degree regular" (*CP* 6.63).

II

Let us now consider in more detail the question of the proper meaning to be attributed to chance in Peirce's scheme. The answer to this will ultimately lead us to a deeper understanding of tychism as a whole. To achieve this, I concentrate on two characteristics attributed to chance viewed as an instance of firstness: freedom and spontaneity. In fact, these are the characteristics of what Peirce calls absolute chance: "Let me here say one word about Tychism, or the doctrine that absolute chance is a factor of the universe. . . . When I speak of chance, I only employ a mathematical term to express with accuracy the characteristics of freedom and spontaneity" (*CP* 6.201).

The characteristic of spontaneity, which is an expression of freedom, is the most important element to help us if we are to arrive at a deeper understanding of Peirce's tychism and also to establish the inner connection of his tychism with the idea of potentiality, both of firstness and of thirdness. This in turn will further indicate that probability is embedded in the very nature of things, physical phenomena and processes making room for qualitative variety, diversification, novelty, growth, and development. "The endless variety in the world," claims Peirce, "has not been created by law. . . . When we gaze upon the multifariousness of nature we are looking straight into the face of a living spontaneity" (*CP* 6.553). "In short, diversification is the vestige of chance, spontaneity" (*CP* 6.267). All the above phenomena not only are related to spontaneity, they are eventually identified with it. "I am a logical analyst by long training, you know, and to say this [the variety in nature] is a manifestation of spontaneity seems to me faulty analysis. I would rather say it *is* spontaneity" (*CP* 1.160).

Having established the relation of chance and spontaneity, it remains to indicate the inner connection of these two concepts with potentiality. What chance, spontaneity, and generality share in common is that they are all manifestations of the same mode of being, namely, of *firstness*. This is nicely illustrated in Peirce's cosmological theory as expounded in "The Logic of the Universe" of 1898.

> The very first and most fundamental element that we have to assume is a Freedom, or Chance, or Spontaneity, by virtue of which the general vague nothing-in-particular-ness that preceded the chaos took a thousand definite qualities. The *second* element we have to assume is that there could be accidental reactions between those qualities. The qualities themselves are mere eternal possibilities. (*CP* 6.200)

In fact, in this picture that Peirce draws of the universe, there is a continuous evolution along the lines of the three modes of becoming. "Chance is First, Law is Second, the tendency to take habits is Third" (*CP* 6.32).[13] The first stage is connected with the *indeterminate*, the nothing-in-particular, the vague, the indefinite; the second with the *determinate*, the particular, the concrete, the definite; finally, the third with the *law* that makes possible the passage from the one to the other. Through this procedure, "the indeterminate future becomes irrevocable past. In Spencer's phrase the undifferentiated differentiates itself. The homogeneous puts on heterogeneity" (*CP* 6.191). It is a highly interesting point that a few lines below this comment, Peirce proceeds to establish the connection of all the above-mentioned characteristics of firstness and chance with potentiality. The process of diversification, he claims, began in "the utter vagueness of completely undetermined and dimensionless potentiality" (*CP* 6.193).

Furthermore, the established interrelation of chance, indeterminateness, and potentiality can be even better illustrated in the following passage: "The evolution of forms begins, or, at any rate has for an early stage of it, a vague potentiality. . . . It must be by a contraction of the vagueness of that potentiality of everything in general, but of nothing in particular, that the world of forms comes about" (*CP* 6.196). Peirce goes on to remark, "The original potentiality is the Aristotelian matter or indeterminacy from which the Universe is formed" (*CP* 6.206).

We therefore return, in following the path of Peirce's cosmological speculations, to a point that has already been stressed in the foregoing analysis of the first mode of being. Firstness, and consequently chance, expresses an indeterminacy or spontaneity which has a marked similarity to the Aristotelian definition of matter, which is also identical with the indeterminateness of freedom and potentiality. "By matter I mean that which in itself is neither a particular thing nor a quantity nor designated by any of the categories which define Being" (*Metaphysics* 7.3.1029a20–22). There is textual evidence that Peirce was indeed well aware of Aristotle's connection of matter with potentiality. "The embryonic being, for Aristotle," remarks Peirce, "was the being he called matter" (*CP* 1.22).

On the basis of the above analysis, I wish to suggest that chance as spontaneity corresponds mainly to the Aristotelian term *automaton*,[14] not *tychē*. I thus read spontaneity as self-determination, taken as a type of intrinsic causality grounded on potentiality and opposed to extrinsic mechanistic determinism. Such an interpretation also has the merit of rendering the ideas of probability and indeterminacy explicable in terms of the habit-taking tendency of the third category. Habit, claims Peirce, viewed as a generalized tendency "must have its origin in the original continuity which is inherent in potentiality" (*CP* 6.204).[15]

We are thus led to the following synoptic description of Peirce's tychism, in his own words:

> By thus admitting pure spontaneity or life as a character of the universe, acting always and everywhere, though restrained within narrow bounds by law, producing infinitesimal departures from law continually, and great ones with infinite infrequency, I account for all the variety and diversity of the universe, in the only sense in which the really *sui generis* and new can be said to be accounted for. (*CP* 6.59)[16]

In this picture we have the two crucial elements which will offer the ground for the establishment of the conceptual relation of Peirce's tychism with quantum mechanics and microphysics, namely *indeterminacy* of law and *qualitative variety* of nature.

III

So far as quantum mechanics is concerned, my argument is based on the realist interpretation of Heisenberg's principle of indeterminacy.[17] The Copenhagen interpretation, attributing uncertainty to our ignorance, dominated the scene around the 1930s. But very soon, attempts at a realist interpretation appeared from various quarters. Their common characteristic is the acceptance of the idea of potentiality, very much like the one elaborated in Peirce's scheme, which in turn is based, as has been shown, on the Aristotelian ontology. Strangely enough, it was Heisenberg, one of the most prominent exponents of the Copenhagen interpretation, who started this idea. In his *Physics and Philosophy*, he described the probability wave of Bohr, Slater, and Kramers in the following remarkable way: "The probability wave of Bohr, Kramers, and Slater . . . meant a *tendency* for something. It was a quantitative version of the old concept of '*potentia*' in Aristotelian philosophy. It introduced something standing in the middle between the idea of an event and the actual event, a strange kind of physical reality just in the middle between *possibility* and *reality*" (Heisenberg 1971:42).

Furthermore, in referring to the peculiar fact of "coexistent states" in quantum mechanics, Heisenberg (ibid.:159) suggested that "if one considers the word 'state' as describing some *potentiality* rather than a reality . . . then the concept of 'coexistent potentialities' is quite plausible, since one potentiality may involve or overlap other potentialities." He drew the conclusion that the atoms or elementary particles themselves form "a world of potentialities or possibilities rather than one of things or facts" (ibid.:160). Or, in Peirce's words, we would say they belong to the domain of firstness rather than of *secondness*, *thisness*, or *haecceitas*.

Along the same line of thought is Popper's propensity interpretation of probability in quantum mechanics. He repeatedly emphasizes the fact that "propensities" are "physically real," they are "physical propensities, abstract relational properties of the physical situation" (Popper 1959:28). Thus, Popper arrives at an important conclusion which Peirce would be very happy to endorse: Propensities "are not merely logical possibilities, but physical possibilities." In the context of his realist interpretation, Popper describes his theory as a theory of matter or particles, according to which they are considered as realizations of potentialities. In this picture, waves are described as "mathematical representations of *propensities* or of dispositional properties, of the physical situation . . . interpretable as propensities of the particles to take up certain states" (Popper 1982:126). A further implication of this position is the emphasis posited by Popper on the process of actualization, that is, the process of transition from possible to actual, taking place in nature. Heisenberg, though not Peirce, is referred to by Popper as a forerunner of the "propensity interpretation," in virtue of his comment on the concept of probability relating closely to the concept of Aristotelian "potentia." What better meeting point could one wish to find for Popper, Heisenberg, Peirce, and Aristotle![18]

Finally, Popper (1982:160) arrives at a description of physical realism that is very Aristotelian as well as Peircean: "We may describe the physical world as consisting of *changing propensities for change.*" In this respect it is interesting to see how Popper summarizes his own "metaphysical research programme": "Everything is a propensity. Or in the terminology of Aristotle we might say: To be is both to be the actualization of a prior propensity to become, and to be a propensity to become" (ibid.:205).

Heisenberg's expressed realist interpretation is also a meeting point for a theory proposed quite recently: the pilot-field theory of Henri Krips. Krips refers to Heisenberg's position, expressed in 1930, as one of the first attempts toward a realist interpretation of indeterminacy in quantum theory. In fact, in his *Physical Principles of Quantum Theory* (Heisenberg 1930:14), Heisenberg interpreted the indeterminacy "as an essential characteristic of the electron and not as evidence of the inapplicability of the wave picture" (Krips 1987:26). Claiming that he follows a line in many respects analogous to the realist attempts of Heisenberg, Krips propounds his own pilot-wave theory (ibid.), which is also similar to Popper's propensity interpretation. He thus interprets the system as "a particle (albeit one with indeterminate values for some of its physical qualities)" which is always "guided by a probability field. Thus its wave and particle aspects are always co-present" (ibid.:3). So far as the field is concerned, Krips points out that "it is not a field of probabilities or even of propensities. . . . Rather it is a categorical basis for propensities, the presence of which is displayed by the probabilities of certain hypothetical experiments to have certain results" (ibid.:67).[19]

Attention must be confined here to two points: (1) the *objective* and *ontological* character attributed to probability described as propensity, and (2) the *indeterminate* character attributed to the particle which at the same time is both a particle and a wave.[20] These two coexistent states of *particle* and *wave* could be described in Heisenberg's words as coexistent potentialities, which brings us right back to Peirce's idea of ontological possibility. Furthermore, it can serve as a very plausible connecting link with the definition of matter in microphysics.

IV

To be more specific, as science approached deeper levels of reality, it was shown that particles cannot be considered as the rigid building blocks of physical reality; rather, they should be viewed as having an unstable existence. "Not even the quality of being (if that may be called a 'quality')," as Heisenberg (1971:67) remarked, "belongs to what is described. . . . It is a possibility for being or a tendency for being"; or, one might say, in both Aristotelian and Peircean terminology, it is a potentiality, yet undeveloped but capable of development.

Moreover, as the pursuit of particles continued, a great variety of qualitatively different particles has been discovered,[21] such as protons, neutrons, positrons, hyperons, mesons, pions, and leptons. All of these particles have been shown to be undergoing processes of transmutation and even of creation and disintegration. For instance, a neutron-proton transmutation of hydrogen and helium builds up into nuclei of heavier elements such as carbon, oxygen, and iron.[22] Similar experiments have shown that nuclei can be transformed into one another or that matter and antimatter can mutually annihilate each other and convert into radiant energy; for example, high-energy electrons and positrons annihilate to produce radiant energy, which in turn produces new varieties of matter and antimatter.[23] Even more interesting is the fact that, in radioactivity, a neutron in the nucleus *spontaneously* converts into a proton and ejects an electron without energy being supplied.

Such incredible phenomena bring to light two fundamental features of the microcosm: (1) the qualitative variety in the physical world, which was emphatically pointed out by Peirce; (2) the continuous conversion of forms of matter into one another, which has also been one of the favorable themes of Peirce's tychism, albeit he could not refer to the as yet undiscovered phenomena in microphysics. The issue is now this: How is this variety and mutability of matter to be accounted for? or, What is the substratum of this change? It seems quite plausible, in my view, that an answer could be given by an appeal to the idea of potentiality as a ground of the transition from potential to actual being.[24] The idea of potentiality as a substratum for change can be further illustrated by an appeal to a deeper level of reality, that of quarks.[25] In the realm of quarks potentiality can also be connected with indeterminacy, characterizing both the Aristotelian and Peircean definition of matter.[26]

What is matter, for Peirce? Matter is very similar to that of Aristotle. It is the undifferentiated, indeterminate being, capable of determination, with a disposition/propensity/inherent tendency to be actualized. All the above-mentioned characteristics seem to fit nicely with the characteristics attributed to quarks, the constituent elements of hadrons, such as protons and neutrons. The interesting point connected with quarks is that they could be taken as an instance of potential being, as they have never been found to have an actual existence—that is, they never display a feature of secondness; rather their being is that of firstness in their character of being constituents of hadrons.[27] They are characterized not by numbers but by qualities such as flavors and colors and are responsible for the properties of other particles. For example, a proton is composed of three quarks, whose electric charges are $\frac{2}{3}$, $\frac{2}{3}$, and $-\frac{1}{3}$. Thus the total electric charge of the proton is 1. But how can an entity (quark) have a fractional charge of $-\frac{1}{3}$? This can be explained only if we consider quarks as existing only potentially. In this case quarks could be accepted as real but not actual, that is, only as quarks-in-the-form-of-a-proton.

V

We are now ready to pull all the threads together. My concluding suggestion is that we may arrive at a unified picture of physical reality by combining into an integral whole both the idea of *qualitative variety* of matter and that of *indeterminacy*. If the definition of matter as an indefinite and indeterminable substratum of change, or as potentiality expressing an inherent tendency to actualization, is accepted, then indeterminacy can be accounted as grounded in the very nature of things. Such a picture has also the merit of not excluding causality. By just loosening the bonds of mechanical determinism, it also leaves room for self-determination, which, in Peirce's terminology, is an expression of spontaneity or chance.

Notes

I am indebted to my colleague of the physics department, Argyris Nikolaidis, for valuable discussions and helpful comments on issues connected with the field of microphysics.

1. See, for example, F. M. Hamblin 1945, Peter T. Turley 1969, Charles Hartshorne 1973, and J. Jauhari 1985.

2. Some of the arguments of this paper are based on the ideas I have elaborated earlier in Sfendoni-Mentzou 1980.

3. I have presented my arguments on the essential connection of Peirce's thought with the Aristotelian-Scholastic realism in Sfendoni-Mentzou 1991.

4. One of the first Peirce scholars to call attention to the importance of Peirce's metaphysics and in particular to the idea of potentiality in relation to Aristotle's ontology is E. Moore (1950), as I have earlier pointed out (Sfendoni-Mentzou 1980). See also Moore 1968.

5. "We must, however, distinguish when a particular thing exists potentially, and when it does not; for it does not so exist at any and every time. . . . And in all cases where the generative principle is contained in the thing itself, one thing is potentially another when, if nothing external hinders, it will of itself become the other. E.g., the semen is not yet potentially a man; for it must further undergo a change in some other medium. But when, by its own generative principle, it has already come to have the necessary attributes, in this state it is now potentially a man, whereas in the former state it has need of another principle" (Aristotle, *Metaphysics* 9.7.1048b43–1049a20).

6. "For while uniformity is a character which might be realized, in all its fulness, in a short series of past events, law, on the other hand, is essentially a character of an indefinite future" (*CP* 8.192).

7. Cf. "As *general*, the law, or general fact, concerns the potential world of quality, while as *fact* it concerns the actual world of actuality" (*CP* 1.420).

8. The connection of probability with the concept of "would-be" has been analyzed in detail in Sfendoni-Mentzou 1980:64ff.

9. Here Peirce means that "the person or thing that has the habit, *would* behave . . . in a certain way *whenever* a certain occasion should arise" (*CP* 8.380).

10. "The signification of the phrase 'final cause' must be determined by its use in the statement of Aristotle that all causation divides into two grand branches, the efficient, or forceful; and the ideal, or final. If we are to conserve the truth of that statement, we must understand by final causation that mode of

bringing facts about according to which a general description of result is made to come about, quite irrespective of any compulsion for it to come about in this or that particular way; although the means may be adapted to the end" (*CP* 1.211).

11. "So far as I know Peirce was the first post-Newtonian physicist and philosopher who thus dared to adopt the view that to some degree *all clocks are clouds;* or in other words, that *only clouds* exist, though clouds of very different degrees of cloudiness" (Popper 1975:213).

12. For instance by P. Turley (1969). Turley claims that there are three conceptions of chance: (1) the first holds chance to be a matter of human ignorance; (2) the second attributes to chance an objective character; (3) the third conception of chance is represented, according to Turley (1969:246), by "Peirce's Tychism, the hypothesis of absolute chance. Chance is here taken to mean real violations of the laws of nature, in a word, lawlessness."

13. Cf. "Mind is First, Matter is Second, Evolution is Third" (*CP* 6.33); "but what I mean is, that all that there is, is First, Feelings; Second, Efforts; Third, Habits" (*CP* 6.201).

14. I share this view with Feibleman (1946:190 n. 108). For a detailed analysis of this account, see Sfendoni-Mentzou 1980:172ff.

15. "The original potentiality" for Peirce "is essentially continuous, or general" (*CP* 6.205).

16. Cf. "But my hypothesis of spontaneity does explain irregularity, in a certain sense; that is, it explains the general fact of irregularity, though not, of course, what each lawless event is to be. At the same time, by thus loosening the bond of necessity, it gives room for the influence of another kind of causation, such as seems to be operative in the mind in the formation of associations, and enables us to understand how the uniformity of nature could have been brought about" (*CP* 6.60).

17. According to this principle, neither the location nor the momentum (the product of mass and velocity) of a particle can be determined with an unlimited precision. As a result, we cannot make accurate predictions of events as we could in Newtonian mechanics. We could very well speak of the position or of the momentum of an electron as in Newtonian mechanics, and we could very well observe and measure those qualities separately, but never simultaneously with an arbitrarily high accuracy. For there is a reciprocal relationship between the possible precision of definition of the momentum p and that of the position x of a particle. The more accurately the position is determined, the less accurately the momentum can be measured, and vice versa.

18. For a more detailed analysis of the connection of Popper's propensity

theory with Peirce's tychism and Aristotle's ontology, see Sfendoni-Mentzou 1989.

19. It is suggested that there is a field $f(S, t)$—what M. Born calls a probability field—associated with the q-system S at time t. According to Krips (1987:2), "q-systems are associated with fields, indeed probability fields, which pilot them."

20. Krips (1987:70) describes the situation: "On the pilot-field model however, a q-system is always already a bricolage of both wave, *qua* field, and particle. Second, when Bohr refers to 'wave' and 'particle' he means classical waves and particles. But in the pilot-field model for q-systems, the q-system is not so much a (diachronic) combination of a classical wave and a classical particle, as a (synchronic) combination between what I have called 'a probability field' and a particle."

21. In cloud chambers and bubble chambers, where the production and properties of particles are studied, and in several other cases, qualitatively new particles have been produced.

22. Some more examples: When a neutrino converts into an electron, a W^+ particle is emitted. When a π^+ decays, it usually produces a positively charged particle called a muon ("antimuon") and a neutrino. Less often it will produce an antielectron (positron) and a neutrino.

23. Such as a muon and an antimuon or quarks and antiquarks. The quarks and antiquarks cluster together, forming the familiar nuclear particles such as protons and pions, which are what is detected.

24. In this respect motion regains its internal dimension, in the Aristotelian sense. This has been pointed out nicely by D. Bohm. According to Bohm (1957:147), motion is expressed not only by the displacement of bodies in space but also by "all possible changes and transformations of matter, internal and external, qualitative and quantitative, etc."

25. I have profited from Eugene Schlossberger's (1979) account of the analogy between Aristotelian prime matter and quarks; I also owe a lot to the discussions in Suppes 1974 and Capek 1979.

26. " . . . nor does everything contain matter, but only such things as admit of generation and transformation into each other" (Aristotle, *Metaphysics* 8.5.1044b27–30); "For what I mean by matter is precisely the ultimate underlying subject, common to all the things of Nature, presupposed as their substantive, not incidental, constituent." And again, "The destruction of a thing means the disappearance of everything that constitutes it except just that very underlying subject which its existence presupposes, and if this perished, then the thing

that presupposes it would have perished, with it by anticipation before it came into existence" (Aristotle, *Physics*, 1.9.192a32ff.; cf. *Physics*, 1.7.191a7ff.; 5.2.226a11ff.).

27. It is extremely interesting to note that Aristotle not only connects matter with potential being, as has already been pointed out in this chapter, but also defines matter in terms of constituent parts of an actual being: "For it would seem that the formula which involves the differentiae is that of the form and the actuality, while that which involves the constituent parts is rather that of the matter" (*Metaphysics* 8.2.1043a19–21).

References

Aristotle. 1968a. *Metaphysics*. Loeb editions. Trans. H. Tredennick. Cambridge: Harvard University Press.

———. 1968b. *Physics*. Loeb editions. Trans. P. H. Wicksteed and F. M. Cornford. Cambridge: Harvard University Press.

Bohm, David. 1957. *Causality and Chance in Modern Physics*. London: Routledge and Kegan Paul.

Capek, Milic. 1979. Two Views of Motion: Change of Position or Change of Quality? *Review of Metaphysics* 33:337–46.

Feibleman, James. 1970. *An Introduction to the Philosophy of Charles S. Peirce*. Cambridge: MIT Press.

Hamblin, Francis Murphey. 1945. A Comment on Peirce's Tychism. *Journal of Philosophy* 42:378–83.

Hartshorne, Charles. 1973. Charles Peirce and Quantum Mechanics. *Transactions of the Charles S. Peirce Society* 9:191–201.

Heisenberg, Werner. 1930. *The Physical Principles of the Quantum Theory*. New York: Dover Publications.

———. 1971. *Physics and Philosophy*. London: George Allen and Unwin. 1st ed. 1959.

Jauhari, Jamila. 1985. Peircean and Quantum Generals. *Transactions of the Charles S. Peirce Society* 21:513–34.

Krips, Henry. 1987. *The Metaphysics of Quantum Theory*. Oxford: Clarendon Press.

Moore, Edward C. 1950. Metaphysics and Pragmatism in the Philosophy of C. S. Peirce. Ph.D. diss., University of Michigan.

———. 1968. On the World as General. *Transactions of the Charles S. Peirce Society* 4:90–100.

Murphey, H. F. 1945. A Comment on Peirce's Tychism. *Journal of Philosophy* 42:378–83.

Popper, Karl R. 1959. The Propensity Interpretation of Probability. *British Journal of the Philosophy of Science* 10:25–42.

———. 1975. *Objective Knowledge*. Oxford: Clarendon Press. 1st ed. 1972.

———. 1982. *Quantum Theory and the Schism in Physics*. Vol. 3 of the *Postscript* to *The Logic of Scientific Discovery*. London: Hutchinson.

Schlossberger, Eugene. 1979. Aristotelian Matter, Potentiality, and Quarks. *Southern Journal of Philosophy* 17:507–21.

Sfendoni-Mentzou, Demetra. 1980. Probability and Chance in the Philosophy of C. S. Peirce. Doctoral diss., University of Thessaloniki.

———. 1989. Popper's Propensities: An Ontological Interpretation of Probability. In *Imre Lakatos and Theories of Scientific Change*, ed. K. Gavroglou et al., 441–55. Norwell, Mass.: Kluwer.

———. 1991. Towards a Potential-Pragmatic Account of C. S. Peirce's Theory of Truth. *Transactions of the Charles S. Peirce Society* 27:27–77.

Suppes, Patrick. 1974. Aristotle's Concept of Matter and Its Relation to Modern Concepts of Matter. *Synthese* 28:27–50.

Turley, Peter T. 1969. Peirce on Chance. *Transactions of the Charles S. Peirce Society* 5:243–54.

18 Aristotle and Peirce on Chance

Philip H. Hwang, Dongguk University, Seoul, Korea

Aristotle believes that all events are caused, or negatively speaking, no events are uncaused. But he also admits that some changes are made by chance (*tychē*) or spontaneity (*automaton*). But this does not mean that those changes are uncaused. Rather, they are produced by an "unintending" cause.

Peirce, for his own tychism, accepts Aristotle's theory of chance. But he goes one step further than Aristotle and argues that, to a certain extent, some events happen absolutely—by pure chance, without any cause. He not only points out empirical facts to support his theory of pure chance but also tries to prove it.

In this paper I compare the position of these two philosophers on chance and show that Peirce misinterprets Aristotle's concept of chance. Specifically, Aristotle rejects a *mechanical* or *logical* determinism, the view that propositions about states of affairs must be truth-functionally determinate. At the same time, he defends a *causal* determinism, the view that all events are susceptible of analysis in terms of his four causes. He is anxious to show that his causal theory is powerful enough to account for events which are generally attributed to chance. Peirce's interpretation of Aristotle's chance as an absence of cause is therefore a misconception of Aristotle.

I

Aristotle believes that all events in the past, present, or future are not *causa sui*, or "self-determined." If there is a change (an effect), it is necessary that there is a cause, or it is impossible for change to come about. Consequently, if there is an effect, it is necessary that there is a cause. This position is well expressed in two famous dicta: "If cause is present, the effect is present, and if absent, absent. For by proving the cause you at once prove the effect, and conversely nothing can exist without its cause" (*Rhetoric* 2.24.1400a28–31); and "When the effect is present the cause is also present—for instance, if a plant sheds its leaves or the moon is eclipsed, there is present also the cause of the eclipse or of the fall of the leaves" (*Posterior Analytics* 2.16.98a36–38).

Aristotle distinguishes four different kinds of causes and contends that his predecessors used only some of the four and neglected the rest. They mentioned some of the four causes "only to touch on them, and then forget them" (*Physics* 2.8.198b15–16). "All these thinkers, then, as they cannot settle on another cause, seem to testify that we have determined rightly both how many and of what sort the causes are. Besides this is plain that when the causes are being looked for, either all four must be sought thus or they must be sought in one of these four ways" (*Metaphysics* 1.7.988b17–20).

The question is whether it is contradictory to hold that all events have their causes and, at the same time, to maintain that some events are caused by chance. How is it possible to say that we have to look for causes in chance events? In order to answer this question, we need to consider Aristotle's own explanation of the nature of chance events.

A point Aristotle makes in his discussion of chance in the *Physics* is that all changes in nature can be classified in three ways: (1) some changes "always come to pass in the same way" by absolute uniformity; (2) some changes happen "for the most part" by regularity; and (3) some changes are those of which "chance is said to be the cause" (2.5.196b10–15). That chance events happen neither always (necessarily) nor normally (regularly) may be more clearly understood by his own example of "accident." Suppose that a farmer found a valuable treasure while digging a hole for planting seeds. Finding the treasure is an accident for him, for he neither always nor regularly finds it while planting. In this sense, an accident is said to have "an indefinite chance cause."

> There is no definite cause for an accident, but a chance cause, i.e., an indefinite one. Going to Aegina was an accident for a man, if he went not in order to get there, but because he was carried out of his way by a storm or captured by pirates. The accident has happened or exists—not in virtue of the subject's nature, however, but of something else; for the storm was the cause of his coming to a place for which he was not sailing, and this was Aegina. (*Metaphysics* 5.30.1025a23–29)

Thus Aristotle offers his first explanation of chance events in the negative form. They happen as an exception, neither always nor usually. That is, they happen neither of uniformity (necessity) nor of regularity (normality).

In the *Metaphysics* Aristotle offers a second explanation of chance events in his discussion of three kinds of "becoming," or generation, in the world: by nature, by art, and by chance or spontaneity.

The "becoming by nature," or "natural production," needs four causes, and so does the "becoming by art," or "making from thought." But in the latter case the efficient cause contains two elements, namely, an "idea" as an immanent element and a "process" as a transient element. For example, an artist as an efficient cause has a certain idea or thought about the object he or she is going to make (an immanent element) before starting to make a painting (a transient element). The artist has a certain artistic idea whose "form is in his or her soul" and then proceeds "from this point onward." In other words, the immanent idea is what proceeds "from the starting point" as a "thinking," whereas the transient process is what proceeds "from the final step of the thinking" as a "making" (*Metaphysics* 7.7.1032a26–1032b23).

The becoming by "spontaneity" also involves all four causes, but it is different from the "becoming by art" in that the efficient cause lacks its immanent element. If something happens by art, there is "a form in the soul" or a certain deliberate thinking of the causal agent, whereas if by spontaneity, it simply happens without such thinking. This is why it is often called a "becoming by luck" in the sense that it happens, as natural change does, naturally or spontaneously. In short, the spontaneous change happens without the agent's deliberate thought (*Metaphysics* 7.7.1032b5–29).[1] This does not mean that the "becoming by spontaneity" and the "becoming by nature" are the same, just because both can happen spontaneously or naturally. The former involves, or requires, an efficient cause (artist), whereas the latter is produced strictly by nature.

Aristotle concludes that the "becoming by chance" is neither by nature nor by art.[2] It is produced by an efficient cause, only without the agent's deliberate thought. The agent did not think "this effect," and the events are actually other than what the agent thought. This means that chance is *not* an absence of cause, but an efficient cause without the agent's thought of it. It is an "unintended" efficient cause, a cause without intention.

But here is a difficulty: Aristotle wants to extend chance events not only to human activities but also to the realm of natural activities. Since all changes in nature happen by necessity, there is no reason, so it seems, why we should not conceive that they happen without an end. It seems absurd to say that all natural changes are "for the sake of something," thus involving the agent's intention. Rather, it seems correct to say that all natural changes happen necessarily, not by chance.

Suppose that a man's crop is spoiled by heavy rain. The rain did not fall for the sake of this purpose (spoiling the crop), but "the result just followed." Thus, it can be called a chance event, but it did fall by necessity. For "what is drawn up must cool, and what has been cooled must become water and descend" (*Physics* 2.8.198b19–21). "A difficulty presents itself: why should not nature work, not for the sake of something, nor because it is better so, but just as the sky rains, not in order to make to corn grow, but of necessity" (198b17–19).

After presenting this line of difficulty, however, Aristotle contends that this is not the "true view" to explain natural phenomena. Roughly his argument runs as follows: We do not ascribe, for example, a heavy rain in the rainy season to chance, but only in the dry season. We do not ascribe heat in summer to chance, but we do that in winter. Now such natural phenomena—rain in the dry season and heat in winter—can be either the result of a certain purpose or the result of "mere coincidence," that is, that which lacks any intention. But since, Aristotle holds, nothing can be the result of mere coincidence, it follows that there must be a certain end or intention in natural phenomena such as rain and heat, and the lack of a cause is only an appearance (see *Physics* 2.8.198b33–199a5).

In order to understand this argument, it is necessary for us to accept Aristotle's view that "nature" as well as "human intelligence" is purposive, intentional, or teleological. According to Aristotle, all natural events, as well as artificial ones, belong to "the class of causes which act for the sake of something"; furthermore, the changes by art partly complete what nature cannot do or partly imitate what nature does. For example, a house built by art "would come to be in the same way as by nature," and conversely, if we imagine a house built by nature, it would be built in the same way as by art (*Physics* 2.8.199a12–15). "If, therefore, artificial products are for the sake of an end, so, clearly, also are natural products" (199a17–18).

In summary, Aristotle "proves" that all chance events are produced by an unintended efficient cause with two assumptions: (1) all events are caused, and (2) all events, natural or artificial, are teleological or intentional, and they are therefore involved in a certain purpose or end. Without these assumptions, it would be impossible to understand his argument on chance and chance events.[3]

In the *Physics* Aristotle offers a third explanation of chance events, this time in terms of a final cause. It is generally thought that we can apply a final cause, or "for the sake of something," only to events which happen always or usually, not to events which happen according to some deliberate intention. But this is not true, for a final cause is applicable not only to events which happen uniformly or regularly but also to incidental or accidental events. Such incidental events, as far as being involved in a final cause, Aristotle contends, are properly said to be by chance or spontaneity. He states: "Hence it is clear that even among the things which are outside the necessary and the normal, there are some in connection with which the phrase 'for the sake of' is applicable. When a thing of this kind comes to pass among events which are for the sake of something, it is said to be spontaneous or by chance" (*Physics* 2.5.196b20–

23). Here chance is not a cause *per se* but a cause *per accidens*. It is not a final cause as such. It is an efficient cause, involving an "incidental" final cause. Let us clarify this claim by examining Aristotle's own example.

A man went to the market to shop but incidentally or accidentally met his debtor, who paid him a debt. Since the man was collecting money for a feast, he would have gone there for the purpose of collecting money, *if* he had known it. But actually he went there for another purpose (shopping), and incidentally he happened to meet his debtor. "It was only incidental that he got his money by going there." His purpose, collecting money, was fulfilled, not because he went there for that purpose and not because he was going there always or normally whenever he was collecting money, but because he met his debtor there accidentally. The final cause of the activity of the agent (shopping) was other than the end of the chance event (collecting money). But we can say in this case that "intelligent reflection and chance are in the same sphere" (*Physics* 2.5.197a6–8), for he might have gone there for the purpose of collecting money if he had known it.

In this section, I have shown that Aristotle explains the nature of chance and chance events in three ways. First, chance events happen neither always nor normally. They happen neither of necessity nor of regularity. Second, chance is an efficient cause which lacks the agent's intention, and correspondingly, chance events are produced by the agent unintentionally. That is, chance events are the effects of an agent who intended something other than the chance events. Third, chance as an efficient cause also involves an incidental final cause, and correspondingly, chance *events* involve an incidental final cause. All these amount to saying that events attributed to chance are not uncaused but are caused by an "unintending" cause.

II

Peirce seems to construct his tychism out of Aristotle's doctrine of *tychē* (chance), for he not only quotes Aristotle as "the prince of Philosophers" but also holds that his indeterminism exhibits a kind of "Aristotelian pattern" (*CP* 6.320; see also 6.5, 11, 206; 7.579). Like Aristotle, he believes that chance happens neither always nor usually and that Aristotle's theory of absolute chance is "the utmost essence of Aristotelianism" (*CP* 6.36).[4] But Peirce also believes that some chance events happen without any cause whatsoever. Such absolute chance has not only "a logical subjective value" but also "an ontological objective value" (*CP* 6.93).

How is this possible? Peirce explains his theory in terms of possibility, fallibilism, and probability. First, according to Peirce, the term "possibility" can mean either something contrary to actuality or something which has an actual existence but lacks a rational necessity. Considered as something which is not actual, it has two meanings. First, taken *objectively*, it means something "as yet undeveloped." It has only a capacity for realization, which can lead "to final completeness of being" at a future time. Second, taken *logically*, it means that there are some grounds for asserting an actuality but not sufficient to justify it "positively." For example, the statement "It will possibly rain tomorrow" has to do with degrees of certainty. "Possibly" here means "may" rather than "can" (*CP* 6.364–65). Considered as something which has an actual existence but lacks rational necessity, it also has two meanings. Third, it means *contingent chance*, which, while necessary causally, is not necessary teleologically; that is, it is something unplanned. "Possibility" here has no purpose or intention. But Peirce argues that this view ultimately leads to the logical sense as defined above. Finally, it means *pure chance*, which is not derivable from and causally explainable in reference to past empirical facts.

Peirce replies to these concerns in his theory of fallibilism, the view that all philosophical reasonings are only tentative and provisional. All positive reasonings are subject to doubt and refutation, for we have to draw our conclusion by the proportion found in a sample.[5] Accordingly, there are three things we can never attain by reasoning: absolute certainty, absolute exactitude, and absolute universality. Peirce says:

> We cannot be absolutely certain that our conclusions are even approximately true; for the sample may be utterly unlike the unsampled part of the collection. We cannot pretend to be even probably exact; because the sample consists of but a finite number of instances and only admits spe-

cial values of the proportion sought. Finally, even if we could ascertain with absolute certainty and exactness that the ratio of sinful men to all men was 1 to 1; still among the infinite generations of men there would be room for any finite number of sinless men without violating the proportion. (*CP* 1.141)

Some "theological" determinists nevertheless claim that a religious revelation is absolutely certain, exact, and universal. But Peirce argues that this is not so, for three reasons. First, since revelation is not established by logical reasoning, we cannot be certain whether any given message by revelation is really inspired, and moreover we cannot even prove it with a high degree of probability. Second, even if inspired, we cannot be certain that it is true, for all inspired matter is still subject to "human distortion and coloring." Third, since inspiration is of a somewhat incomprehensible nature, we cannot be certain that we correctly comprehend it (*CP* 1.143).

Some "rationalistic" determinists claim that a priori truth such as mathematical axioms, first principles in logic, and the maxim of causality are absolutely certain and without exception. But Peirce argues that most positive historical proofs show that all "innate truths" are "mixed up with error" and that there are a fortiori some exceptions. It has been held that mathematical axioms are absolutely exact. But Peirce argues that our experience can never show any truth to be exact. For example, we can say that the sum of three angles equals 180 degrees, but we cannot be certain that it is *exactly* the case.[6]

Finally, Peirce also explains absolute chance in his theory of probability: the view that all positive reasoning of human beings is only "probable" or that we can be certain "only under some circumstances." "Probability" may in fact render a likely thing "practically" certain. This is why we believe that the inference of probability has a higher degree of certainty than, say, those of plausibility and likelihood.

Peirce argues that our statements are only "probable," for they involve an infinite series "in the long run." Let us say that the probability of getting a certain number on a throw of dice is one-third. This means no more than that the dice have a certain "would-be." To verify it, it is necessary for the dice to undergo an endless series of throws. We can be sure that the dice will have a certain character only when the series is endless. This means that no collection of single events can constitute a would-be, nor can any knowledge of single events tell us with certainty of any would-be (*CP* 2.667).

Peirce does not believe that the world is a sheer chaos or that all events are "chance combinations" of independent events, which would make any reasoning from one fact to another utterly impossible. What is important is that there is a "higher universe" to which the conception of probability can have no applicability. This is a universe of absolute chance, producing "uncaused" events. Such events just happen without any cause, and this makes our world more dynamic and diverse (*CP* 2.684).

Peirce sees several difficulties in his theory of pure chance. Some determinists seek to back up their a priori position with empirical argument. They say that the exact regularity of the universe is a natural belief and has generally been confirmed by experience.

But Peirce argues that natural beliefs also require correction and purification in order to be separated from natural illusion, because the adaptation of nature, however beautiful and marvelous, has never been found to be quite perfect. Moreover, those necessitarians have to prove that their determinism is founded on something other than observation. For experience informs us that single objects exist at a single date in a single place. In other words, observation is a posteriori; that is, it has reference to single facts, to particulars, and yet the doctrine of necessity is of an a priori nature.

Another difficulty in Peirce's theory of pure chance can be stated as follows: Even if there exists absolute chance, it is inconceivable to human beings, for we find no observed phenomena with reference to which the hypothesis of absolute chance is needed to provide an explanation. In addition, even if such chance is conceivable, it is unintelligible to reason.

Peirce offers four theses against this theory of the inconceivability and unintelligibility of absolute chance. First, the world is full of real growth and increasing complexity, which presents all kinds of developing diversifications. This general prevalence of growth in nature is opposed to strict mechanical law. From these broad and ubiquitous facts, Peirce argues, we can infer fairly that there is in nature an agent called absolute chance, by which the complexity and diversity of nature can be increased (*CP* 6.36). Mechanical law can explain only one side of the growth, not all the freshness of increasing growth in nature.

Second, there is an inexhaustible and multitudinous variety of the world. This is obviously the most intrusive character of the universe, which the rule of necessity cannot account for in the least. This variety seems to deny the existence of spontaneity, but in fact it emerges from spontaneity. Mechanical law can explain only one determinate result, not variety itself (*CP* 6.63). No mechanism can account for the fresh and new variety of nature. The only way to account for this infinite diversity is by admitting absolute chance as the most important, or at least an indispensable, element of the universe, which produces "infinitesimal departures" from law continually. Without such a hypothesis of chance-spontaneity, the variety of the universe is manifestly inexplicable (*CP* 6.63).

Third, those who hold a deterministic law as "immutable and ultimate fact" do not seek to explain the diversity and irregularity of the universe. They leave the whole specificity of the world utterly unaccounted for. They regard certain laws of the world as ultimate regularities and yet disregard, or neglect, many relationships of "lawlessness" in nature, which demand an explanation (*CP* 6.64).

Furthermore, as we have seen, the deterministic laws require an explanation, and like anything which is to be explained, they must be explained by something else. Since laws in general cannot be explained by any law in particular, the explanation must consist in showing "something lawless." The explanation of laws is possible only by showing how they are developed out of absolute chance (*CP* 6.613).

Determinists would say that absolute chance needs an explanation as much as anything else. However, if absolute chance or irregularity is a mere absence of any particular circumstances necessitating a certain result, Peirce argues that it calls for no explanation. For it is absurd to say that the absence of any definite character should be accounted for. The "non-existent" spontaneity requires no explanation (*CP* 6.605).[7]

Fourth, there is consciousness, a feeling which may be "very inconvenient" to the mechanical philosophers. The whole action of mind cannot be a part of the physical (mechanical) universe as determinists maintain. Our consciousness cannot be "but an inward aspect, a phantom." We have the right to believe that the universe of the human mind is a "rounded system uninterfered with from elsewhere." The state of feeling at any moment is not exactly calculable. A strict mechanical causality leaves no room for the ever-changing states of our feeling (*CP* 6.61).

In conclusion, Peirce's tychistic argument involves two steps: rejecting deterministic mechanism and showing his own demonstration for his tychism. In refuting mechanical determinism, he argues that all reasoning of human beings is only provisional, and therefore the determinists' law of causality is nothing but a simple postulate which is not supported by logical inference. The doctrine of fallibilism leads us to accept a hypothesis of chance-spontaneity, by which we can account for many irregularities in the world. Furthermore, the theory of probability shows that mechanistic law, including the law of causality, is only "probable" or can be certain only "under some circumstances." Peirce also offers four theses to support his doctrine of pure chance, namely, the real growth in nature, the variety of nature, the explanation of the diversity and irregularity in the universe, and finally the ever-changing consciousness of the human mind. His whole argument may be written in simple modus tollens form: If determinism were true, then certain obvious facts, such as the variety and the like, could not be the case. But it is quite clear that these are the case. Therefore, determinism is not true.

III

In conclusion, I will not try to say whether Aristotle's or Peirce's doctrine of chance is more nearly true. I will also not pursue the question of whether it is more correct to hold Aristotle's distinction between chance and spontaneity or Peirce's interchangeable use of both terms. It will suffice to point out that any interpretation of Aristotle's doctrine of chance as an uncaused pure chance is contrary to Aristotle's own position.

As we have seen, Aristotle interprets chance in terms of an efficient cause which lacks any intension and a final cause which occurs incidentally. But the change by chance or spontaneity is for Aristotle by no means a change without a cause. Such changes are still described in connection with a phrase "for the sake of something" (*Physics* 2.5.196b21–22). For Aristotle, it would be absurd to suppose that purpose is not present just because we do not observe the agent's deliberate intention (2.8.199b27–28).

Peirce fails to see that Aristotle's chance or spontaneity does not denote the absence of cause, but rather the absence of intention. He fails to see that, in the events attributed to chance, an *accidental* occurrence may produce a *desired* effect. He fails to see that there are in fact events that "have happened as you could have wished, without your contriving it." In short, he fails to see that chance events are not "self-determined" events (see Heath 1877:114).

A question remains, then, why Peirce and other philosophers misunderstand Aristotle's doctrine of chance.[8] There are, it seems to me, several reasons for this.

Aristotle himself rejects the strict mechanical determinism of the atomists, such as Leucippus and Democritus. According to the atomists, the world consists of an infinite number of atoms which are "tossed about"; that is, the atoms of different shapes move in different ways. They move at random in the void in a number of possible ways until, at last, they fall by chance into a vortex of motion (Robinson 1968:208). However, the atomists also maintain that the atoms "come into being by necessity" and that there are no exceptions. They contend that we ascribe an event to chance not because it has no cause but because we simply do not know its cause. All events rigidly follow from "the law of motion," for nothing occurs out of nothing but for a reason and by necessity (ibid.:112).

Aristotle, however, believes that the atomists, while saying that there is always a certain movement, do not say what this movement is; nor do they tell us why the movement is. Thus, why it moves in this or that way is a question Aristotle regards as requiring an answer. At any rate, Aristotle rejects a mechanical determinism.

Another reason why some philosophers misunderstand Aristotle's doctrine of chance is that Aristotle seems to imply an idea of absolute chance, especially when he presses Empedocles' doctrine of "strife" and "love." Aristotle says: "What is the cause of this? Not fire or earth, surely. But neither is it Love or Strife, for the former is simply the cause of coming together, the latter of separation. . . . 'Chance,' not 'proportion,' is the name one applies in such cases; for things can be mingled at random" (*On Generation and Corruption* 2.6.333b12–16).

Aristotle seems to imply an absolute chance when he states that "things can be mingled at random" in the foregoing quotation. His point seems to be that it is not their mere presence in a "particular ratio." But if so, then what brings them together in this particular ratio? Is it not a "mere chance" (Robinson 1968:166)?

It is evident that Aristotle here maintains that chance events cannot be accounted for by an absolute mechanical law, for "things can be mingled at random." However, his rejection of a mechanical determinism does not exactly mean that he holds an indeterminism of pure chance. Rather, he holds a causal determinism, the view that all events, including chance events, are susceptible of analysis in terms of causes. Therefore, in order to understand the nature of chance events, we have to look for, not absolute chance which entails some uncaused events, but "the cause in question" (*On Generation and Corruption* 2.6.333b13).

It is in this line of argument that Peirce misunderstands Aristotle's doctrine of chance. For Peirce says:

> Already, the prince of philosophers had repeatedly and emphatically condemned the dictum of Democritus (especially in the *Physics* Book II, chapters 4, 5, 6) holding that events come to pass in three ways, namely, (1) by external compulsion, or the action of efficient cause, (2) by virtue of an inward nature, or the influence of final cause, and (3) irregularly without definite cause, but just by *absolute chance*; and this doctrine is of the inmost essence of Aristotelianism. (*CP* 6.36, emphasis added)

Peirce's misconception of Aristotle's doctrine of chance is quite evident; however, this does not entail that his whole tychism is false. It is nonetheless an irony in the history of philosophy that Peirce, who praises Aristotle as "the prince of philosophers," distorted so drastically, in Peirce's words, "the inmost essence of Aristotelianism."[9]

Notes

1. Some effects can come about both by art and by spontaneity. A person's health is produced by art, namely, by a physician's "train of thought." But it can also be produced spontaneously, for the patient may get well "naturally," without a doctor's help. In contrast, a house can be produced only by art, not by spontaneity.

2. For Aristotle, "spontaneity" is a wider term than "chance." Every result of chance is from spontaneity, but not every result of spontaneity is from chance (*Physics* 2.6.197a33–197b37).

3. Two comments are needed here. First, in this chapter I do not pursue whether these two assumptions are conclusively proved, as Aristotle believes. Second, it is not necessarily contradictory to hold that all events are intentional and, at the same time, that some causes are absent in some events. An event which lacks a final cause may be intentional from the viewpoint of "higher beings." If there exists a God or Unmoved Mover, then an event which lacks purpose in itself may fulfill a purpose of God's.

4. "The position of Aristotle in this matter is altogether right . . . but it is a position that nobody can understand" (*CP* 6.93).

5. For Peirce, all statements of matters of fact, whether philosophical or not, are in principle subject to doubt.

6. Peirce's view on mathematics is not perfectly clear. He says in one place that he does not mean by his fallibilism that 2 × 2 is not exactly 4 or that people cannot usually count with accuracy. All he means is that we cannot attain an absolute certainty on any question of fact. But he generally takes mathematics as an inexact science, for example, in passages such as *CP* 2.646 and 1.400–402.

7. I find this somewhat strange, for the argument operates from the deterministic position.

8. According to Heath (1877:115), Mill and Grote are such philosophers. Mill takes Aristotle's chance and spontaneity as "co-ordinate agents with 'Mind' in producing the phenomena of the universe"; Grote, as "an independent agency inseparable from Nature," which is essentially irregular and unpredictable.

9. Peirce contends that it is not correct to suppose that a final cause is necessarily a purpose. A purpose is merely one of the "patterns" of a final cause which is most familiar to our experience. Furthermore, Peirce contends that we cannot think a final cause without an efficient cause, for a final cause is only an "ideal," whereas an efficient cause is "forceful" as an "external compulsion" (*CP* 1.211–12).

References

Heath, D. D. 1877. On Some Misconceptions of Aristotle's Doctrine on Causation and *To Aytomaton*. *Journal of Philology* 7:97–115.

Robinson, John M. 1968. *An Introduction to Early Greek Philosophy*. Boston: Houghton Mifflin.

Part 3

The Life of the Mind

19 Peirce's Definitions of the Phaneron

André De Tienne, The Catholic University of Louvain, Belgium

One of the most fundamental conceptions used by epistemologists of every stripe is that of the *phenomenon*. Since the ancient Greeks coined the word to mean "whatever comes to the light of mind," philosophers have used it for more than two thousand years to designate whatever objects were apprehended by our faculties of perception. For more than two millennia also, the notion of phenomenon has been tainted with the Platonic contempt for what is said to be only an imperfect or inaccurate reflection of true reality. The ideal real cannot be channeled to our consciousness through the senses without undergoing major alterations, so that a mere (sensuous) appearance of something is always considered to be essentially deceptive.

Philosophers have thus made every endeavor to unveil the secret nature of that which was hidden under the phenomenal disguise and to that effect have resorted to different intellectual devices, such as the faculty of intellectual intuition. They were more or less successful until Kant set up his tribunal of pure reason and demonstrated our inability of ever going beyond the phenomenon to catch even a glimpse of the noumenon, or thing-in-itself. For Kant, the thing-in-itself is not cognizable, though it stands as the transcendental source of validity for all that we can know. The phenomenon stands for the noumenon; it is a *Vorstellung*, and most thinkers after Kant, even the early Peirce, took for granted that every phenomenon was precisely that: a *Vorstellung*, or representation.

Peirce was among the first philosophers to question Kant's noumenon/

phenomenon distinction seriously and to reject his transcendentalism. He argued that nothing is uncognizable, everything is knowable, because everything is representable (which is not to say that everything is a representation). If the phenomenon is no longer simply the correlate of an unknown noumenon, one would expect its definition to undergo some significant reform; it can no longer be interpreted in terms of representation (or re-presentation), where the "re-" denotes the Platonic suspicion of appearances. Such a reform did take place in Peirce, especially in his later writings, after the turn of the century, when he made his doctrine of the three categories into a science of the phenomenon in general, a science he named phaneroscopy. A few months before the end of 1904, Peirce decided to replace the old "phenomenon" with the new technical term "phaneron," to make sure that the two conceptions underlying those words would not be confused.

Peirce gave several definitions of the phaneron. Unfortunately they do not always appear consistent; their diversity has in fact more than once been a source of misinterpretation of Peirce. In this paper I survey Peirce's definitions of the phaneron, sort them out into different categories, and seek to make sense of the variety by reducing the manifold to unity.

The Etymological Definition

The basic definition of the phaneron, the one that permeates all the other definitions, is that which Peirce derives from the etymology of the word: "The word *φανερόν* is next to the simplest expression in Greek for *manifest*. . . . There can be no question that φανερός means primarily *brought to light*, *open to public inspection throughout*. . . . I desire to have the privilege of creating an English word, *phaneron*, to denote whatever is throughout its entirety open to assured observation" (*MS* 337:4–5, 7, written in 1904).[1]

A phaneron is that which is manifest, that which appears. Peirce does not present the phaneron as something that unveils itself in a process independent of mental intervention. The manifest is not appearing by itself but is fully apparent, plainly exhibited, evident. The evidence is all-encompassing: nothing in the phaneron can escape the light by hiding in some dark corner. Peirce confers on the phaneron the broadest meaning; it is not only an appearance but more largely anything that just *seems*. The only assertion in phenomenology is "that there are certain seemings" (*CP* 2.197, 1902). Peirce draws a distinction between appearance and seeming because the former has a lesser extension than the latter. We can doubt that something, at some point, appeared; we do not doubt, however, that it seemed to appear. Seeming is enough: something was before the mind, and that is what counts.

We can deny something to be thinkable, but this denying is enough for that something to become thinkable and thus be a phaneron. Consciousness is therefore a key element of the phaneron; it cannot, in fact, be separated from that which appears. The phaneron is not an object that the subject mind can manipulate as it pleases. The cognitive distance that this requires between subject and object does not exist in the phaneron. Peirce is convinced that in the subject-object distinction "there lurks . . . one of the worst fallacies of metaphysics" (*MS:L* 482:27, c. 1904). The *esse* of the phaneron is its *percipi* (*MS* 908:5, 1905), and no *perceptum* is without *percipiens*; Peirce therefore insists that an external reality cannot be a phaneron: "it is not *entirely* open to observation" (*MS* 337s:8). An external object is opaque; sides of it elude our apprehension.

Peirce does not restrict the phaneron to that which appears to our senses. His definition is so broad as to obviate any distinction between the real (in an ordinary sense) and the unreal, or fictitious. What is fancied or hallucinated is just as much an "appearing" as a red flag waved before our eyes. The authenticity of the phaneron is irrelevant; all it needs to do is to seem in order to be. The light to which the phaneron is brought is not external but internal to the contemplating mind; it emanates from it in a way that overlooks any sharp distinction between mind and phaneron. The phaneron is the crucible in which the inner and the outer worlds are conflated. The basic elements of that conflation can be separated only by a special chemical analysis, that of phaneroscopy, to which Peirce at one point gave the alternative name of "phanerochémy—the chemistry of appearances" (*MS* 1338:23, 1905–6).

Phaneron and Awareness

The phaneron is open to mental observation: we may wonder what kind of observation this is, since the two poles of the phaneron—ego and nonego—are conflated. In each act of cognition, of no matter what kind, something is made manifest and appears to the mind with irresistible force. The mind can conduct a variety of operations, all of which must be based upon a common denominator, a "necessary and sufficient" mental condition for a seeming to take place. Peirce calls this common denominator "direct awareness": "I think it will conduce to perspicuity to invent the noun '*Phaneron*' . . . to denote an object of any kind of which a person is aware not merely in being first aware in something else, but directly" (*MS* 612:7–8, 1908).[2]

Awareness is involved in every state of consciousness as its basic ingredient. The slightest vigilance is a source of mental light (however dim) and brings something to the mind. Our awareness of a phaneron is always total and puts it into our "Immediate and Complete possession" (*MS* 645:3, 1909). The most important feature is the immediacy, the directness, with which one is aware of the phaneron. The appearance and the mind are conflated, which means that there is nothing to mediate between the two: there is no intervening sign. We are put *facie ad faciem* before the very phaneron itself, Peirce says (*MS* 645:5). Direct awareness is a face-to-face encounter, which is the same as saying that that which appears to a mind is *not* represented. A seeming is not a representation, at least not in the first place, and thus a phaneron never conveys any cognitive information. Direct awareness is therefore not to be confounded with cognitive intuition, which is a faculty whose existence Peirce denies. It follows, then, that the mode of manifestation of a phaneron must be in some essential respect quite different from that of a sign.

The Particular Definition

Peirce's most frequent definition of the phaneron, though not the most satisfactory, I call the "particular" definition. The following quotation is a reconstruction made up of ten different excerpts:

> I use the word *phaneron* to mean or denote an *ens* or object of no matter what kind, a thing in the widest sense of that word, in short, whatever is present to or can come before the mind, at any time and in any sense or in any way whatsoever, without caring whether it be regarded as real or not, as fact or fiction, whether it be a tinge of feeling, or be imagined, or thought or desired, and whether it be objectified or not.[3]

The particularity (in a logical sense) of this definition originates in the apparent objectification of the phaneron. The formula "*whatever* is present to the mind" may refer to *any* kind of thing, but eventually, in an actual case, to *some* thing. The whole problem lies here, in the determination of what that something can be. If we need a criterion to make such a determination, Peirce's definitions do not appear to offer any. A catalog of phanera that Peirce gives as examples in one manuscript (*MS* 611:17–18) lists such "objects" as those perceived by the five senses, emotions, pleasant or unpleasant feelings, impulses of all kinds, perceptions of abstract relations, recollections and senses of recollecting, and so forth.[4] In all those instances, what is perceived is never detached from the act of perceiving.

The problem could be summed up in the following question: What is it that entitles us to speak of *a* phaneron?—*a* phaneron, that is, a single phaneron among many others, distinguishable as a unity, circumscribable. We know that no object—a table, for instance—taken in itself, independently of our apprehension, is a phaneron. If we say that only that which we directly perceive of the table is a phaneron, we must also recognize that we are in fact aware, at the same time, of a thousand other things, all of which concur to form the phaneron. If such is the case, then it is not possible to liken the phaneron to a haecceity of any sort, since it does not appear to be denumerable or describable with any fullness.

Indeed, the above question amounts just to this one: Is it possible to give any description of a phaneron? In this regard, I quote two of Peirce's definitions of phenomenology: "In Phenomenology there is no assertion except that there are certain seemings; and even these are not, and cannot be asserted, because they cannot be described" (*CP* 2.197, 1902); and "Phenomenology . . . endeavors to describe in a general way the features of whatever may come before the mind in any way" (*MS* 693:46, 1904). An obvious contradiction seems to be lurking here: on the one hand, we are told that the phaneron cannot be described; and on the other, that it can. I do think, however, that it is possible to solve this contradiction and dissipate the confusion.

Lived Phaneron and Objectified Phaneron

Part of the solution consists in our avoiding a dangerous confusion of concepts, and for that an important distinction must be introduced. There are two ways of "observing" a phaneron. The first is already familiar to us and is called direct awareness. It is the act by which the mind acknowledges the manifestation of that which appears without any mediating representation. The phaneron as "object" of immediate awareness is the manifestation itself as it is taking place or as it is being "lived." Let us characterize this type of appearance with the name *lived phaneron.*

There is also another way of observing a phaneron, that of the phenomenologist. The latter, according to Peirce, is chiefly busy describing the phaneron, determining what its formal and indecomposable elements are, and classifying those elements and compounds according to their valency. To achieve this goal, the phenomenologist is forced to create, as it were, an objectifying distance between himself or herself and the phaneron. The *objectified phaneron* (as opposed to the lived phaneron) becomes, in this way, the product of a mental separation or abstraction, as it is indeed isolated from the continuous stream of manifestation and becomes re-presented. Only then can it be considered as an object: when it is no longer genuine, no longer "lived." It is dead, for it has irremediably lost an essential ingredient, that of being a relational knot where the two poles, ego and nonego, are interwoven "confusedly." It has been externalized and thus made partially opaque. Strictly speaking, therefore, the objectified phaneron is no longer a phaneron. Whenever Peirce speaks of phenomenology as a descriptive science, he has in mind the particular definition of the phaneron, which thus leads him to misuse the term. If we return now to the lived phaneron, the question we have to ask is that of its unity: where does it start, where does it end? The next type of Peircean definition brings a most interesting answer.

The Collective Definition

The two following quotations will immediately bring home the main characteristics of the collective definition:

> I propose to use the word *Phaneron* as a proper name to denote the total content of any one consciousness, . . . the sum of all we have in mind in any way whatever, regardless of its cognitive value. (*MS* 908:4, 1905)
>
> By the *Phaneron*, I mean the single entirety, or total, or whole, of that which the reader has in mind in any sense. (*MS* 338:2, c. 1904)[5]

The difference with the particular definition is striking. Peirce is no longer speaking of "whatever is present to the mind"; rather, what he is referring to is the "total content of consciousness," the "sum total" of all appearances. The phaneron is not a particular object present to the mind but the collection of all things that present themselves together to our consciousness. If the phaneron is such a totality, it must then comprehend that by which it is brought to light, namely, the consciousness that apprehends it and lights it up. The collective definition confirms in a way the relevance of our suspicions toward the particular definition. Speaking of the genuine "lived" phaneron, we are unable to describe it as a denumerable unity. The unity of the phaneron is that of cohesion, of togetherness, of globality; we cannot define its limits. But if such is the case, if we are right in our holistic interpretation of the phaneron, is there any sense left in our speaking of one phaneron among many others? Does it make sense to speak of several "totalities"? From the standpoint of an "objectified" phaneron, the mere reference to a totality is meaningless. As to the "lived" phaneron, the idea of a multitude of totalities is absurd, because the phaneron is temporally open-ended and is not definable; it is part of a stream of manifestation and never stops occurring, never stops being present to the mind. Continuity is part of the definition of the lived phaneron; if it was not, we would have the greatest difficulty to account phenomenally for such conceptions as change, motion, and evolution.

But consider the objection that, if a phaneron is an open-ended totality, then there can be only one such totality, and thus only one phaneron: from the day we are born to the day we die, our whole life is a unique and continuous phaneron, one totality containing everything that was, is, and will be present to our mind. But surely Peirce does not speak of only "one" phaneron? However strange it may sound to say so, I think he does. More than once, Peirce refers symptomatically to the Phaneron as a proper name and uses it only in the singular.[6] None of his "collective" definitions contradicts the idea of a unique phaneron. And the following excerpt from a 1904 letter to William James strongly supports it: "My 'phenomenon' for which I must invent a new word is very near your 'pure experience' but not quite since I do not exclude time and also speak of only *one* 'phenomenon'" (*CP* 8.301). Yet, if there is only one phaneron, how can we interpret all those instances where Peirce speaks of the phaneron in an objectifying way?

Tentative Conciliation of the Definitions

Peirce's most satisfactory definition of the phaneron is the collective definition conjoined to the etymological one. They are totally compatible, and this is an advantage which the particular definition does not share. The preceding discussion made it clear that the particular definition, though Peirce's favorite, was the most troublesome. The distinction between lived and objectified phaneron proved very useful in this regard, as it showed that the notion of phaneron could be particularized only at the expense of its integrity or genuineness. Peirce, however, was not unaware of the discrepancy between the particular and the collective definitions. Indeed, he provides himself the means to solve the difficulty: "We call whatever is in the mind, whether as feelings, as stresses, or efforts, as habits, or habit-growths, or of whatever other kind they may be by the name of *ingredients of the phaneron*; . . . whatever we at all know we must know through ingredients of the phaneron" (*MS* 477:10–11, c. 1905).[7]

By "ingredients" of the phaneron Peirce thus understands the detail of the phenomenal whole, which is nothing but the "anything," the "whatever is present to the mind." Whatever the phenomenologist isolates will therefore be an objectified ingredient of the phaneron and not the phaneron itself, which can never be objectified. While examining an ingredient of the phaneron, phenomenologists themselves are involved immediately in a lived phaneron from which they are unable to detach themselves.

This inability stems evidently from the unicity of the phaneron. If there were more than one, phenomenologists would have to check constantly whether or not they let elements of their own lived phaneron impinge on the phaneron they are observing, to make sure their description is not distorted. But of course such a checking is impossible, since the act of observing a phaneron is itself part of the phaneron. Edmund Husserl did not see this, as he thought it possible for phenomenologists to bracket all their preconceptions and natural beliefs so as to be able to get an immediate intuition of the intentional essence of the phenomenon. Peirce, however, does not believe in the existence of a faculty of cognitive intuition (his "immediate awareness" is different), and he rejects the need of bracketing one's beliefs: as long as we have no reason of doubting them, we should keep and trust them. One does not need, in other words, to be concerned about interferences between different phanerons, and this is another way of justifying the claim that there is only one phaneron, understood as a manifest collective totality.

Notes

The research summarized in this paper has been done under the auspices of the Belgian National Fund for Scientific Research (FNRS), for which I am a senior research assistant. I wish to thank the Peirce Edition Project, Indiana University at Indianapolis, for their generous support.

1. Other references for this type of definition include the following: written in 1902: *CP* 1.280; 2.84, 120, 197; in 1903: *CP* 5.43; in 1904: *MS* 336:9; 337s:5, 6, 7–8; in 1905: *MS* 284:41; 908:5, 8; in 1905–6: *MS* 1338:23; c. 1907: *CP* 7.36 n. 13; in 1908: *MS* 612:7; c. 1908: *CP* 1.288; in 1909: *CP* 8.303; in 1910: *MS* 655:24. (Throughout this chapter, the date of composition appears with each Peirce reference.)

2. Other related references include, in 1908: *MS* 611:16, 19; 612:9, 13; in 1909: *MS* 645:3, 5, 26–27.

3. The excerpts come from the following texts: written in 1903: *CP* 1.186; in 1904: *MS* 336:4; 693:46, 62; *L* 107:19; in 1905: *CP* 8.213; *MS* 1338:21; *PW* 189; in 1908: *MS* 612:8, 14; in 1909: entries in *Century Dictionary* supplement: phaneron, phenoscopy; n.d.: *L* 482:27. Other references for this type of definition include, in 1903: *MS* 478:144; *CP* 8.265; in 1904: *MS* 336:6; 337:8; 337s:7–8; *L* 427:12; in 1905: *PW* 190; in 1908: *MS* 611:19; in 1909: *CP* 8.303; *MS* 645:3, 7.

4. Peirce's use of the word "object" can be puzzling. In at least two places (*MS* 693:33 and *L* 482:27), he explains the meaning he attaches to it by tracing it back to its thirteenth-century definition, which he distinguishes from the German *Gegenstand*. An object, he says, is that "creation of the mind in its reaction with a more or less real something, which creation becomes that upon which cognition is directed" (*MS* 693). "Object" is thus defined in the same terms as a phaneron (this is even clearer in *L* 482), which implies that it is not to be understood as the external correlate of a subject. The "objectification" is thus only apparent in the above definition. It may be added that Peirce uses several phrases to describe the mental manifestation of the phaneron: "whatever *comes before* the mind," "whatever is *present to* the mind," "whatever is *before* the mind," "whatever may *present itself to* the mind," "whatever is *open to* mental observation," "anything that can *emerge in* knowledge or in fancy." Those phrases are purposively broad and vague; the mode and the time of the appearance do not matter, as far as the phaneron is concerned. But it remains that the phaneron is the object studied in phaneroscopy and is thus an "object" of scientific investigation. The next subsection of this chapter should help clear up the matter.

5. Other references for this type of definition include the following: written in 1904: *CP* 8.301; in 1905: *MS* 284:41 (twice), 42, 48; 998:5; *L* 224:35; *CP* 1.284 (or *MS* 1334:35); c. 1905: *MS* 477:4; in 1906: *MS* 292:72; c. 1906: *MS* 293:23.

6. See *MS* 338:2 (c. 1904) and *MS* 908:4 (1905).

7. See also, written c. 1904: *MS* 998:5; in 1905: *MS* 908:5, 9; c. 1905: *MS* 477:4; c. 1906: *MS* 293:23–24.

20 An Application of Peirce's Valency of Relations to the Phenomenon of Psychological Dissociation

Martin Lemon, Northwestern University Medical School

The purpose of this chapter is to outline some hypotheses regarding a personality structure. While many personality structures have been proposed by various theorists, empirically testable hypotheses in this area are, to my knowledge, unprecedented. I will begin with several broad statements for the purpose of orienting the reader. Some form of the search for self-understanding—or at least self-identity—is surely one of the few human endeavors common to all cultures and societies throughout history. Until the past one hundred years or so, persons typically relied upon their religious traditions as the sole means of exploring and seeking to identify the most basic aspects of their identities. By the late nineteenth century, however, in the minds of many Westerners, the authority of science had eclipsed the authority of religion, and it was around this time that a relatively new type of inquiry emerged which would come to influence a large segment of humanity. The fields of psychology and psychiatry—and particularly the writings of persons such as Sigmund Freud and Carl Jung—yielded the first thoroughly articulated secular accounts of human nature ever to attract widespread attention. And while such writings have often been categorized as "scientific," there has been considerable controversy regarding whether they, in fact, satisfy any strict definition specifying the characteristics of scientific constructs or theories. Even so, the ideas expressed by persons such as Freud and Jung have undoubtedly added clarity to the human quest for self-understanding. That is, at the very least, their theories have been beneficial heuristically.

The greatest single contribution to human knowledge on the part of psychology and psychiatry may well be the concept of personality in the sense of a relatively enduring intrapersonal structure which embodies several identifiable tendencies. Jung's archetypes and Freud's *das Es*, *das Ich*, and *das Über-Ich* (translated literally, "it," "I," and "over-I," or more commonly, "id," "ego," and "superego") are surely the most famous personality structures. Interestingly, since around 1950 virtually all personality theorists have avoided employing structural concepts. The only exception to this trend is the popularity of the concept of the "self," which is common to several theories of recent vintage.

Presumably, one reason structural concepts have become scarce is because positivism (the epistemological posture which holds that science must be exclusively concerned with observable phenomena) has so thoroughly infected the behavioral sciences throughout the twentieth century. And even though proponents of "cognitive psychology" (which deals with unobservable mental processes) have risen to hegemony within the field of psychology in the past twenty-five years or so, their theories generally pertain to processes and not structures. Cognitive psychology attempts to study what the mind *does* and is only peripherally concerned with what the mind *is* or what a human being is in the most basic sense. If asked what the mind is, a cognitive psychologist would likely respond that it is "an information processor" and then go on to say that his or her research investigates how the mind processes information. For instance, a cognitive psychologist might study the relationships between "pattern recognition" and "reaction time" in order to test some set of hypotheses regarding how information is stored and retrieved. And while his or her hypotheses would be scientific in the sense that they are disprovable, they would not be about causal structures and would therefore fail to address adequately basic questions such as, What is a human being?

The concept of a "causal structure" will be of central importance to the arguments which I will later present in support of my hypotheses. Therefore, I wish to explain at the outset the nature of a causal structure. As I am using the phrase here, "causal structure" refers to the most elementary unit, entity, or system with which a given science will be concerned. In physics, for example, the atom qualifies as a causal structure. Moreover, in describing causal structures as elementary, I am not endorsing reductionism; instead, I am assuming that nature is stratified so that causal structures may exist on a variety of levels (Manicas 1987). I further assume that knowledge of the properties possessed by causal structures on one level does not necessarily explain the nature of causal structures on higher levels (Manicas 1987). To illustrate, consider that knowledge about the atomic positions of an unknown protein does not allow us to infer the protein's causal properties as part of a larger system. I am assuming, in other words, that biology is not reducible to chemistry and/or physics, and likewise that the social sciences are not reducible to any combination of the physical sciences. Furthermore, a central premise of this essay is that the process of hypothesizing and studying causal structures is one of the best ways any science can generate the most powerful explanatory concepts applicable to a given level.

For present purposes, and following the work of Manicas (1987), a causal structure shall be defined as the embodiment of a theorized mechanism which gives rise to several dispositional properties. A causal structure's dispositional properties or tendencies are hypothesized causal powers which are triggered when a particular set of circumstances impinges upon the structure. The nature of the theorized mechanism is such that when certain circumstances emerge, the causal powers *invariably become manifest* unless prevented by some event or circumstance whose origin is external to the causal structure. In this way, the theorized mechanism *defines* the causal structure.

For example, the molecular structure of water, H_2O, gives rise to the tendency, or causal power, such that when temperature is reduced to the freezing point, water freezes. To advance a hypothesis regarding this tendency is not to suggest merely that when temperature is reduced to the freezing point, water *usually* freezes. Instead, the nature of the hypothesis is such that if the hypothesis is correct, when temperature is reduced to the freezing point, water *must* freeze. If it does not, and if there were no other mitigating factors which might explain why the causal property or power was not expressed, then the hypothesis must be incorrect (Manicas 1987). The essence of a causal structural hypothesis is precisely that the mechanism which defines the causal structure *will be activated* under certain conditions unless it is prevented by events whose origin is external to the causal structure.

While there are many examples in theories from the physical sciences of constructs which satisfy the above definition for causal structures—molecules and magnetic fields, for example—there is no theory in any of the social sciences which hypothesizes entities which qualify as causal structures, including those of Freud and Jung (Manicas 1987). Instead of developing hypotheses about causal structures, the social sciences have attempted to illuminate the relationships among sets of variables. Recent research in the behavioral sciences has studied, for example, the relationships between physical punishment received in childhood and aggressive behavior expressed in adulthood (Eron 1987), between assertiveness and substance abuse in adolescence (Wills, Baker, and Botvin 1989), and between interspousal aggression, marital discord, and child behavior problems (Jouriles, Murphy, and O'Leary 1989). Such research is typical of the behavioral sciences today, yet it fails to address such basic questions as, What is a person?—questions which can be addressed by employing causal structural hypotheses. A recent book by Manicas (1987) has suggested several historical reasons why the social sciences now seem unaware of the importance, or even the possibility, of developing causal structural hypotheses.

Returning to the notion of a personality structure, it seems that a construct which might lend itself to elaboration and augmentation so as to assume the form of a causal structure is that of a *dynamism*. Harry Stack Sullivan (1956) selected "dynamism" as his primary structural term, which he defined in two ways: first, as "a relatively enduring configuration of energy which manifests itself in characterizable processes in interpersonal relations," and second, as "the relatively enduring pattern of energy transformations which recurrently characterize the organism in its duration as a living organism" (quoted from Rychlak 1981:329). As Rychlak (1981) has noted, the first definition focuses upon social or interpersonal qualities, while the second definition applies to living organisms in general and could therefore be employed in biology or physiology as well as psychology.

Sullivan was influenced by the pragmatic philosophies of William

James, George Herbert Mead, and John Dewey (Rychlak 1981). It is not likely, however, that Sullivan ever came in contact with the writings of Charles Peirce. And it is Peirce's philosophy—particularly his ideas on the valency of relations—which might have provided Sullivan with the conceptual tools to clarify his concept of a dynamism further. Through such clarification, the dynamism might have come to possess the requisite qualities of a causal structure as defined earlier.

Regarding his three categories of the valency of relations, Peirce wrote, "The little that I have contributed to pragmatism (or, for that matter to any other department of philosophy), has been entirely the fruit of this outgrowth from formal logic, and is worth much more than the small sum total of the rest of my work, as time will show" (*CP* 5.469). Briefly, the categories of relations consist of what Peirce called firstness, secondness, and thirdness, and the relations corresponding to these categories he called monads, dyads, and triads. These three relational concepts can be applied to the analyses of entities and actions of both a physical and mental nature, as well as to the analysis of verbal concepts. Peirce wrote extensively about the nature of monads, dyads, and triads, but a comprehensive definition of each is beyond the scope of this present chapter (see *CP* 1.284–353). Thus, I will only sketch each of these concepts briefly.

Peirce wrote that a monad was a single quality considered or experienced without regard to degrees of that quality and without comparison or contrast with other qualities (*CP* 1.303). The notion of a monad is, perhaps, the most difficult of the three relations to define, and since it is of little direct relevance to the hypotheses I am about to present, I will not dwell on its nature here.

A dyad, according to Peirce, was the most familiar and most easily described relation. A dyad is the embodiment of a causal act, in the sense of mechanical causation, wherein two subjects come into relation such that the first may be said to have caused the second, and further, where there was no intervening process or event between the causal act and its effect (*CP* 1.326–28). A dyadic relation is the essence of a mechanical or efficient cause. For instance, consider the following. Suppose I stand two dominoes on end and arrange them side by side and then push over the first so that it in turn knocks over the second. This set of events amounts to a pair of dyadic relations—the first, involving my finger (ignoring, for the moment, any consideration of the intent which preceded my movement) and the first domino, and the second, involving the first domino and the second domino. Peirce characterized dyadic relations as "brute actions of one subject or substance on another, regardless of law or any third subject" (*CP* 5.469). Verbs such as "push," "throw," "kill," "expel," and "repel" reflect dyadic relations. Furthermore, adjectives that reflect absolute qualities such as "true," "false," "good," "evil," "ambitious," and "lazy" are dyadic in that they each imply an absolute relation to their polar opposites. The word "true," for instance, implies only two alternatives: true and false.

In contrast, the essential nature of a triad, as defined by Peirce, was that of mediation. The nature of a sign, or of any representational act or cognition, is triadic. Peirce wrote:

> A sign or *representamen*, is something which stands to somebody for something in some respect or capacity. It addresses somebody, that is, creates in the mind of that person an equivalent sign, or perhaps a more developed sign. That sign which it creates I call the *interpretant* of the first sign. The sign stands for something, its *object*. It stands for that object, not in all respects, but in reference to a sort of idea, which I have sometimes called the *ground* of the representamen. (*CP* 2.228)

According to Peirce, the sign has three elements—the interpretant, the object, and the ground. Furthermore, Peirce states that the essential character of the triadic relation is that it is not reducible to combinations of monadic and/or dyadic relations (*CP* 3.82). An example of a triadic concept is the predicate "gives," in that it implies an irreducible triadic action such as expressed in "Joe gives the book to Sarah" (*CP* 5.469). Triadic notions or ideas are relative rather than absolute in nature and often require phrases or sentences rather than single words to be expressed. The sentence "John is ambitious" implies an absolute, dyadic relation between John and any person who is not ambitious. In contrast, the sentence "John is ambitious sometimes" implies a relative, triadic relation in that the nature of John's character seems to be conditional and perhaps mediated by particular factors or circumstances. All intelligent action—as opposed to mechanical action—is of a mediated, triadic nature. A triadic event is a transcendent event in the sense that mere efficient causes are transcended, mediated, and thereby fundamentally transformed.

Returning to the notion of a personality structure and to the concept of a dynamism, I will now borrow another of Sullivan's key terms: "dissociation." Sullivan described dissociation as the most extreme form of selective inattention. As such, the process or act of dissociation involves the separation, isolation, or expulsion of some cognitive element or elements from awareness or from any integrated mental structure or dynamism. Note that the relational character of a dissociative act is dyadic. When an idea is expelled from a dynamism, this amounts to a purely dyadic event, even though it has been preceded by triadic events.

Now by employing the set of concepts which I have outlined, including causal structures, transcendent triadic events, and dissociative dyadic events, it becomes possible to reformulate significantly the concept of a dynamism in terms of several empirically testable hypotheses. I shall redefine a dynamism as an encampment of interests and activities which characterizes every living organism and every social aggregate, which is organized and organizing with respect to particular purposes or objectives, and which comes into existence only in relation to such purposes and objectives. The organizing is accomplished through a combination of dyadic and triadic capacities and acts which are *both* essential for the dynamism to maintain a stable existence. If only dyadic or only triadic capacities are present, then a dynamism will not form, and if an existing dynamism forfeits either its dyadic or triadic capacities, then it will disintegrate.

The dyadic and triadic elements assume different roles pertaining to the development and maintenance of the dynamism. The particular dyadic capacities of every dynamism include the various forms and degrees of dissociation which serve to defend against substances or ideas which threaten the dynamism's core interests. To be sure, while the act of *identifying* an alien substance or idea is interpretive and, therefore, triadic, the act of *removing* such a substance or idea from the dynamism is dyadic. In this way, the dyadic aspects are actually prompted by, and under the control of, the triadic aspects. Just as a cell relies upon its membrane to repel foreign substances and upon particular organelles, known as lysosomes, to expel waste products, so must a dynamism rely upon its dissociative capacities in order to maintain its integrity.

The triadic activities of the dynamism include diverse forms of cognition and volition. Their essence is their organization in relation to objectives or purposes. And thus, in psychology even the triadic activities of a dynamism are not adequately accounted for by the information-processing model which cognitive psychology has borrowed from computer science. Whereas ordinary computer programs will remain unchanged until altered by someone or something (e.g., lightning or clumsily discarded chewing gum), a dynamism, in contrast, will fail to develop, or will disintegrate, if its relations to particular objectives or contexts are not established and maintained. Once created, a computer program exists in isolation, while a dynamism exists at all times only in relation to contexts which are at least in part external to the dynamism.

Recalling the definition of a causal structure, we see that there are two aspects of the theorized mechanism which define a dynamism. First is the transcendent, triadic aspect. It is hypothesized that a dynamism will cease to exist if it is not organizing in relation to particular contexts or objectives or if the relation between the dynamism and its objectives is invalidated. Second is the dissociative, dyadic aspect. And here it is hypothesized that a dynamism will cease to exist if it loses the capacity to repel things which threaten the maintenance of its relation to particular contexts or objectives.

Perhaps one of the best illustrations of a dynamism is that of a committee of persons. Committees are organized with respect to purposes, and likewise dynamisms come into existence in relation to contexts or objectives. Suppose a city government was to select a committee to investigate the construction of a new fire station. The committee members would inquire as to why, where, when, and how the fire station should be built. Now suppose further that this committee was opposed by a group protesting the tax increases which the committee had recommended as a means of funding construction. The committee must then either adjust its objectives so as to accommodate the group's concerns, or it must somehow dissociate the opposing group's input. The dissociation could be accomplished in many ways, including (1) ignoring the protests, (2) moving committee meetings to locations unknown to the protestors, or (3) somehow censoring the protestors. Also, recalling my hypothesis that a dynamism will disintegrate if its relations to contexts or objectives are somehow invalidated, we see that if the city's population was to shrink significantly, so that a new fire station was no longer necessary, then the committee (or dynamism) would likely disband.

Turning to how my hypotheses might be structured operationally for study, recall that Peirce addressed the valencies of *concepts* (*CP* 5.469). Words and ideas can be categorized according to their valencies and thereby become objectively observable variables that may facilitate empirical research. For instance, as mentioned earlier, ideas can be categorized as dyadic and absolute, or triadic and relative, and by focusing on this distinction, my hypotheses can be stated with empirical precision.

Since my hypotheses concern the stability of dynamisms, they can be applied to psychopathology in that psychopathology concerns the stability of personalities. My hypotheses state that a dynamism will disintegrate when the relation between the dynamism and its contexts or objectives is invalidated. More specifically, by focusing on the role of dyadic and triadic ideas, hypotheses regarding personality decompensation can be formulated. When a personality or dynamism becomes unstable, such as in schizophrenia, it may begin seeking input which might help redefine its identity or, in other words, its relations to contexts or objectives. Also, at this time of instability, the dynamism will have less capacity to defend against ideas (the dyadic capacity is diminished) which threaten the dynamism's core interests. Now suppose a schizophrenic on the brink of decompensation encountered a family member who said to him or her, "You are lazy." The nature of this input is dyadic in that it is absolute. This input is obviously quite different from triadic statements such as, "You didn't take out the trash today. You are lazy *at times*." Furthermore, the input is a negative evaluation of the person or dynamism, implying that the person is worthless or undesirable. If the person or dynamism accepts this evaluation—and the likelihood of its acceptance is increased by the person's current vulnerability—then, seeking to be responsible or worthwhile (i.e., seeking to maintain his or her relation to essential contexts or objectives) rather than irresponsible or lazy, the dynamism may turn its dissociative capacities against itself. The dynamism, thus, begins depleting its own triadic, mediational capacities until only dyadic capacities remain, and at this point, the dynamism becomes disorganized and disintegrates.

In fact, several recent studies have indicated that when someone who is emotionally close to a schizophrenic repeatedly criticizes the schizophrenic's identity, the likelihood of psychological decompensation increases significantly (Brown, Birley, and Wing 1972; Vaughn and Leff 1976; Vaughn et al. 1982). As I have shown, the process of decompensation can be hypothesized in terms of dynamisms and dissociative acts as defined earlier. Furthermore, hypotheses about the types of comments which might prove either troublesome or beneficial can be specified according to the distinction between dyadic and triadic statements.

In conclusion, it is worth reemphasizing that research in psychopathology—and for that matter all research in the behavioral sciences—could be improved if hypotheses regarding personality structures were incorporated. By combining Peirce's ideas on the valency of relations with Sullivan's notions of the dynamism and dissociation, development of such causal structural hypotheses becomes possible.

References

Brown, G. W.; Birley, J. L. T.; and Wing, J. K. 1972. Influence of Family Life on the Course of Schizophrenic Disorders: A Replication. *British Journal of Psychiatry* 121:241–58.

Eron, L. D. 1987. The Development of Aggressive Behavior from the Perspective of a Developing Behaviorism. *American Psychologist* 5:435–42.

Jouriles, E. N.; Murphy, C. M.; and O'Leary, K. D. 1989. Interspousal Aggression, Marital Discord, and Child Problems. *Journal of Consulting and Clinical Psychology* 57 (3):453–55.

Manicas, P. T. 1987. *A History and Philosophy of the Social Sciences.* Oxford: B. Blackwell.

Rychlak, J. F. 1981. *Introduction to Personality and Psychotherapy: A Theory-Construction Approach.* Boston: Houghton Mifflin.

Sullivan, H. S. 1956. *Clinical Studies in Psychiatry.* New York: Norton.

Vaughn, C., and Leff, J. 1976. The Measurement of Expressed Emotion in the

Families of Psychiatric Patients. *British Journal of Social and Clinical Psychology* 15:157–65.

Vaughn, C. E.; Snyder, K. S.; Freeman, W.; Jones, S.; Falloon, I. R. H.; and Liberman, R. P. 1982. Family Factors in Schizophrenic Relapse: A Replication. *Schizophrenia Bulletin* 8 (2):425–26.

Wills, T. A.; Baker, E.; and Botvin, G. J. 1989. Dimensions of Assertiveness: Differential Relationships to Substance Use in Early Adolescence. *Journal of Consulting and Clinical Psychology* 57 (4):473–78.

21 Knowing One's Own Mind

Gerald E. Myers, Queens College and Graduate Center, City University of New York

Donald Davidson has recently defended the claim that sincere, first-person, present-tense reports about beliefs and doubts in one's own mind, although neither infallible nor incorrigible, are authoritative in a way that first-person other-tense reports and second- and third-person reports are not. "Because we usually know what we believe (and desire and doubt and intend) without needing or using evidence (even when it is available), our sincere avowals concerning our present state of mind are not subject to the failings of conclusions based on evidence" (Davidson 1987:441).

First-person authority needs defending, according to Davidson, because it apparently conflicts with views held by Hilary Putnam and other contemporary philosophers. It also apparently conflicts with views held by Charles S. Peirce (although this is outside Davidson's concerns). I want here to examine Davidson's arguments while monitoring their relevance for Peircean thinking.

Peirce, as is well known, declared, against Cartesianism, that one's knowledge of one's mind, or internal world, is derived entirely from observing external facts, that "man possesses no infallible introspective power into the secrets of his own heart, to know just what he believes and what he doubts" (*CP* 5.498). Whether Peirce's was an earlier version of Gilbert Ryle's theory, that one knows one's own mind quite as one knows another's, is not for me to decide here, although, for reasons given later, I doubt that it is; but that Peirce's recurrent disclaimers about introspection and intuition seem to compromise first-person authority is surely evident.

Peirce's requirement of "external" factors in any process that we would call introspective is strikingly similar to what Davidson finds troubling in Putnam's (1975:227) argument that what words mean depends on factors "external" to "what is in the head." The external factor for Putnam is the way one learned the uses of words, the learning process itself affecting their meanings.

If the meaning of the words that determine the content of propositional attitudes is a function of how they were learned, of the learner's history, then two people might be in identical physical and psychological states, yet mean different things by the same word because they learned to use the word in different environments. They may differ in what they mean by it, even though their "inner" states are identical. Hence, they cannot tell what they mean (believe, doubt, etc.) merely by consulting (introspecting, intuiting, etc.) their physical/psychological inner states. And they may not know what they think until they have identified what their words mean as determined by how they learned them. This, absent the various and often ingenious arguments offered in its support, is the conclusion of Putnam and others that troubles Davidson (although, to his surprise, it does not similarly vex Putnam and others), precisely because it seems to compromise first-person authority. Given the preceding, it appears that another person might be better positioned than myself for knowing what I am thinking.

For Davidson, however, first-person authority is so clearly a fact that either there must be something amiss in any argument that threatens it or the appearance of threat is only illusory. First-person authority is not infallible, so incorrigibility of belief is not an issue. But what is undeniable, Davidson insists, is that while we can err about what we think, such cases are necessarily exceptions but never the rule. The why of this tends to elude philosophical explanation, so Davidson's task is to explain it while simultaneously eliminating any apparent threat, such as Putnam's, to first-person authority.

On Davidson's analysis, we need to ask what is responsible for the idea that first-person authority evaporates simply because meanings are partially identified via relations to events outside one's head (the conditions under which one learned them). The culprit responsible for what is in fact only an empty threat to first-person authority, Davidson (1987:453) says, is "a picture of the mind which has become so ingrained in our philosophical tradition that it is almost impossible to escape its influence, even when its worst faults are recognized and repudiated."

The culpable picture is the familiar one that depicts the mind as a kind of theater in which, when one thinks or believes something, one is somehow related to objects or entities that are thought or believed. If one's attitude is propositional, then, as this picture demands, there is a mental object or target for that attitude. The undermining effect of this picture of the mind on first-person authority is immediately discernible.

If thinking involves having an object before the mind, and if the object's identity makes the thought what it is, then errors about one's own thoughts are always imminent. Why? Because some feature of the object may always elude detection. No philosophical attempt to relate oneself to the object, such that knowing one's own thought is ensured because all errors are precluded, has succeeded or will succeed. This is all the more certain if, with Putnam and others, this identification of the object requires the identification of relations external to one's mind or head.

Abandoning this picture of the mind and its assumption that thoughts must have mysterious objects, Davidson argues, allows us to see how the genetic/social determination of meaning (emphasized by Putnam and others), rather than forcing us to jettison first-person authority, actually helps us to explain it.

Once we appreciate that what we mean by our words depends on the kinds of things that have caused us to use them as we do, we can understand why the asymmetry between first- and third-person reports exists. Davidson (1987:456) writes:

> An interpreter of another's words and thoughts must depend on scattered information, fortunate training, and imaginative surmise in coming to understand the other. The agent herself, however, is not in a position to wonder whether she is generally using her own words to apply to the right objects and events, since whatever she regularly does apply them to gives her words the meaning they have and her thoughts the contents they have.

> Of course, in any particular case, she may be wrong in what she believes about the world; what is impossible is that she should be wrong most of the time. The reason is apparent: unless there is a presumption that the speaker knows what she means . . . there would be nothing for an interpreter to interpret. To put the matter another way, nothing could count as someone regularly misapplying her own words. First-person authority, the social character of language, and the external determinants of thought and meaning go naturally together.

Davidson's explanation suits Peircean theory in relying not on introspection or intuition but on genetic/social factors. The explanation, on my reading, holds that once a language game with its genetic/social determinants is accepted, any move to initiate skepticism about first-person authority occurs too late in the game. Such authority is not established within the game as one more fact but is rather invoked as one of the game's preconditions. Any given first-person present-tense report, absent relevant evidence, is presumed authoritative, not on intuitive grounds but on the fact ("external" to the mind/head) that, in general, such reports must be true.

Whether Peirce would have granted Davidson's claim that a basic asymmetry between first- and third-person reports is also explained is questionable; such asymmetry is arguably hazardous for a Peircean "communal" concept of knowledge (*CP* 5.356). But how plausible is such asymmetry, even on Davidson's analysis? It strikes me that Davidson connects speaker and interpreter, as mutually involved language users, so intimately that the latter's getting the former right must be the rule and not the exception, just as it is for the former's getting it right on her own. The genetic/social factors seem to require, if the language game is not to break down, that, in general, speaker and interpreter comprehend each other. If so, the alleged asymmetry is missing.

Davidson finds the alleged asymmetry because he supposes that we, unlike himself, must utilize evidence for inferring that he has a certain thought. We know on evidence that he thinks something, whereas he knows it noninferentially. Like Peirce, as we saw above, he avoids any appeal to intuition or introspection. He may be mistaken occasionally in claiming to know what he thinks but not generally, since that would wreck the language game. Thus, although explanations for mistaken first-person reports are presumably forthcoming, apparently none are available on Davidson's analysis for specific, successful, first-person reports. First-person authority is a statistical fact that confers a presumption in favor of any specific report.

Consequently, it seems that, for Davidson, there are no specific explanations for specific, successful, first-person reports. How do I now know that I think this and not that? is for Davidson, I judge, a muddled question. If so, is it pointless to insist that I nevertheless "know" what I think? Peircean theory might well answer this affirmatively.

A plausible alternative to Davidson's account agrees that we generally get our first-person reports right but denies that we know them, since the usual considerations of evidence and inference that make reference to "knowing" significant rather than pointless are absent here. Why we generally report our thoughts correctly, I suggest, no more invites philosophical explanation than does our tendency to recognize familiar objects. Credit goes to such human intelligence and its relevant features as are scientifically and routinely explicable.

If the general fact is not that we know our reports but rather that we get them right, and if that creates no special puzzles, then what interests philosophy, I submit, are those instances where asserting our knowledge of such reports is significant. Do we know them through "private access"? Putnam supplies fresh reasons for denying infallible acquaintance with one's own mind, and Davidson also dismisses infallibility, so contemporary opinion joins Peircean fallibilism. But is a limited concept of "private access" salvageable? Oddly, Davidson may intend to answer this affirmatively. Although he defends first-person authority through linguistic "presumptions," thus indicating no need for psychological explanations such as private access, nevertheless he follows (but without elaboration) Putnam in referring to mental contents as identifiable inner states. Davidson (1987:444) writes:

> I think such states are "inner," in the sense of being identical with states of the body, and so identifiable without reference to objects or events outside the body; they are at the same time "non-individualistic" in the sense that they can be, and usually are, identified in part by their causal

relations to events and objects outside the subject whose states they are. . . . First person authority can, without contradiction, apply to states that are regularly identified by their relations to events and objects outside the person.

Davidson does not elaborate what he means by "identifying" mental states. This is regrettable, because it now appears that, besides appealing to the inescapable "presumption" of first-person authority, he presumes that some type of inner state identification establishes specific, successful first-person reports of propositional attitudes. His main brief against Putnam and others is that "mental states as we commonly conceive them are identified in part by their natural history," but this neither touches "the internal character of such states" nor compromises first-person authority about them (ibid.:455–56).

If Davidson supposes that avowals are correct by virtue of correctly identifying mental states, two problems arise. First, to repeat, he owes an explanation of the "identification" involved. Second, if inner states are "objects" to be identified correctly in reporting our thoughts correctly, this resembles the same picture of the mind, attributed to Putnam and others, that Davidson earlier declared to be wrongheaded. He may intend, however, merely that to report a thought is itself to identify it; but, if so, one wonders why he reiterates references to inner states and their identifications. And would such references still occur if first-person authority was conceived as generally getting our avowals right instead of "knowing" them when and as we utter them?

If it remains uncertain whether "private access" is salvageable in Davidson's account, can we find it and first-person authority in Peirce's philosophy of mind? Perhaps, because Peirce's disposal of certain concepts of introspection was intertwined with a complex picture of inner states and one's knowledge of them.

Peirce asserted repeatedly that what passes as introspection is always inferential and never directly observational. An example: "Now the truth is that the data of introspection are . . . altogether analogous to those of external observation" (*CP* 7.418). Another: "In a certain sense, there is such a thing as introspection; but it consists in an interpretation of phenomena presenting themselves as external percepts" (*CP* 8.144). Even feelings and emotions are inferred targets, "because, as for feelings, they are always referred to some object, and there is no observation of feelings except as characters of objects" (*CP* 7.376).

Why was Peirce led to the implausible claim that even reports of feelings are inferences from "external" factors such as our outer behavior? Because, as Christopher Hookway suggests, he probably assumed that, if a genuine introspective capacity exists, so there would exist (as there does not in fact) a range of exclusively introspective predicates. Hookway (1985:26) remarks, after noting that concepts in fact can play both introspective and perceptual/inferential roles: "So, it is rather surprising that Peirce does not offer an account of our ordinary first-person avowals."

Although Peirce did not offer a full account of first-person avowals, what he said about the "dialogic" nature of thinking is suggestive. "All thinking is dialogic in form. Your self of one instant appeals to your deeper self for his assent" (*CP* 6.338). The relevance of this for Peirce's theory of self-knowledge has been noted by W. B. Gallie (1952:82). I want, however, to adapt it for our topic of first-person authority along lines indicated in Myers 1986:206.

We typically claim to *know* our first-person present-tense avowals when putting to rest any prior doubt about them. Whatever the provocation, the self-addressed query, "Do I really believe that?" is remarkably effective. It produces in me special mental states that are expressed as hesitant or confident answers. Although fallible, such answers figure significantly in deciding whether I know what I say I believe.

Such answers, I submit, correlate reliably with eventual decisions about what are the actual contents of one's mind. They correlate much more reliably than do answers to questions asked of other persons about the contents of their minds. There is generally more room for error and deception on my part when you answer my queries about your thoughts.

We are therefore justified, absent contrary evidence, in accepting our own answers to self-addressed questions about what we believe and desire. To this extent, first-person authority is established by a kind of "private access," because the dialogic process to which Peirce referred is revealed to oneself as it is not to another. The impulse to answer my own query with "Yes, I do believe that" is strong evidence that I do believe it and is of a sort that I cannot have with your answers.

Locating first-person authority for the knowledge we have of our own thoughts in Peirce's dialogic process is advantageous. It does not pretend at the same time to explain why we generally get our first-person avowals right, and surely that requires a very different kind of explanation. And it avoids the difficulties, encountered in discussing Davidson's and Putnam's views, of trying to retain first-person authority while simultaneously making it depend on "object" identification, whether the object be an internal state or the target of a propositional attitude. No object identification independent of the answer given to the self-addressed query is mandated.

This is shown by the dialogic process itself. Asking myself whether I believe something produces an impulse to declare, deny, or doubt that I do believe it; the impulse, being a literal part of the assertion or the assertive process that it induces, is no separable object requiring discrete identification (along with the opportunities for error that this would involve). In answering my own query, in feeling the impulse to say that I do believe it, I have the strongest possible evidence, barring anything to the contrary, for inferring that I do believe it.

No doubt, as Peirce insisted, some process of inference is always implicated. But the dialogic process yields as much certainty and first-person authority as we need. Where better to discover it than, in Peirce's words, in the appeal of one's instant self to one's deeper self for assent? First-person authority, after all, implies some special relation that one has to oneself. First-person authority depending on self-assent alone is admittedly blind, but a concept of first-person authority that is deprived of self-assent altogether is simply empty.

References

Davidson, Donald. 1987. Knowing One's Own Mind. *Proceedings and Addresses of the American Philosophical Association* 60:441–58.

Gallie, W. B. 1952. *Peirce and Pragmatism.* Middlesex: Penguin.

Hookway, Christopher. 1985. *Peirce*. London: Routledge and Kegan Paul.

Myers, Gerald E. 1986. Introspection and Self-Knowledge. *American Philosophical Quarterly* 23:199–207.

Putnam, Hilary. 1975. The Meaning of "Meaning." In *Mind, Language, and Reality*, 224–47. Cambridge: Cambridge University Press.

22 Peirce's Psychophysics: Then and Now

Peter J. Behrens, The Pennsylvania State University

The birth of American experimental psychology can be popularly dated with the founding of the first laboratory devoted to research and training at The Johns Hopkins University in Baltimore by G. Stanley Hall in 1883. At Hopkins the first generation of experimental psychologists—notably Dewey, Cattell, Sanford, and Jastrow—was trained in psychological methodology and problems. The problems were primarily the measurement of sensory and perceptual processes, such as the tactile, visual, kinesthetic, and temperature senses and the judgments of rhythm (Boring 1942). Surprisingly, however, the very first study from the Hopkins laboratory was not authored by Hall. The senior author was none other than Charles Sanders Peirce, who, in collaboration with Joseph Jastrow, at that time a first-year graduate student, published a paper in 1884 entitled "On Small Differences in Sensation" in the *Memoirs of the National Academy of Sciences*. This psychophysical investigation of weight discrimination was grounded in Peirce's philosophical system, one that would have a significant impact on Jastrow and other experimental psychologists into the twentieth century.

This chapter will set forth some major philosophical and scientific propositions of Charles Peirce which influenced early American psychology and show the longer lasting contributions of his thinking to so-called modern experimental psychology.

Primarily from his famous mathematician father, Benjamin Peirce, Charles received a science education based upon mathematics and laboratory activities (Gallie 1966). As a result, Peirce's interests in the problems of psychology, such as sensory awareness, sensory discrimination, and cognition, had a decidedly empiricist perspective. It should be recalled that his graduate degree was in chemistry and that he often referred to himself as a chemist, even later in life (Ketner 1987).

Early in his career Peirce became acquainted with the writings of Wilhelm Wundt and Gustav Fechner. Their work exemplified the applications of mathematics and laboratory methodology to psychology. By 1868 Peirce referred to the "facts of physiopsychology," and later he called for psychology to follow the lead of Wundt and Fechner (Cadwallader 1974). He became well versed in both Wundt's *Physiologische Psychologie* and Fechner's *Elemente der Psychophysik*.

Peirce's interest in experimental psychophysics came to fruition in a series of observations on color sensations published in 1877 in the *American Journal of Science and Arts* and in his 1884 study on weight discrimination with Jastrow. These two studies, Peirce's only laboratory work, represent a profoundly rich and influential contribution to both nineteenth- and twentieth-century psychology. But before proceeding with a description of these works, I wish to present some aspects of Peirce's epistemology developed in several essays between 1868 and 1879 which provided the principal foundation for Peirce's empirical psychology.

To begin with, Peirce was highly critical of both nineteenth-century positivism and romanticism. He developed the argument that knowledge arises out of experience in the first of two 1868 essays, titled "Questions Concerning Certain Faculties" in the *Journal of Speculative Philosophy* (Peirce 1868a). Judgment, he said, arises out of the subject's experience of sensation and is constructed from sensation. Sensations, moreover, such as dreams, texture, pitch, and even self-consciousness, are partly derived from the internal conditions of the subject but primarily from observation of the external world.

A second component of Peirce's theory of knowledge follows and was developed in a second essay of 1868, "Some Consequences of Four Incapacities" (Peirce 1868b). That component was his conclusion that Cartesian *doubt* is an improper starting point for knowledge. To make individual consciousness begin with doubt—as Descartes did—is an error.

Belief represents the fundamental starting point of knowledge. In the psychophysical experiment, for example, this means that the subject will begin with a particular belief about the presence or absence of a sensation or the difference between sensations and base a judgment on this belief.

A third important formulation of Peirce's theory of knowledge was his view of the continuity of thought, also developed in his first essay of 1868. Thought is a process which cannot be intuitive or immediate, because experience is not an instantaneous affair. Rather, thought is an event occupying time. In addition, past thoughts do not cease instantaneously but fade slowly. Ideas, therefore, are conveyed by other ideas, which in turn convey information to subsequent ones. The implication of this proposition is that consciousness resides on a continuum, and the distinction between consciousness and unconsciousness is in the immediateness of the experience. Psychophysical judgments are instances of unconscious, uncontrolled inferences, yet they are habitual and are guided by previous experience. A judgment might very well be rendered, yet the basis of the judgment not understood. For Peirce, this was unconscious inference, which was a valid subject of study in the laboratory (Gallie 1966).

A fourth and final component of Peirce's theory of knowledge is the nature of judgment itself, developed in essays between 1877 and 1879 under the general title "Illustrations of the Logic of Science." There are judgments which are indubitable because they are forced upon us. Psychophysical judgments (seeing something as red, for example) are uncontrollable and cannot be altered. What can be altered is our tendency to act, think, or speak in a certain way. Judgments are therefore fallible and prone to error, but they can be corrected by other judgments. The implication of this position is the relevance of the laws of probability to psychology. Peirce believed the subject's responses to minimal stimulation follow the laws of probability and that this could be demonstrated in the psychophysical experiment.

The four propositions for science here sketched formed the basis of Peirce's empirical psychology and enabled him to inquire into the inner world of human thought. The responses of subjects, particularly those responses at or near minimal values of stimulation, could be significant for understanding cognition, the basis of knowledge. Peirce used the concept of threshold, or *Schwelle*, from Leibniz and Herbart, and given quantitative meaning by Fechner. As a mathematician, Peirce was particularly taken with Fechner's measurement formula. "The intensity of a sensation is proportional to the logarithm of the strength of the excitation, the barely perceptible excitation being taken as the unit of strength" (Peirce 1877:247).

Peirce's first psychophysical study concerned the phenomenon of change in hue with change in intensity. This study was probably undertaken between 1875 and 1877 while he was in Europe for the Coast Survey (Cadwallader 1974; Fisch 1981). He observed from the Young-Helmholtz theory of color, which postulates green, red, and blue processes, that if indeed there are three independent chromatic processes, it is most likely that hue will vary with increased intensity not in a linear fashion but according to a law of diminishing returns. To quote Peirce once more: "Now I find, in fact, that all colors are yellower when brighter. If two contiguous rectangular spaces, illuminated with the same homogeneous light, uniformly over each other, but unequally in the two, they will appear of different colors" (Peirce 1877: 248). For example, if both were red, the brighter appeared "scarlet"; if both were green, the brighter appeared "yellowish."

Peirce performed twenty-eight comparisons for each of twenty-one colored disks and calculated the probable error for a single comparison. He concluded that Fechner's law is at least a good approximation as an explanation for the apparent change in hue.

The shift in hue reported by Peirce has come to be called the Bezold-Brücke phenomenon, after von Bezold, who described the phenomenon in 1873, and von Brücke, in 1878. But experimental psychophysics after Peirce has not always supported him or the Young-Helmholtz theory. Purdy, for example, made careful observations and concluded that the phenomenon cannot be reconciled with Peirce's theory because three wave lengths in the spectrum appear to be invariable in hue with respect to intensity. These three points Purdy found to be yellow, green, and blue (Purdy 1931, 1937). More recently, Walraven (1960, 1962) revived Peirce's main ideas in connection with the Bezold-Brücke phenomenon and countered the objections to it raised by Purdy and others.

Briefly stated, the difficulty between Peirce's original conception of the phenomenon and the objections to it may be overcome by a distinction between the information about luminosity and the information about chromaticity conveyed in the visual system. According to Walraven (1960), the assumption can be set forth that the information about brightness is transmitted along different channels than hue from each of the three color systems. The nonlinear relationship between input and response which results in the Bezold-Brücke shift may reside in the chromaticism channel, not the brightness channel. The brightness channel can therefore be regarded as an independent piece of information. With this modification of Peirce's original conception, his theory is suitable to explain the facts of the Bezold-Brücke phenomenon.

The views of Peirce, presented in several of his essays, were also the foundation for the 1884 psychophysical study undertaken at Hopkins, in collaboration with Jastrow, on weight discrimination. In 1882, while Peirce was lecturing and Jastrow was enrolled in Peirce's "Algebra of Logic," Peirce introduced Jastrow to experimental psychophysics: method, procedure, and statistical analysis (Jastrow 1916). Subsequently, at Wisconsin, Jastrow published more on psychophysical methodology, and between 1885 and 1895 minor studies from Wisconsin and other university laboratories appeared regularly in the issues of the *American Journal of Psychology*. Thus, in relation to an important focus of early American experimental psychology, the Peirce-Jastrow study may be viewed as a significant first in at least two ways.

One way the study may be regarded as significant is in relation to Fechnerian methodology. Up until the 1884 study, psychophysical investigations of the *Schwelle* and *Unterschiedsschwelle* were conducted by the presentation of stimuli near the point of minimum detectability, and the observer was given three response alternatives, including a category of "equal" or "cannot say." The "equal" category permitted the subject to remain indecisive with regard to a perceived sensation or difference in sensation. This arrangement, however, did not fit with Peirce's conception of the nature of consciousness. If sensation resides on a continuum of consciousness, then very small differences between stimuli may be correctly judged. If this be the case, then there is no true *Schwelle*.

In the 1884 study the subjects—Peirce and Jastrow themselves—

were required to judge without the "equal" category which of two weights applied to the finger was heavier. Jastrow in his 1930 autobiography summarized their findings in relation to the threshold this way: "If the threshold were a physiological limit below which there would be no registry whatever, there should be no difference in the proportion of right and wrong guesses. . . . But if the proportion continued to diminish below the threshold, the fact would argue for the subconscious registration. Such proved to be the case" (Jastrow 1930:136). Jastrow also offered a strong defense of Peirce's epistemological view in 1888 when he wrote: "Sensation and stimulation each forms a continuum, and it leads to hopeless confusion to apply discrete conceptions to them" (Jastrow 1888: 277). Peirce and Jastrow thus took the bold step of rejecting the "equal" category for psychophysical judgments which Fechner had adopted in his *Elemente*, thereby altering the conception of the threshold in American experimental psychology (Boring 1942).

The Peirce-Jastrow position was reflected later in the work of Fullerton and Cattell in 1892, and into the twentieth century by, for example, G. E. Müller, S. S. Stevens, R. D. Luce, and J. A. Swets. The 1884 paper began a movement away from a rigidly defined threshold in terms of Fechner's steplike function toward a theory of excitability based upon a sensory continuum (Corso 1963).

No more ambitious modern approach to the question of threshold exists and yet is consistent with Peirce's "radical" formulation than the work of Swets and others in signal detection theory (SDT). One assumption of SDT is that a signal has an effect which varies from presentation to presentation, the same conception given to detection by Peirce and Jastrow. A significant addition made by SDT, however, is that the basis for variability is physical and neural "noise." Noise is random central activity which is always present in the detection situation (Treisman and Watts 1966). Thus SDT posits two distributions—one noise (N), and one signal plus noise (SN)—in the psychophysical experiment. Since the signal is added to the noise, the average sensory observation magnitude will always be greater for SN than for N. As the signal strength is decreased, however, the difference between the means becomes smaller and smaller, thus making the decision more difficult (Gescheider 1985).

A second important contribution of the 1884 paper is Peirce's expla-

Table 1. Results from Fourth Set of Experiments Conducted in 1884 by Peirce and Jastrow

Ratio of pressures	Number of sets of 50	Average number of errors	
		Observed	Calculated
1.060	1	0	0.8 ± 0.6
1.030	1	5	4.8 ± 1.4
1.020	6	10.0 ± 0.5	9.6 ± 0.8
1.015	1	14	12.8 ± 2.1
1.010	6	16	16.5 ± 0.9
1.005	6	20.8 ± 0.4	20.6 ± 1.0

nation of the psychophysical law in terms of the normal probability curve. A series of 50 experiments was conducted each day with three different weights, 25 in increasing and 25 in decreasing pressures by the method of right and wrong cases. Thus 150 experiments total were undertaken at one sitting. The number of errors in judgment in the 50 experiments with each differential weight was recorded. One set of experiments conducted on Jastrow in March and April 1884 appears in table 1. A correspondence of observed errors to the probable errors calculated from probability theory appeared. According to the authors, "The number of errors follow, as far as we can conveniently trace them, numbers assigned by the probability curve" (Peirce and Jastrow 1884:82).

We may turn to a second assumption of SDT to provide a modern formulation of Peirce's psychophysics. SDT posits that probability theory applies to sensory discrimination. That is, the detection problem is solved by a statistical decision theory in terms of a criterion selected by the subject. This means that in order to respond "yes" or "no" in the psychophysical method of right and wrong cases, the subject employs a decision procedure to select a criterion. Sensory performance is thus determined by the position of the criterion chosen in addition to the separation of the distributions (Treisman and Watts 1966). Moreover, many modern studies incorporate a "confidence" scale in order to establish the degree of confidence a subject has for each decision. In a typical SDT study on auditory discrimination, Treisman and Watts (1966) asked subjects to decide if changes occurred or did not occur in the loudness of a 500 cps tone. The yes-no decision was augmented by a four-point scale of confidence that there has been a change in loudness. Peirce and Jastrow had also employed a four-point scale, ranging from 0 to 3, for their judgments. And they used these scaled responses to develop the following mathematical expression for the degree of confidence:

$$m = c \log \frac{p}{1 - p}$$

Here *m* denotes the degree of confidence on the scale, *p* denotes the probability of a correct answer, and *c* denotes a constant which may be called the index of confidence. So, for example, for Peirce when the ratio of weights was 1.060 grams, the average confidence, *m*, was calculated to be 0.70; the observed *m* was also 0.70. But when the ratio was 1.015, *m* was calculated to be 0.10 and observed to be 0.14. The judgments associated with any degree of confidence were more likely to be correct with greater differences between weights than with smaller differences.

Signal-detection theory has been credited for a significant shift in focus from the traditional emphasis on the stimulus values and threshold sensitivity to characteristics of the observer in the psychophysical experiment, including the decision criterion for judgment and the so-called receiver operating characteristics (Falmagne 1985). The latter refers to a function which summarizes the possible performances of an observer faced with detecting a signal in a noise. This function relates the probability of a response of "yes," given a signal occurrence, to the probability of the same response of "yes," given a noise alone (Egan 1975). However, as I have attempted to show here, judgment criteria and observer characteristics were at least implicit attributes of Peirce's psychophysical work. This should not be surprising, considering Peirce's understanding of the nature of consciousness, judgment, and inference, in addition to his innovative experimental methodologies. Taken together, then, there is much in Peirce's psychophysical studies that deserves recognition as a nineteenth-century foreshadowing of twentieth-century "light."

However, this is not to say that Peirce's contribution to psychology is limited only to psychophysics. Indeed, a close reading of many of Peirce's papers points to a considerable creative contribution to and significance for such diverse topics as psycholinguistics, psychoaesthetics, psychopathology, social psychology, and cognitive studies. These remain largely unelucidated with regard to Peircean formulation.

No more appropriate tribute to Peirce can conclude this paper than that from Joseph Jastrow, who not only was profoundly influenced by Peirce in his short association with him at Johns Hopkins but also became a prominent and prolific psychologist in his own right, thereby implicitly and explicitly disseminating Peirce's thinking. In his 1914 eulogy on Peirce, Jastrow (1914:571) reflected on the Hopkins environment enriched by Peirce in these words:

> The impression that I retained of his analyses of logical and philosophical problems is that of observing a plummet line descending through troubled waters foot by foot, sounding the depths, avoiding the weeds and the shoals, and reaching an undiscovered bottom. . . . If, in addition, it be remembered that these logical, psychological, and philosophical pursuits were in a sense avocational . . . the scope of his attainments will be more truly perceived. In a sense he represents the American Helmholtz.

References

Boring, E. G. 1942. *Sensation and Perception in the History of Psychology.* New York: Appleton-Century-Crofts.

Cadwallader, T. C. 1974. Charles S. Peirce (1839–1914): The First American Experimental Psychologist. *Journal of the History of the Behavioral Sciences* 10:291–98.

Corso, J. F. 1963. A Theoretico-historical Review of the Threshold Concept. *Psychological Bulletin* 60:356–70.

Egan, J. P. 1975. *Signal Detection Theory and ROC Analysis.* New York: Academic Press.

Falmagne, J. 1985. *Elements of Psychophysical Theory.* New York: Oxford University Press.

Fisch, M. H. 1981. Peirce as Scientist, Mathematician, Historian, Logician, and Philosopher. In *Proceedings of the C. S. Peirce Bicentennial International Congress*, ed. K. L. Ketner, J. M. Ransdall, C. Eisele, M. H. Fisch, and C. S. Hardwick, 13–34. Lubbock: Texas Tech Press.

Gallie, W. B. 1966. *Peirce and Pragmatism.* New York: Dover Publications.

Gescheider, G. A. 1985. *Psychophysics: Method, Theory, and Application.* Hillsdale, N.J.: Lawrence Erlbaum.

Jastrow, J. 1888. A Critique of Psycho-physic Methods. *American Journal of Psychology* 1:271–309.

———. 1914. The Passing of a Master Mind. *Nation*, May, 571.

———. 1916. Charles S. Peirce as a Teacher. *Journal of Philosophy, Psychology, and Scientific Methods* 13:723–26.

———. 1930. Joseph Jastrow. In *History of Psychology in Autobiography*, ed. C. Murchison, 135–62. Worcester, Mass.: Clark University Press.

Ketner, K. L. 1987. Charles Sanders Peirce. In *Classical American Philosophy*, ed. J. J. Stuhr, 13–31. New York: Oxford University Press.

Peirce, C. S. 1868a. Questions Concerning Certain Faculties Claimed for Man. *Journal of Speculative Philosophy* 2:103–14.

———. 1868b. Some Consequences of Four Incapacities. *Journal of Speculative Philosophy* 2:140–57.

———. 1877. Note on the Sensation of Color. *American Journal of Science and Arts*, 3d ser., 247–51.

Peirce, C. S., and Jastrow, J. 1884. On Small Differences of Sensation. *Memoirs of the National Academy of Science* 3:75–83.

Purdy, D. 1931. Spectral Hue as a Function of Intensity. *American Journal of Psychology* 43:541–59.

———. 1937. The Bezold-Brücke Phenomenon and Contours for Constant Hue. *American Journal of Psychology* 49:313–15.

Treisman, M., and Watts, T. R. 1966. Relation between Signal Detectability Theory and the Traditional Procedures for Measuring Sensory Thresholds. *Psychological Bulletin* 66:438–54.

Walraven, P. L. 1960. On the Bezold-Brücke Phenomenon. *Journal of the Optical Society of America* 51:1113–16.

———. 1962. *On the Mechanics of Colour Vision.* Soesterburg, The Netherlands: Institute for Perception.

23 Peirce and Self-Consciousness

Antoni Gomila, University of the Balearic Islands, Spain

Self-consciousness, as a philosophical concern, characterizes the early modern era. Nowadays, though, it seems as if we have lost interest in it. Once the early modern project of founding objective knowledge within the limits of subjective consciousness failed, the psychological structure itself of self-consciousness appears abandoned from almost every point of view. Paradigmatic examples of this situation can be found both within the analytic tradition and among outstanding voices of the so-called Continental philosophy. Think, for instance, of Wittgenstein's claim in the *Tractatus*: "The subject that thinks, that has representations, does not exist" (5.631), or Foucault's (1969:28) statement of the self's death (which he applied to himself) writing in *L'Archéologie du savoir*: "Do not ask me who I am nor ask me to continue being invariable: it is a moral of the civil state which governs our documents."

But the phenomena uncovered by rationalists and empiricists with respect to self-consciousness cannot be so easily eliminated. As Tugendhat (1986) has argued, these phenomena can be grouped along two main axes: the epistemic one, insofar as self-consciousness constitutes a domain of direct knowledge; and the practical one, having to do with the possibilities of action as an agent.

Meanwhile, the lack of a renewed theoretical approach to self-consciousness has the consequence that the empirical research on these phenomena becomes rather bizarre, ranging from the reductionist materialism that considers consciousness, in itself, as an anomaly requiring to be explained away, analyzed in respectable physicalist terms,[1] and the naturalism that purports to characterize the neurological mechanisms involved.[2] Neither approach takes into account the central features of self-consciousness: the capability of setting, assessing, and modifying one's own goals, and the ability to express one's thoughts linguistically.

Certainly, some philosophers continue to adopt a traditional stance, either the Cartesian view that the self is a special substance (Swinburne, Eccles) or the Kantian position that the self is transcendental (Z. Vendler, J. Rosenberg, T. Nagel, J. Searle).[3] To my view, such conceptions have been submitted to convincing critiques (by, for example, Hume and Husserl) that undermine their assumptions.

In this context I believe Peirce's contribution is of great value. He provided some very interesting cues to a conception of self and self-consciousness that are in agreement with an evolutionary view of human nature. By means of his semiotic conception of thinking and the role he attributed to society and language, he tried to develop an account of objectivity and self that could overcome the Kantian a priori and that went beyond the solipsistic approach in philosophy.

Peirce on Self-Consciousness

It is not my intent to gather here everything Peirce said about this subject, let alone to present it in a coherent way—a rather difficult task, given the evolution of his thinking.[4] I shall concentrate upon the most suggestive of his writings on the subject, which are contained in his papers during 1868–69. There we find an explicit effort to refute the views of Descartes and Kant on objective knowledge through Peirce's theory of thinking as a semiotic process. The development of his own epistemological view required Peirce to offer a new account of self-consciousness, not as the starting point as regards knowledge, but as the outcome of a semiotic process, linguistically mediated, that takes place during the socialization of children.

His main ideas on this theme appeared in the 1868 article "Questions Concerning Certain Faculties Claimed for Man." The topic of this paper was a critical examination of Descartes's views on subjectivity. The principal features of Descartes's position were his concept of intuition as providing direct and immediate access to a private reality and the feeling of certainty that accompanied this kind of knowledge. Peirce's criticisms consisted of a series of arguments. In the first place, Peirce argued that founding objective knowledge upon the inner criterion of evidence entails the rejection of the possibility of finding public, positive criteria of evidence, within experience (*CP* 5.214). For Peirce, the inner criterion adds nothing to the public ones. In the second place, in a quite ironic vein, Peirce equated this appeal to one's "inner authority" to the medieval practice of resorting to authorities for certainty (*CP* 5.215), in contrast with the scientific method of considering exclusively experimental evidence.

In consequence, the Cartesian notion of self-consciousness is rejected. For Descartes, self-consciousness consisted in a kind of self-acquaintance, of immediate knowledge of one's self. But if Descartes's notion of intuition, the faculty through which this sort of knowledge is achieved, is no longer acceptable, then self-consciousness must be conceived within some other framework.

Toward this goal, two main ideas were available to Peirce. On the one hand, his attack against intuition was not something of secondary interest; on the contrary, his opposition to this "claimed" faculty stemmed from his own account of thinking of mental life as a semiotic process. On the other hand, Peirce had already adopted the "active" notion of belief, in the sense of the psychologist Alexander Bain, according to whom belief and action were conceived as intrinsically connected. In this vein, mental contents are no longer to be assessed in solipsistic terms but on the grounds of behavioral evidence.

With respect to the first point, I shall only outline the main traits. Peirce's semiotic approach resulted from his "critical realism" as applied to meaning, according to which the Kantian notion of "incognizable things-in-themselves" is semantically incoherent and therefore can be eliminated. Cognizability, understood as symbolic meaningfulness, constituted the domain of possibility: "Over against any cognition, there is an unknown but knowable reality; but over against all possible cognition, there is only the self-contradictory. In short, *cognizability* (in its widest sense) and *being* are not merely metaphysically the same, but are synonymous terms" (*CP* 5.257).

Objective knowledge of reality, then, can be achieved, not because of the structure of our consciousness, as in Kant, but "in the long run," through a process semiotically mediated of inferences. Every thought, because of its semiotic character, brings another, even unconsciously (*CP* 5.253). What must be established is the right pattern of these inferences, so that the advancement to truth can be guaranteed. It is in this context that Peirce's interest in logic becomes paramount. He conceived the study of the different types of valid inferences (deductive, inductive, and abductive) as the study of the right patterns of reasoning. Moreover, its semiotic character implied that their meanings (or "interpretants") were a matter of a public process of interpretation. Actually, Peirce understood thinking through the metaphor of "inner dialogue," as internalization of the public process of interpretation.[5]

As for the second point, Peirce accepted and brought to its limit Bain's idea of the intimate relation between belief and action, without blurring the first within the second, as behaviorism did. For Bain, what one believes is what guides one's actions and directs one's "volitions." Accepting such a conception, which was to develop into his notion of "habit," Peirce provided himself with an important source of public evidence to use in assessing mental content: "If belief is taken in the active sense, it may be discovered by the observation of external facts and by inference from the sensation of conviction which usually accompanies it" (*CP* 5.242). The "sensation of conviction" was also understood by Peirce in a behavioristic manner, so that its validity can be ascertained through observable criteria.

This sketch of the conceptual resources available to Peirce will help us to understand the kind of account he gave of self-consciousness. To begin with, the thought of one's self as a "self" was conceived by him as an inference from the thoughts of others, not an exceptional thought, as in Descartes. Therefore, no special faculty must be postulated to account for it. Furthermore, this thought must also exhibit a practical dimension; that is, it must be, to some extent, empirically observable in behavior. Finally, and as a consequence of this last point, he is required to give a developmental account of self-consciousness, since public evidence—as in the use of personal pronouns, particularly the first-person "I"[6]—makes it obvious that infants lack such self-acquaintance.

Thus, what is required is an account of the semiotic process through which such a thought develops in the child, an account that could explain the role played by the society in which the process takes place. Moreover, it is an account "from outside," that is, without appealing to criteria of subjective evidence, but appealing to observable behavior, thus providing a basis for ascription of self-consciousness.

Peirce articulated such an account through two main stages, one prelinguistic, the other linguistic. The first moment in this process would be the discovery of one's own body, by means of perceptions and volitions, which could make possible the appearance of a rudimentary notion of "one's self": "A very young child may always be observed to watch its own body with great attention. There is every reason why this should be so, for from the child's point of view his body is the most important thing in the universe. Only what it touches has any actual and present feeling: only what it faces has any actual color; only what is on its tongue has any actual taste" (*CP* 5.229).

Peirce considered what could be inferred by children from causal interactions with their environment as a result of their will. He thought the thinking powers in children were sufficient to establish a connection between their actions and the changes such actions could provoke:

> The child, however, must soon discover by observation that things which are thus fit to be changed are apt actually to undergo this change, after a contact with that peculiarly important body called Willy or Johnny. This consideration makes this body still more important and central, since it establishes a connection between the fitness of a thing to be changed and a tendency in this body to touch it before it is changed. (*CP* 5.231)

Thus, a first sense of one's self springs as an inference from the bodily interactions with the environment. But then the child learns to understand the language, and a more advanced stage in the process begins. What is most remarkable in this moment is that the child receives information through a new source. Up until that moment, the child's own body and senses were the only channel of information input, but language makes communication possible. This influence is so strong, Peirce observed, that linguistic information enjoys a preeminent status for the child, as "the best evidence of fact." The upshot, as Peirce envisioned it, will be a conflict between what is perceived and what is heard, solved in favor of linguistic information, that is, information provided by other people. The importance of such conflicts is that they give rise to new experiences of error and ignorance, whose role in the appearance of the self is determinative.

Peirce described the experience of ignorance in the following way:

> A child hears it said that the stove is hot. But it is not, he says; and, indeed, that central body is not touching it, and only what that touches is hot or cold. But he touches it, and finds the testimony confirmed in a striking way. Thus, he becomes aware of ignorance, and it is necessary to suppose a self in which this ignorance can inhere. So testimony gives the first dawning to self-consciousness. (*CP* 5.233)

And as regards error, Peirce introduced in the description of this experience a new range of phenomena: the emotions. What is peculiar about them is that they are related to the body which experiences them in a different manner than to those people who merely observe them. As a result, their attribution to somebody is frequently denied. This is the origin of experiencing error:

> These (emotional) judgement[s] are generally denied by others. Moreover, he has reason to think that others, also, have such judgements which are quite denied by all the rest. Thus, he adds to the conception of appearance as the actualization of fact, the conception of it as something *private* and valid only for one body. In short, *error* appears, and it can be explained only by supposing a *self* which is fallible. (*CP* 5.234)

Peirce concluded his account with a rather cryptic remark: "Ignorance and error are all that distinguish our private selves from the absolute *ego* of pure apperception" (*CP* 5.235).

We can understand such a remark, in accordance with the general approach adopted by Peirce, as an expression of the rejection not only of

Cartesian dualism and its claimed special mental faculties but also of Kantian transcendentalism—the unexperienceable self, which is presupposed by every experience—but saving what is most original in Kant's theory: the self as a formal condition, as a structure, the principle of the unity of consciousness. From this perspective, moreover, we can shed light on the mechanisms by which error and ignorance give rise to a "self"-thought. The principle of unity of consciousness requires that contradictory inputs be reconciled, unified. The conflicting thoughts that originate in communication and perception cause the dawning of self-consciousness, because this way there is no contradiction in reality but in the representation of it that the self builds: within the domain of representation the conflict is located and solved.

Spreading the Theory

Peirce did not claim to have said the last word on the subject. He was well aware that his account was only one attempt to explain the process involved and, as such, could merely be considered as a plausible hypothesis: "The only essential defect in this account of the matter is, that while we know that children exercise *as much* understanding as is here supposed, we do not know that they exercise it in precisely this way. Still the supposition that they do so is infinitely more supported by facts, than the supposition of a wholly peculiar faculty of the mind" (*CP* 5.236).

Interestingly enough, nevertheless, recent empirical findings in social psychology offer an unexpected support both to the general approach to the subject matter—the self as a social product, linguistically mediated—and also (though only partially) to the sort of stages in the development of self-consciousness Peirce argued for. No doubt, the influence of symbolic interactionism, especially the work of G. H. Mead in *Mind, Self, and Society*, which concerned itself with an account of the social character of the self in its origins, is not alien to this outcome—and may be a linkage between Peirce's insights and current research. My purpose, in the remainder of this paper, will be to expand Peirce's views by presenting this new evidence. In so doing, I expect his view on self-consciousness and self to be improved and brought to the foreground of philosophical discussion.

First of all, as Peirce claimed, the presence of a sense of "one's self" in the prelinguistic child has been experimentally demonstrated. As Peirce thought, this rudimentary "self" is closely related to the sense of one's body, both perceptively and by means of the ability to influence the environment through one's movements.

> The first stage in the awareness of the self is somatic, organized around kinesthesis, proprioception, and other bodily activity. The infant soon discovers that external objects—especially those moving, vocalizing, variable ones that are persons—are instrumental in relieving bodily stress. Re-afferent feedback provides the child with its first evidence that an event may be contingent on its own performance. When it closes its eyes the world disappears. When it cries a face appears. (Crook 1985:251)

This primitive, prelinguistic self-awareness has been detected using behavioral tests, as Peirce required. One of the most important experimental designs to this goal has been the use of mirrors (Lewis and Brooks 1979). The procedure consists in confronting the infants with their mirror images, after having marked their noses surreptitiously with rouge, and observing subsequent behavior. Self-directed behavior—that is, behavior directed to one's own body, particularly the face—is considered a sign of a rudimentary self-concept. As a matter of fact, such a behavior involves not only self-recognition, in assessing the mirror image as an image of one's self, but also an ability to show surprise in discovering something unexpected. The results confirm what Peirce claimed: before mastering the language (which occurs at about twenty-four months), children reach a concept of self. "Below 12 months almost no infant responded to the red mark as a result of observing itself in the mirror, but beyond this age there was a rapid development in nose-directed behavior (25 percent of 15–18 month-old infants, 75 percent at 21–24 months)" (Crook 1985:251–52).

On the contrary, if the mark was placed in other parts of the body, apart from the face, the results were completely different, without such self-recognition. This could indicate the adaptive pressure in a social species to solve the problem of identification of its members and the important function played by the capability to discriminate different faces—one's own included.[7]

Similar phylogenetic reflections arise in considering the results of these same experiments performed with rhesus monkeys and chimpanzees. In an identical experimental situation as that of children, rhesus monkeys showed no sign of self-directed behavior, while chimpanzees touched their noses, evidencing a degree of self-recognition and the ability to detect a change in the normal appearance equivalent to that of children (Gallup 1972).

All this suggests that this prelinguistic concept of self, this rudimentary self-consciousness, shared both by chimpanzees and children below two years old, is to a large extent phylogenetically fixed, as a consequence of the social nature of anthropoids. At the next stage of the development of self-consciousness—the linguistic one—where human self-consciousness in its whole richness appears, we can find the clues for an understanding of the phylogenetic process that culminated with the dawning of humanity in evolution. In other words, to deal with the problem of the origin of self-consciousness in evolutionary history, in its full, specifically human sense, we must be concerned with the problem of the origin of language, since, as Peirce also suggested and as human ontogeny shows, it is through the mediation of language that self-consciousness arises. I do not mean it to be an instance of the principle that ontogeny recapitulates phylogeny, at least without qualification. This is the subject of another study. Let me simply say here that any nativist view of the origin of language is inadequate, because language develops only through linguistic stimulation in social interaction.

It is time, therefore, to address the linguistic stage in the process to self-consciousness. Peirce relied upon the experiences of error and ignorance to account for the appearance of a full notion of self at this moment. I want to present the account developed by Harré (1987), which, though placed at a more abstract and sophisticated level, strongly resembles Peirce's remarks. Although his work takes into account recent findings in social and developmental psychology, Harré himself acknowledges his debt to Mead and Vygotsky (another psychologist concerned with the social character of the human mind) in considering language as a tool that, in addition to the function of communication, serves the function of thinking, through a process of "internalization." Harré follows a similar tactic as that used by Peirce: the idea of "self" is not immediate and direct but is a social product, the result of inferences, motivated by the social relations in which others are involved, carried out frequently through language. Harré contends that the unifying notion of "self" stems from the social concept of "person." He tries to show how the concept of person can appear without presupposing an inner self. He distinguishes three types of situations in that process: "First, people treat babies as persons from the moment of their first appearance. . . . Second, by copying every word and gesture as best he can, a baby seems to be treating those around him as persons. Third, among the ways of speaking and acting that a baby imitates is the way in which other people treat him as a person" (Harré 1987:101).

The key notion in this account is that of "treating babies as persons." Harré spells it out in three modes of linguistic interaction. First, there is the practice of naming. Second, there is the "psychological symbiosis" phenomenon, studied by developmental psychologists (Shotter and Newson 1974), which consists of adults, particularly mothers, who "interact with their offspring in terms of psychological attributes that they assign to the infant" (Harré 1987:101), including cognitive capabilities, preferences, character, and sensibility. Although the scope of this phenomenon is greater, it involves an obvious linguistic dimension: adults "talk" to infants, attributing to them such psychological traits, long before their mastering of language. Third, treating somebody as a person involves judging his or her actions, preferences, and decisions. The outcome of being treated in such ways is that the child becomes a member of the group, a unique individual, and recognizes himself or herself as a person.

How is it possible, then, to go from the mastering of the concept of person, through the different situations just sketched, to that of a self? In Harré's view, the use of proper names and pronouns to refer to one's self—the "I," in particular—learned in the process of being treated as a person and used by the child as well, gives rise to a special kind of problem: that of "epistemic warrant." When a child says, "I'm tired," he or she must face the possibility of being contradicted or disbelieved; doubts about what was said may arise among the child's interlocutors.

Once again, we can recognize the similarity with Peirce in the fact that the conflict is placed completely at the linguistic level, instead of between the perceptual and the linguistic. At this point, Harré presents an explicit account of how this confrontation leads to the concept of self.

In effect, in being contradicted, Harré argues, children are forced to reflect on, and justify, the basis for their avowals; and to express the results of their deliberations, they must use a language game in which, instead of one pronoun, there appear two: "I'm sure I'm tired," or "I have no doubt I'm tired." The problem now is that while the first pronoun refers to the child as a public person, what about the second? Harré (1987:105) describes the consequences this question may have for the child in the following way:

> I am quite used to keeping my thoughts and feelings, sensations and opinions, to myself. For me, the distinction between public face and private being is routine. No wonder, then, that I am inclined to slip into following the model all the way, and am likely, therefore, to fall prey to the idea that there is an inner self to which my private thoughts and feelings are annexed.

Self-consciousness, accordingly, is the result of becoming competent in the use of certain grammatical forms, that of "direct knowledge," required in the normal practice of linguistic interaction and consisting of a certain unifying perspective. As for Peirce, though less cryptically, the conception of the self that emerges for Harré opposes the traditional ones—Cartesian and Kantian—in favor of the notion of "self" as inner entity. It is conceived as "a centered organization of thought, feeling, actions, and memory" (Harré 1987:105); in other words, the "unity of consciousness" principle again but without any transcendental dimension.

In my view, Harré's argument can be paraphrased for the other aspect of self-consciousness, the practical one. It is not only in questions of epistemic warrant that such "special" judgments take place; they also take place in justifying wantings, decisions, desires ("I wish I were a better player," "I have decided *X* because I want *Y*," "I did such-and-such because I believe so-and-so to be good," etc.). Acknowledging this can be the first step toward the development of a nonreductionist theory of action, according to which behavior is not just a mechanical consequence of "efficient mental states"[8] but the outcome of a deliberative process of critical assessment of opportunities, values, and goals.

The fact that the self is a social product does not entail a kind of determinism of the subjective sphere, in the sense that whatever is thought is externally determined. As Peirce remarked, thinking is a semiotic process, that is, a process driven by the meaning of the symbols involved, so that external facts can influence thinking only through the meanings the subjects attach to them. What is socially determined, accordingly, is the symbolic world characteristic of a society, and socialization can be understood as the process of acquiring competence in such a world—of learning the rules, the norms, the rites, and the roles that constitute the society.

Therefore, we do not need a metaphysical notion of "free will" (parallel, for Kant, to that of "trascendental self," both being "truths of the practical reason") to guarantee the personal dimension of action we are so familiar with. Our becoming persons brings about our development of a self, of a personal perspective, which allows us to decide, to some extent, who we want to be and what we want to do.

To sum up, Peirce's views on self-consciousness as a mental structure resulting from the social dimension of mind and the semiotic character of thought can still shed light on the nature of our conceptions of self. As a matter of fact, his conception has recently received support from the unconnected work of developmental and social psychologists. My contention, therefore, is that the study of Peirce's views of the self should not be seen as merely "backward history"—honoring thinkers of the past for ideas that have been developed only recently. On the contrary, considering the unsatisfactory situation concerning the current views on the consciousness of the self, I have tried to give credit to the originality and insight of Peirce's views on this subject.

Notes

1. See Jackendoff 1987, a recent representative of this attitude, where the main "solutions" to this "problem" are presented and criticized and his own alternative account is developed.

2. A distinguished example of this approach is Underwood and Stevens 1979.

3. An interesting review of the claims of these thinkers is Hannay 1988.

4. Such an attempt has been made by Singer (1982).

5. See Singer 1982:137–41.

6. Peirce took this observation from Kant, but he failed to develop the role of pronouns in the development of self-consciousness.

7. Chomsky (1980:chap. 5) argued for the existence of an innate faculty designed for this specific purpose.

8. This is the standard notion of behavior explaining in cognitive science, what is called folk psychology: "*X* did *p* because he or she desired that *q* and believed that if *p* then *q*." One of its most influential defenders is Davidson (1963).

References

Chomsky, N. 1980. *Rules and Representations*. Oxford: B. Blackwell.

Crook, J. H. 1985. *The Evolution of Human Consciousness*. Oxford: Oxford University Press.

Davidson, D. 1963. Actions, Reasons, and Causes. *Journal of Philosophy* 60:685–700.

Foucault, M. 1969. *L'Archéologie du savoir*. Paris: Gallimard.

Gallup, G. 1972. Self-Recognition in Primates. *American Psychologist* 32:329–38.

Hannay, A. 1988. The Claims of Consciousness: A Critical Survey. *Inquiry* 30:395–434.

Harré, R. 1987. Persons and Selves. In *Persons and Personality*, ed. A. Peacocke and G. Gillet, 98–112. Oxford: B. Blackwell.

Jackendoff, R. 1987. *Consciousness and the Computational Mind*. Cambridge: MIT Press.

Lewis, M., and Brooks, J. 1979. *Social Cognition and the Acquisition of Self*. New York: Plenum Press.

Mead, G. H. 1934. *Mind, Self, and Society*. Chicago: University of Chicago Press.

Shotter, J., and Newson, J. 1974. How Babies Communicate. *New Society* 29:345–47.

Singer, M. 1982. Personal and Social Identity in Dialogue. In *New Approaches to the Self*, ed. B. Lee, 129–78. New York: Plenum Press.

Tugendhat, E. 1986. *Self-Consciousness and Self-Determination*. Cambridge: MIT Press.

Underwood, G., and Stevens, R. 1979. *Aspects of Consciousness*. 2 vols. London: Academic Press.

24 The Relevance of Peirce for Psychology

Clyde Hendrick, Texas Tech University

I was fascinated a few years ago to read Peirce's review of William James's *Principles of Psychology* in the *Nation* (*N* 1:104–10). I was also intrigued by Thomas Cadwallader's (1974) argument that Peirce was actually the first American experimental psychologist. It seems clear that Peirce was relevant to psychology in its development as a discipline during his own day. Unfortunately, attribution to him appears to have become lost early in this century. It would be a fascinating project for a historian of psychology to go back to the nineteenth century and develop the threads of Peirce's thought as it influenced and wound its way into psychology as it matured in this century. However, in this chapter I discuss the relevance of Peirce for current psychology. I will necessarily be selective because I have found Peirce to be relevant to almost all areas of contemporary psychology with which I am familiar. As just one minor example, many years ago I read a small volume by Karl Weick entitled *The Social Psychology of Organizing* (1969). Weick made the strange statement that all meaning is retrospective. I should say that statement was very strange then, but since reading Peirce, the meaning now seems natural enough. Peirce said many times that the present moment is something that we cannot cognize; we can only cognize it after it has gone. For example, "The only thought that is really present to us is the thought that we can neither think about nor talk about" (*CP* 7.425). Following this logic, it emerges that meaning, however we define it, is something that must come from our past experience; thus it must be retrospective.

In my opinion, the relevance of Peirce for contemporary psychology stems from the relationship of his ontological categories of firstness, secondness, and thirdness to the phenomena dealt with by psychological science. As I understand the matter from a psychological point of view, Peirce started his analysis from whatever appeared before the mind, the Phaneron (e.g., *CP* 1.284). The phenomena of experience seem to be a good jumping off place for psychology. Indeed, despite some of the rhetoric of my discipline, I do not believe there really is another starting place. Peirce discussed his three categories in many different ways. I will simply accept them as metaphysical categories in all their pristine glory. But I do want to apply them specifically to psychological phenomena, and in so doing I must interpret and liberalize the categories slightly, but I hope in that process not to distort them too much. I will first explicate Peirce's categories from a psychological point of view, and then I will apply the categories to several concrete instances relevant to psychology.

Peirce's Categories from a Psychological Point of View

First, I will examine, somewhat informally, Peirce's three categories. I will then describe a hypothetical experiment that attempts to give some empirical validity to a distinction among the categories. According to Peirce, all three categories intrinsically inhere in any intelligible phenomenon. Presumably, they can be separated conceptually but not empirically. In fact, it is clear that Peirce cognized these categories on the basis of ongoing, everyday phenomenal experience. I maintain in my thought experiment that under a certain limited condition the categories can in fact be empirically separated.

Firstness

During my adult life, on peaceful occasions, I sometimes have an image of the "now" in which all the wonderful diversity of nature just is, out there, simultaneously manifesting itself in a collage of qualities. The rose garden, grass, spray of water—all are just there. In this moment, causation is suspended; there is just a manifold of qualities present for me. I suspect that it was from this kind of experience that Peirce abstracted his concept of firstness. Peirce's label of *presentness* fits the experience quite well. It seems to result from a receptive mood when the rest of the mind is idling in "neutral." Clearly the experience of presentness in this sense cannot be cognized or described. It always puzzled me that when I became aware of myself being aware of presentness, the presentness was gone. After reading Peirce, it is easy to understand why presentness cannot be cognized in the present but only remembered from the "just past."

Psychologically, the sense of presentness seems to be preconceptual. The experience of firstness requires suspension of the time sense. An event may be experienced as presentness across duration, but there can be no awareness of duration, or the sense of presentness is gone. For example, the lazy flight of a hawk might be experienced as it "just is." However, the swoop of a hawk toward prey probably cannot be so experienced. In general, I would guess that the sense of presentness is possible mostly from environmental constancy or change that occurs at a constant rate.

As the first category of phenomenology, rooted in Peirce's notions of common sense, my guess is that presentness is an evolutionary development. It stems from that unconscious part of our brain that keeps us at home in our world. In other words, presentness, or firstness, is the fit between the nervous system and environment. It had to be the first organic development of mind. It probably exists in all living species that have nervous systems. It is not enough for survival of mobile creatures, so it is overlaid with secondness. But firstness *is* first. As a matter of fact, when one thinks about it in a certain way, secondness and thirdness do not make sense except against a background of firstness. Therefore, firstness is not just a logical null category. It is as real as the other two categories.

Secondness

We all understand secondness. We live in a constant flux of action and reaction. Secondness is what Marine boot camp is all about. It is the notion that if you take an eighteen-year-old boy and subject him to six weeks of barely endurable secondness, somehow you make a real man of him! On particularly aggravating days—when the computer is down, I slip on the sidewalk, and a telephone call is cut off—I am fond of saying that "my world was full of secondness today." It helps me keep my sense of humor and prevents an attribution of thirdness, such as "God was out to get me."

Psychology as it has so far developed is a discipline that mostly studies secondness. The focus on observable behavior, with the language of stimulus and response, action and reaction, testifies to this assertion. Beginning around the time of John Watson, somewhere about 1915, psychology made a strong behavioral turn, with a focus on behavior in an overt, external sense. The ethos was that only observable existents could be studied profitably. The further assumption was made that only what exists is real. Thus, subjectivity and mind were banished from the realm of psychology and treated as epiphenomena (i.e., as essentially meaningless). Language was difficult to handle but was eventually construed as "verbal behavior."

To some extent, Peirce would have applauded this development. After all, every serious hypothesis should be pursued for all it is worth. And Peirce had a heavy hand in banishing introspection as a psychological method because of his insistence that we do not have immediate access to our mental states but arrive at them only through inference.

Nevertheless, I expect that Peirce would have been unhappy with the one-sided emphasis on the study of overt behavior as bodily movement and action. There is, after all, another world of experience, the symbol world of signs and meaning. As a matter of fact, psychology, along with several other disciplines, has discovered this world during the past fifteen years. Mind is back, under the code phrase of "the cognitive revolution." The cognitive turn has been so strong, in fact, that I sometimes fear that we are in danger of losing the solid world of secondness in which psychology so long dwelt. But there is no doubt that the discipline has at long last discovered thirdness.

Thirdness

I will not say much about thirdness. The world of thirdness has not yet been much explored within psychology. Although, as I noted, the cognitive world has been discovered, *it is approached mostly from a methodological standpoint of secondness*. That is to say, the methods of research tend to use fairly simple dyadic models of cause and effect for the study of the triadic world of symbols. I do not expect much progress to be made until there is a methodological revolution that develops tools of thirdness to deal with the phenomena of thirdness.

There are many suggestions and substantial discussions in Peirce's writings for the development of a triadic psychology. For example, the notion that an individual person is both a general and a partially indeterminate being is rich in potential conceptual development. Peirce's notions of self (Colapietro 1989), personality, person, and community (Singer 1984) beg for extended development, along with many other examples.

Speaking of Peirce's categories as a psychologist necessarily requires a certain looseness of reference. We live in a world in which all three categories come in a bundle in all intelligible phenomena. We cannot live the categories in purely abstracted metaphysical terms of thirdness. But if we cannot distinguish the categories in the empirical world, the hardheaded experimental psychologists will wonder what good they are. Categories distinguishable only by reason but not by experiment are apt to be construed as no categories at all, or at least not very useful categories.

In pondering this situation, it occurred to me that there is a way of demonstrating the separability of Peirce's categories in the empirical world. Let me illustrate this method by a simple hypothetical example, although it is based on research tradition in experimental psychology. The example involves a simulated perceptual experiment.

A Hypothetical Perceptual Experiment

Cognitive processes take time. The more elaborate and complicated the cognition, the more time is required for it to occur. According to Peirce, all three categories—firstness, secondness, and thirdness—are involved in each occurrence of a phenomenon. But what is a phenomenon? Assume the occurrence of some event susceptible to one of the senses, such as vision. In order to perceive the event, it must have some minimal duration. But "perceive" is a global term, and many different things can occur under this general label, depending on the length of the time interval permitted.

I want to show that the categories of first, second, and third emerge in visual perception as a function of time intervals. At very short intervals, I claim, there can be sheer presentness but without experienced secondness or thirdness. As the interval is lengthened, secondness and then thirdness enter into experience. If my claim is true, there is an experimental basis for asserting conceptual independence of the three categories.

Consider a common type of perceptual experiment, such as determining thresholds for various types of visual stimuli. My example is a moderately fictionalized experiment in which simple stimuli of alphabet letters are presented to a subject by a tachistoscope, an instrument that projects images on a screen for brief durations. Shutter speed can be varied to control precisely the duration of a presentation. Given a fixed-energy light source, an experimenter can determine the durations at which something, such as the letter *A*, can be detected (was that flash a blank or was something there?). Durations for actual identification of a stimulus can also be determined. It takes longer to identify a stimulus than to detect that it simply occurred.

Now, consider the hypothetical range of presentation speeds shown in figure 1 below. Stimulus duration ranges from almost instantaneous to about one second. Suppose a "psychological instant" is 0.000001 second, too short for any effect on our sensory apparatus. Nothing perceptual happens. As duration is increased to approximately 0.001 second, bare awareness that "something happened" enters. A sense of *presentness* with respect to the stimulus field occurs. The duration would need to be lengthened to perhaps 0.01 of a second for there to be a definite sense of an object on the screen. At this point the sense of otherness, or secondness, of the stimulus impinges on the subject's eyeballs. At about 0.1 second, full identification (the letter *A*) is possible, and thirdness of the symbol world enters into the situation.

Thus, psychologically speaking, it takes time for the categories of experience to register in the organism. The same stimulus letter *A* is

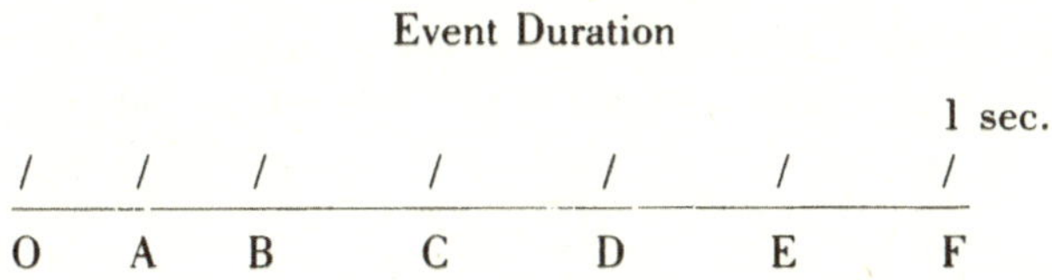

O = an instant of time, for example, 0.000001 sec. Interval is too short for any awareness, but physiological activity is in process.

A = duration required for bare awareness, perhaps 0.001 sec. No content to awareness, which may be based on a physiological interval marker. Embryonic firstness in the sense of vague presentness.

B = duration required for vague awareness of a general environmental surround, perhaps 0.005 sec. Sense of presentness.

C = duration required for detection of object flashed on screen, perhaps 0.01 sec. Sense that "something happened." Firstness (as presentness) and beginning of secondness.

D = duration required for identification of simple object (for example, the letter *A*), perhaps 0.02 sec. Sense of familiarity, but recognition is only preverbal. Firstness, secondness, and beginning of thirdness. Beginning of what Peirce called a percept (*CP* 1.253).

E = duration required for symbolic identification of an object (e.g., "That is an *A*"), perhaps 0.1 sec. Operation is at the level of Peirce's perceptual judgment (*CP* 5.151); thus thirdness is now full partner with firstness and secondness.

F = duration required for complex symbolic identification (e.g., "That *A* is oddly shaped" or "I don't like its color"), perhaps 1.0 sec. The perceptual judgment is elaborated, and thirdness may be predominant.

Figure 1. Hypothetical example of tachistoscopic presentation of an object (the letter *A*) to an experimental subject at varying shutter speeds. This experiment illustrates the manner in which firstness, secondness, and thirdness can enter into consciousness as a function of temporal duration.

presented at 0.001 second as at 0.1 second. But firstness emerges first, then sheer otherness of the object, and finally its symbolic identification. The categories are in this way separated as a function of time duration.

I suggest that such standard perceptual experiments provide direct evidence for the separability of the categories. I do not claim more than shown. Metaphysically, all three categories occur in all phenomena. But with respect to human experience of phenomena, at the very borderline between perception and nonperception, an experiential distinction can be made.

Applications of Peirce's Categories

Please keep in mind that I have liberalized the usage of the categories. Strictly speaking, firstness and secondness can be thought and described only from the standpoint of thirdness. However, it is clear that firstness and secondness must enter into experience in some way, otherwise they could not be thought or described. I assume that we do have awareness, as I discussed previously, of qualities and of the brute reactions of the world around us. These must enter into our experience, and, on the basis of our recollections of and inferences about such experiences, we are then able to describe them with symbols. I further assume that any given experience may have a relative predominance of one category or another. My previous example of firstness seems to be an instance in which we do experience a degree of qualitative immediacy without much of the apparatus of secondness or thirdness at that moment. Other situations involve secondness more clearly. For example, soldiers in warfare in direct combat are in a situation fraught with secondness. Sitting at my desk composing this paper seems, on the other hand, to be living primarily in the symbol world of thirdness, with secondness and firstness more secondary. It is in this sense that I mean that there can be a relative emphasis in the psyche at a given time on one or the other of the three categories.

I will simply describe several different phenomena in which use of Peirce's categories makes sense to me. I begin with an easy one.

Mysticism

Psychologists have long studied religious experience. Mystical experience is one of the more esoteric facets of religion. From my readings on mysticism, it occurred to me that mystical experience may be related in a direct way to Peirce's categories of firstness and secondness. There are at least two general varieties of mysticism: one which seems to be a kind of merging of self with being, with a stress toward something called Oneness; the other type of mysticism seems to be a radical disjunction of the person or self from the universe at large. Nature mysticism is very often the attempt to achieve oneness with the universe. It may be an apprehension of presentness "writ large" to all of nature. In such a mystical experience the sense of presentness of all that there is would be almost overwhelming. The object quality, or secondness, and any cognition about the world would be either stripped away or exist at a very minimal level. In short, the person experiencing such a mystical state would be merged into a total qualitative world of immediacy.

In the other kind of mysticism the "wholly other" may be an apprehension of the terrible secondness of the world with respect to self. This kind of experience of the absolute otherness of the world can inspire states of terror or dread. Because both firstness and secondness are preverbal and precognitive experiences of the world, there would not be appropriate language forms to capture such experiences when they loom up to fill the mental horizon. It may be for this reason that mystics around the world have always complained that it is impossible to capture such mystical experiences in written or spoken form.

Past Me and Present Me

The older literature in psychology sometimes distinguished between the I and the Me. In reading Peirce, it occurred to me that his definition of the three categories automatically implies at least two Me's. One Me, the present Me, lies in the spontaneous now that is incessantly unfolding, and the past Me perpetually reflects on it. The past Me is the Me of memory systems, inferential machinery, reasoning processes, and the like, which incessantly assimilates the material from the immediate Me to create the past. As a matter of fact, this is the only way in which experience can be made consistent. As it ceaselessly unfolds, it is also ceaselessly absorbed into the cognitive apparatus which is by definition always in the past.

The past Me might be partitioned into two components, one a strict past Me of memory systems and the second a future Me that controls the mental modules of expectation and anticipation of future events. To make an inference is to make a prediction that can be assessed only against future experience. Therefore the machinery of inference is oriented toward the future. However, as a physical process occurring temporally, the inference process emerges from the "just past." In this sense, memory systems and anticipatory systems must both be part of the structure of the past Me.

This conceptual structure creates interesting possibilities. For example, psychologically speaking, how long does the immediate now last? Do people differ in the time span in which the spontaneous Me is active? Do all of us differ at different times in the time span of the spontaneous Me? Do people differ in terms of the smoothness of assimilation of immediate experience into the reasoning and memory apparatus?

Some preliminary observations may be made. People do certainly differ in terms of the spontaneity they seem to manifest in ongoing interaction. Some people appear to keep their immediate behavior under rigid self-control. In fact, there is a research condition within social psychology called self-monitoring that strikes me as very relevant to this possibility that people do differ in the degree to which the spontaneous Me is active as opposed to the degree to which it is immediately pulled into the structure of past experience.

There are several other possibilities. To the extent that there is a disjunction between the immediate Me and the past Me, there is a strong entry point for inconsistency of behavior and lack of self-understanding. The simple statement "I don't know why I did that" suggests that the spontaneous actions of the immediate now are interpreted and perceived as inconsistent with the structure of the Me that encompasses all my pasts, reasoning machinery, and the like.

Most likely, there is a trend in the development of concept of self. To a certain extent, education and, more generally, socialization are training in the impoverishment of the now. Conceptual work, which is the occupying of the mind with thirdness, requires relative inattention to the world around us. That is, the experience of firstness and secondness must be subdued if we are going to engage in the serious thirdness of thinking. It appears that infants and young children live relatively more in the immediate now than do adults. In other words, the spontaneous Me is the primary Me, and the past Me is relatively undeveloped in infants and children. As experience accumulates, the past Me should gain greater control over the immediate Me. However, in addition to individual differences in the degree of that control, it can sometimes go awry. I believe that there are individuals who are consistently "out of sync" with regard to present Me and past Me. That is, the experience of the immediate present is not assimilated, or at least not very well, into the structure of the past Me. If lack of assimilation continues over time, the past Me becomes a series of snapshot frames disconnected from each other and from the ongoing Me. Such a disjuncture is one way to define psychopathology. As a matter of fact, this approach would provide a good generic description of the meaning of "borderline personality" and is an opening wedge into defining other kinds of psychopathology.

The Unconscious

Taking the notions of the present Me and the past Me seriously, several psychological implications can be derived. One implication is the concept of the unconscious. Classical psychoanalytic theory views the unconscious as something "deep," emotional material existing somewhere in the far recesses of the mind and held there by a motivational force called repression. Repressed materials are usually considered as emotionally significant memories from one's past, often from childhood.

Based on the distinction between a present Me and a past Me, I propose that the locus of the unconscious is not in the recesses of one's past but that it exists in the present. More precisely, the locus of the unconscious is in a disjunction of connection between the spontaneous behavior of the present moment and the orderly inferential machinery of the past Me. Please keep in mind that what we are, our personhood, is our past Me. It contains the stores of memory material, future anticipations, and the logical thinking apparatus that binds the flow of signs coming into creation in the immediate now with our past store of signs, including our future anticipations of self. When the duration of immediate experience is in some sense too long or the physiological machinery cannot integrate it, there is a failure in the assimilation process. Such a failure leads to two different behavioral processes that have been called "unconscious behavior."

In the first place, if the present Me is too long unconstrained by the past Me, our animal urges as biologic beings relatively control our behavior. Such a person is said to be impulsive, or "acting out," or "dominated by unconscious processes," such as sexual and aggressive urges. In Freudian terms, such a person is "id dominated." We see that the locus of this causal attribution is not something deeply buried in the psyche but rather is the nature of our being in the immediate present. Thus, the nature of this type of unconsciousness is actually our spontaneous presentness in the world when we are relatively cut off from the spiritual heritage of our own psyches. That heritage is all the past store of signs, and especially symbols of self, that constitute who we are. Conscience and morality reside in the past Me. Nonrationality is the lack of a proper connection between our onflowing inference machinery and the immediacy of the now.

The second kind of unconscious process lies in a disjunction of the present Me and past Me in which the past Me does not draw the correct inference for future behavior implied by immediate behavior. For example, suppose a person is in a situation in which any objective observer would assume sexual arousal to be the expected response. However, if the person does not draw that inference, ongoing behavior cannot be labeled as sexual, and the experience of sexual arousal is not assimilated to the past Me. The person denies sexual arousal.

With a bit of development of this line of thinking, one can soon start discussing repressed sexuality. When the person eventually ends up on the analyst's couch, a great deal of effort will be expended helping the person "dredge up" this ostensibly repressed material. In fact, what the analyst actually does is teach the client how to draw correct inferences. When the task is complete, the person will in future experiences be able to label and integrate such ordinary occurrences the way the rest of us do. He or she is "cured."

Therefore, unconsciousness, in this quasi-Freudian sense, is quite simply a type of failure of the ongoing inferential sign process for present and future behavior. And in that sense, unconsciousness is a phenomenon of the present, not of the dark and buried past. Much more elaboration is needed, of course. For example, the role of dreams in this context is very interesting, but that must be developed on another occasion.

Romantic Love

My current area of empirical research as a social psychologist concerns romantic love. It occurred to me that Peirce has something to say about the general structure of this situation. Romantic love is a peculiar institution in Western society, one from which a great many of us enter into that quasi-permanent bond of relationship called marriage. Love is fragile and, as a basis for marriage, might well be questioned, even though romantic love has grown in ideology to the point it is today considered as almost the only legitimate basis for entering into marriage.

Let us examine the structure of romantic love. In its simplest form, "A loves B" suggests a dyadic relation. "A loves B" is directional, an emotional force directed from A to B. But it says nothing about B with respect to A. We all know that reciprocity of love is strongly desired when we are in love with someone else. Thus we hope that another relation will emerge, namely, "B loves A." When both dyadic relationships are constituted, then we have the basis for a triadic relationship binding two persons together. In fact, my wife and I once considered writing a book entitled *One Plus One Equals Three*, meaning you, me, and our relationship.

We see in this situation that a human relationship built on romantic love is necessarily very fragile. It is ultimately based on two distinct dyadic relations, and if either one of those relations fails, then the structure of the joined human relationship is in danger of failing also. In the real world the enduringness of a couple does not rest solely on the enduringness of romantic love. There are a great many other joint endeavors that the couple develops over time, to the point where romantic love may fade but the couple is bound together so strongly that it would never occur to them to part simply because they are no longer in the heat of romantic love.

This example is embryonic in development, but for students of love, I think it has great potential for developing a generic foundation that gives a deeper insight into the joys and tribulations of love than has heretofore occurred in the relevant literature.

Sexuality

Sexual intercourse may well be the best example of the combination of all three of Peirce's categories at maximum intensity in a life's experience. During the sex act the sensual qualities of sexuality constitute a complex and varied series of feeling states. The presentness of the other person is very intense and tends to fill the entire phenomenal world. The actual contact and resistance of bodies is an ideal example of secondness in operation. In addition, signs that accompany sexuality will, in the ideal state, consist of various symbolic operations, such as sentiments of love, intimacy, perceived reciprocity, recognition of the personhood of the other, and so forth. Sexual behavior is indeed a complex symbolic state, and perhaps that is what is meant when somebody says that "sex is mostly in your head."

The ideal of a sexual encounter clearly embodies all three categories, and in great intensity. Because our evolutionary success as creatures is tied to our reproductive ability, it is probably no accident that sexuality embodies all three categories at high levels of intensity. Many of the problems that people have with sex have to do with treating sex as only categories one and two, without sufficient symbolic accompaniments of category three. Without the thirdness, sexuality is merely another behavioral act, and expressions such as "I was used," "I was only a sexual object," "he did not treat me as a person," and the like reflect only firstness and secondness in the act. We implicitly expect to be treated as signs, or symbol creatures, and to be loved and cherished for our "symbolhood" as well as for our bodies. Perhaps this implicit assumption explains why most cultures are very uneasy about prostitution. By definition, it is an act of sexuality that includes firstness and secondness. That is, prostitution is a behavioral exchange for pay. The rare prostitute who understands the importance of the symbolic world of thirdness in the sexual act can make a fortune! In conclusion, it seems to me that theorizing about sexuality from either a psychological or philosophical approach can benefit greatly from considering it from the vantage point of Peirce's three categories.

Life-Style Variables

As illustrated by the above examples, using Peirce's categories from a psychological point of view requires considering the relative prominence of the three categories at any given point in the flow of life. Taking the categories in this way, a great many interpretations can be made, a few of which I illustrate briefly here. First, the study of individual differences in personality is a traditional area within psychology. One variable, first identified by Carl Jung many years ago, is called introversion-extroversion. It is presumed to be a basic divide in terms of people types, and there has been much research on this particular variable over the years. An extrovert is a person who tends to be oriented toward the world out there, the world of things, objects, and actions. We also have an informal notion of the extrovert as happy-go-lucky, but that is not a necessary part of the definition. In contrast, the introvert, who may be equally active in the social world, is a person who is more concerned with his or her ideas of that world. This person deals in concepts and is much more concerned with the relative fit of concepts to each other than with the mesh of actions and persons out there in the external world. While this category does not fit exactly, it seems relatively clear that an extrovert lives in and is more concerned with the world of secondness, whereas the introvert is more concerned with the world of thirdness. I suggest that this distinction in terms of secondness and thirdness could undergo profitable development in the study of personality variables.

As another example, the study of occupational choice is a long-standing tradition of counseling psychology. Career choices are some of the more important choices that people make, and yet the research and tradition of study of such choices came late and is still not as well respected, perhaps, as studying rats pressing bars in a Skinner box. But if people do have dispositions to prefer a life of relative firstness, secondness, or thirdness, then this dispositional difference should be taken into account in occupational counseling. There is no research or theorizing about Peirce's categories with respect to occupational choice, but again I suggest that it is an area ripe for development.

Another example might assume that we need balance in the experience of the three categories in ongoing life. For example, we know that a certain degree and variety of sensory stimulation is important for our welfare. About three decades ago there was a large amount of research on the effects of impoverished sensory environments. A subject would be isolated and the sensory environment would be reduced to an absolute minimum. After some period of time many subjects began to hallucinate, lose track of time, and have other experiential distortions. As I noted previously, occupations may differ in terms of the relative requirement for categorical mode. For example, professoring is in large part a life of thirdness, constantly working and dealing with symbols. In fact, after several years of professoring, it is actually quite difficult to empathize with the mode of life that deals with the world of secondness. For example, driving a truck or operating a jackhammer clearly involves relatively more secondness in life than does professoring. I have noted in my own life that if I am too long at a symbolic job I get bored, anomic, depressed, and I need to play. At the same time, when I have been too long at some physical task, such as digging in the garden or, in my early life, working in sawmills, my brain virtually screams for the stimulation of thinking. I suspect that the use of drugs, for example, may be related to an imbalance of experience. Living in a sensory world of presentness such as television, with its relative impoverishment of vigorous thinking, can lead to the desire for the kinds of feeling states that come with hallucinogenic drugs and perhaps cocaine. The world of poverty and the secondness of ghetto life would seem to dispose one more to a drug such as heroin. Too much thirdness to the point of boredom might dispose one to the minor highs of nicotine and caffeine as a way to obtain a sense of immediacy and contact with the physical world. But sometimes thirdness from the excitement of new ideas provides too much stimulation, and the brain needs slowing down. In that case, a stiff bourbon may be the drug of choice to quiet the noisy circuits of the mind.

Conclusion

In conclusion, I find Peirce to be the most seminal thinker of relevance for psychology that I have discovered in my twenty years of professoring as a psychologist. Carl Jung is one other great thinker who is underrated. I suspect there are deep and pervasive connections between Jung and Peirce, although on the surface, two more different thinkers could not be imagined. However, that will have to be another story for another time. I will close by saying that I now believe that if a sufficient number of psychologists would engage in a sustained study of Peirce's writings, in time that study would lead to a veritable revolution in the way we do psychology.

References

Cadwallader, Thomas C. 1974. Charles S. Peirce (1839–1914): The First American Experimental Psychologist. *Journal of the History of the Behavioral Sciences* 10:291–98.

Colapietro, Vincent M. 1989. *Peirce's Approach to the Self: A Semiotic Perspective on Human Subjectivity*. Albany: State University of New York Press.

Singer, Milton. 1984. *Man's Glassy Essence: Explorations in Semiotic Anthropology*. Bloomington: Indiana University Press.

Weick, Karl E. 1969. *The Social Psychology of Organizing*. Reading, Mass.: Addison-Wesley.

25 Peircean Benefits for Freudian Theory: The Role of Abduction in the Psychoanalytic Enterprise

Matthias Kettner, Frankfurt University, Germany

In a comprehensive logic of science, Peirce thought, a prominent place would have to be accorded to what he came to call "abduction," a mode of reasoning not captured by the inductive/deductive dichotomy as traditionally understood. It took a long time for Peirce's seminal ideas about abduction to be appreciated. With the waning of the once-widespread belief that the hypothetic-deductive account of scientific reasoning exhausted most of what was logically interesting about science, Peirce's pragmatic notion of scientific inquiry and his concept of abduction as reasonable hypothesis-formation made their late but momentous impact. Norwood Russell Hanson, in a series of papers starting in 1958, revived the quest for an abductive "logic of scientific discovery." The debate which ensued over Hanson's revival of Peircean ideas has served to bring out many difficulties which beset the concept of abduction. It has, however, also underlined the epistemic relevance of that concept in the philosophy of the natural sciences (see Hanson 1958, 1965; Cutting 1973; Harris and Hoover 1980; Nickles 1987). In what follows, I suggest that an account of abduction on Peircean lines can provide the centerpiece of a methodology of hermeneutical method and thus can figure prominently in a comprehensive logic of science able to include the cultural sciences. However, I will only have time to argue for that general point by taking psychoanalysis as my example of a cultural science. I begin in section I by discussing Peirce's concept of abduction; the aim of this discussion will be to introduce a concept of generalized Peircean abduction (GPA) and to highlight its connection with hermeneutical methodology in section II. I will then outline the role of GPA in psychoanalysis in section III; finally, I will show in section IV how an abduction account of psychoanalytic interpretation can meet an important charge made by Adolf Grünbaum in his philosophical critique of the foundations of psychoanalysis.

I

How can we intelligently look for that of which we are as yet ignorant? How do we rationally conjure up plausible new ideas in the face of problems which challenge our orthodox habits and expectations? These are the leading questions which fuel Peirce's various and somewhat divergent attempts at spelling out the rationality of hypothesis formation in the sciences (see Burks 1946; Fann 1970; Frankfurt 1958). According to Peirce, hypothesis formation in the sciences proceeds by way of a logically distinct mode of inference which he calls abduction. "All the ideas of science come to it by way of abduction. Abduction consists in studying facts and devising a theory to explain them" (*CP* 5.145). "Abduction is the process of forming explanatory hypotheses. It is the only logical operation which introduces any new idea" (*CP* 5.172). But what is the relationship between facts to be studied and an explanatory theory or hypothesis to be devised or accepted on probation? Are there logically interesting differences between, on the one hand, devising or accepting on probation some explanatory hypothesis and, on the other hand, confirming or testing it? And what is the scope of the originating character of abduction which Peirce emphasizes so much? If abduction is an inference, what is its canonical logical form, and how is it logically warranted? I will address these questions, starting with Peirce's specification of the canonical logical form of abduction and then pointing out some of its weaknesses. The point of the discussion will be to outline a refined concept of GPA.

The logical form of abduction, Peirce says, can be represented by the following schema of two premises and a conclusion:

The surprising fact *C* is observed.
But if *H* were true, *C* would be a matter of course.
Hence, there is reason to suspect that *H* is true. (see *CP* 5.189)

The gist of abduction, then, seems to consist in inferring the reasonableness of a hypothesis *H* from its ability to incorporate, into that which we rationally ought to expect, facts which we found surprising. Let us call this the surprise formula of abduction. Note, first, that according to the surprise formula, it is not the conceptual content of the hypothesis *H*, or *H*'s being true, that is inferred. Rather, what is being inferred is a plausibility warrant for *H*. The content of *H* must already be present in the second premise. So if *H* is a new idea, and if supposedly we get new ideas by abduction, the surprise formula of abduction fails to represent how these originate. What it does bring out, however, is a conceptual link in rational processes between the emergence of novelty and the demand for a prima facie, justified account of it. It captures the kind of reason giving that would ensue if a proponent of a would-be hypothesis was discursively challenged to defend the plausibility of his or her suggestion. For such a challenge to control the outcome of thinking, no actual discourse is necessary; merely the capacity of a self-critical thinker to anticipate it is required. The surprise formula, then, can be taken to represent a regulative idea, either operative in the imaginative processes at the intuitive end of hypothesis invention, by intrinsically constraining such invention, or operative on imaginable hypotheses already at hand by constraining rational choice between them. The surprise formula is indifferent as to whether explanations get introduced into the realm of the reasonable by selective-eliminative procedure among explanation candidates or by genuinely cooking them up from hitherto nonexplicit potentially explanatory resources.

What is the logical standing of the surprise formula? Taken as a purely formal principle, abduction does not seem to be a valid inference because any hypothesis *H*, meeting the demand expressed in the problematic second premise, would then allow the inference that *H* is reasonable. And this is absurd. Let the surprising fact be, for example, that a letter, which I presumed to have put on my hotel-room table the other day, has disappeared from the table. On purely formal grounds, the following three thoughts will indistinguishably satisfy the abductive schema:

> that I must have inadvertently swept it off the table;
>
> that someone from the hotel purloined the letter;
>
> that the letter underwent spontaneous combustion last night.

From each of those three thoughts, and in fact from a number of others, if true, my letter's absence from where I had previously placed it would deductively follow and would thus be made "a matter of course." But obviously, assuming normal background conditions, only the first abduced explanation (that I must have inadvertently swept the letter off the table) can be considered reasonable, whereas the second and the third are a far cry from being reasonable. Of course, by varying the relevant background conditions, we can think of possible worlds in which it would be a thoroughly reasonable first guess to abductively infer a theft or, more exotic, spontaneous combustion. However, that reasonableness is relative to relevant background conditions at all demonstrates that the validity of abduction depends on the specifiability of background assumptions relevant for the explanandum at hand. Such assumptions can be required by a formal principle, but they are not themselves formal principles. Hence, a refined notion of GPA will have to include (1) material background assumptions such that different background assumptions differ in the plausibility ordering which they impose on the logically possible explanations.

Peirce himself occasionally emphasized the dependence of abduction on a set of constraints regulating the choice of the explanatory hypothesis. One such principle which he defends on a priori grounds is that (2) we must not frame hypotheses which will subsequently block inquiry. Another constraint on GPA stems directly from Peirce's "pragmatic principle": (3) our hypotheses must have pragmatic meaning, that is, they must admit of being made evidentially sensitive to observations bearing on their content. Plainly, then, they must be open to suitable empirical confirmation. Finally, Peirce even included (4) considerations of scientific economy (verification costs, theoretical fertility) to constrain abduction.

GPA, then, can validly operate only against a background of normalcy expectations and constraints which Peirce's logical form of abduction fails to represent. Let us now consider another, but related, weakness of the surprise formula. Abduction, according to the surprise formula, looks very much like a directly data-driven inference from problem to explanatory conjecture. However, notwithstanding Peirce's nice metaphor, the facts by themselves do not at all "suggest an explanation." Rather, mere surprising facts hopelessly fail to determine suggestible explanations. Instead of mistakenly construing abduction as an inference from data to explanatory conjecture, then, we should conceive of GPA as an inference from phenomena plus given explanatory resources (e.g., theories) to corrected variants of such resources under the guidance of discrepant facts (cf. Blachowicz 1987). The pragmatic point of abduction does not so much consist in the creation of novelty for novelty's sake but in the preservation of coherence.[1] Discrepant facts strain explanatory resources which have proved fairly reliable so far but now have to be accommodated to new facts in order to recover overall coherence.

GPA, I suggest, represents such accommodation by inference to the most natural coherence-saving explanation. What will count as "most natural" will depend on features of the surprising fact and on the precise aspect under which it is found surprising. They will determine, or help to determine, which explanatory resources we should suitably tap, and these in turn will determine (or help to determine) material principles of coherence (i.e., principles stronger than logical consistency) and standards of evaluation for coherence saving. Let us call this feature of abduction its *domain specificity*. The domain specificity of abduction is what distinguishes abduction from formal logical principles (which hold in all possible worlds) and what marks off the respective interpretative processes of, for example, a detective, an astronomer, and a psychoanalyst as distinct, although all three abductively infer explanatory hypotheses.

Kepler's astronomical abduction that Jupiter's orbit might be of the noncircular type, for instance, employed other material principles of coherence and other standards of evaluation for coherence saving than, say, police-inspector Dupin guessing his way through the murder in the Rue Morgue, or Sigmund Freud's psychological abduction that neurotic symptoms may be conflicting expressions of unconscious phantasies, or, for that matter, my pedestrian abduction that my letter may incidentally have been swept to the ground.

II

I now outline briefly the relevance of generalized abduction to a comprehensive logic of science that would deal with methodologies championed in the natural sciences as well as with those found in the social sciences, without reductively assimilating social sciences to natural science.

From the domain specificity of generalized abduction, it follows that generalized abduction is relevant to any rational activity aiming at explanation. Generalized abduction does not privilege some particular kind of explanans or explanandum. For those who would claim, for example, that the only genuine form of explanation is deduction from laws of physical nature plus initial conditions, no support can be drawn from generalized abduction. Generalized abduction works in the domain of everyday action-explanation no less than in the domain of physics or, say, neurophysiology, for all the difference that exists between the explanatory resources available in each of these different disciplines. Traditionally, contrasts between the so-called natural sciences and the social or moral sciences have been wildly exaggerated. Such over-used dichotomies as "understanding versus explaining" or, for that matter, the more recent alleged dichotomy of "reasons versus causes" are now happily going out of business. One methodologically important contrast, however, has emerged which proves quite helpful for grouping together large sectors within the social sciences and setting them over against the rest of science. This is the contrast between nonhermeneutical and hermeneutical methods of data assessment. There are a number of important distinctions to be drawn here which, unfortunately, I must leave aside. Some clear-cut examples of the contrast I have in mind must do: take, for example, statistical macroeconomics and discourse analysis, or take genetic screening and Rorschach testing. Hermeneutics deals with inferring meanings and meaning connections. To the extent that inferable meanings and meaning connections figure in the specification of certain kinds of explanans vis-à-vis suitable explananda, hermeneutics can be part and parcel of an explanatory enterprise.

Now there are two potential contributions of GPA to a comprehensive philosophy of science. In the first place, inferring meanings and meaning connections has often been belittled as a purely intuitive, logically uninteresting, or, even worse, rationally unconstrained activity.[2] However, a good deal of the business of hermeneutics can be shown to conform to GPA. By employing a GPA frame for analytic purposes, interpretation as the ascription of meanings to some text or text analogue can be seen as a rational activity prompted by surprising discrepancies, constrained by relevant background assumptions, geared to knowledge about relevant explanatory resources in terms of meaning, and directed at rendering surprising facts a matter of course. Thus the first contribution of GPA is a general one: GPA can help us to account for interpretive explanation generally.

A second major contribution of GPA concerns specifically the explanation of a broad range of human behavior, owing to the fact that human beings are interpretants, that is, they themselves possess and exercise certain abductive capacities and act accordingly. For those inferring them, meanings are not causally ineffective. By specifying abductive capacities and dispositions and by attributing them to some person *S*, we can predict and explain that portion of the person's behavior in which inferred meanings play a causal role. This pattern of everyday explanation, explanation by abductive sets, in fact permeates our everyday interactions. If Peirce was right in describing perceptual judgments as abductions, the explanatory relevance of abductive sets goes right down even to the level of the perceptual behavior we expect people to exhibit, for instance, when we reasonably expect that someone will come to believe that a telephone is ringing upon hearing the sound of a telephone ring. Thus GPA provides the centerpiece for an account of interpretative explanation of self-interpreters.

III

Let us now look specifically at psychoanalysis and what GPA can do for it. The methodological core of psychoanalysis is psychoanalytic interpretation. (The outworn distinction between the clinical and the metapsychological, on which many philosophical discussions of psychoanalysis thrive, is of secondary importance. Freud believed that the clinical practice of psychoanalysis is but one application of the interpretative method—here, specifically, to benefit neurotic patients; and he also believed that metapsychology is a theory which explains how that method can, in fact, be successfully so applied.)

Many psychoanalysts have fallen prey to a comfortable intuitionist rhetoric which depicts their interpretative business as simply "listening with the third ear." Aside from the narcissistic gains of having a third ear where others only have two, this rhetoric is detrimental to the rational credentials of psychoanalysis.

Anybody who studies consensus formation within the profession, especially supervision and control-analysis, will notice that the interpretations of psychoanalysts, far from being intramentally insulated, are susceptible to rational criticism from the start (see Fleming and Benedek 1966; Hess 1980; Klüwer 1970, 1971; Kettner and Schulz 1985; Thomä et al. 1967). Psychoanalytic interpretations are proposed on grounds which can be intersubjectively evaluated and criticized. By employing a GPA framework, patterns of discovery could be discerned and rational criticism of such patterns could be extended to and supplemented by external experts, thus increasing the rational transparency of the key psychoanalytic method. Any attempt to assimilate psychoanalysis to some natural science would be well advised first to take into account the importance which Freud himself attached to the hermeneutical component of the psychoanalytic enterprise. Most metapsychological statements, it is true, make causal claims apparently similar to those made in the natural sciences. Such metapsychological claims (e.g., that paranoia has a homosexuality etiology), however, are really high-level abductions generalized in methodologically not very convincing ways from hermeneutically generated low-level abductions. Hermeneutically generated low-level abductions are methodologically at the very foundations of psychoanalysis. And here analogies with the natural sciences should no longer appear illuminating, because observation statements in the natural sciences, although being abductions, are not hermeneutically generated abductions.[3]

A look at such low-level abductions reveals that psychoanalytic interpretations are conceptually continuous with folk psychology.[4] That is to say, such interpretations do not type psychological states or episodes in terms of neurophysiological, syntactic, or other "scientific" properties but, rather, in terms of content, as specified in the rich intentional vernacular of our commonsense conception of the mental. They do not replace but rather presuppose and extend that conception. When Freud, in his 1915 paper entitled "The Unconscious," rehearsed his reasons for introducing the notion of unconscious, yet intentional, mental states into psychology, he emphasized that "all the categories which we employ to describe conscious mental acts, such as ideas, purposes, resolutions and so on, can be applied to them. . . . Thus we shall not hesitate . . . to deal with them in the most intimate connection with conscious mental acts" (Freud 1953–:14:168). Other of Freud's remarks that can be taken as remarks on abductive rules highlight the conceptual continuity with folk psychology from a different angle: psychoanalytic interpretations use folk psychology to exhibit the surprising facts which interpretations have to address and account for.

According to Freud (1953–:2:293), if

> we examine with a critical eye the account that the patient has given us without much trouble or resistance, we shall infallibly discover gaps and imperfections in it. . . . [The] train of thought will be visibly interrupted and patched up by the patient as best he may, with a turn of speech or an inadequate explanation; at another point we come upon a motive which would have to be described as a feeble one in a normal person. The patient will not recognize these deficiencies when his attention is drawn to them. . . . [We] make the same demands for logical connection and sufficient motivation in a train of thought, even if it extends into the unconscious, from a hysterical patient as we should from a normal individual.

What, then, does distinguish the meta-abductions of the analyst and the folk-psychological abductions of the person analyzed? What do they purport to explain? Psychoanalytic interpretations do not strive to explain how something works (e.g., the refrigerator maintaining a certain low temperature) or how someone manages to do something (memorizing 250 telephone numbers); rather, they typically explain why some person

S does something *A*, by ascribing to *S*'s doing *A* some latent meaning not obvious to *S*, and by claiming that the latent meaning of *S*'s doing *A* together with the fact of its not being obvious to *S* is, or has been, causally relevant for *S*'s doing *A* (or for the events which constitute *S*'s doing *A*).[5] Such explanations crucially involve reference to idiosyncrasies in someone's making sense of his or her own behavior; that is, they crucially involve reference to a person's abductive sets. At a high level of abstraction from individual episodes of a person's exercising his or her abductive sets, the explanations by abductive sets that psychoanalysts are particularly interested in will usually be found to conform to functional explanation: when all the rich details are dropped from psychoanalytic interpretations, one is typically left with "defense" in order to avoid "hidden feelings" because of the "anxiety" they would create or have created once.[6]

IV

In the concluding section of this paper, I show how an abductive account can meet Adolf Grünbaum's critique of psychoanalytic interpretation, especially Grünbaum's charge of "thematic affinity fallacy." In his powerful attack on the foundations of psychoanalysis, Grünbaum (1983) has claimed, among other things, that the clinical method of psychoanalytic investigation (the method of free association in therapy) altogether lacks the epistemic resources necessary for validating causal claims about the psychodynamics of unconscious mental states. Now, for some of Freud's etiological hypotheses, linking certain experiences as pathogens (traumatic experiences) to certain nosological entities (neuroses), I think that Grünbaum's diagnosis is exactly right, and so is his conclusion, namely, that therapy cannot be a self-sufficient research context but must be supplanted by well-designed, extraclinical studies.

Take Freud's case of the Rat Man (Freud 1953–:10:153–249). That a certain experience of the child's being punished by his father should have been the "original pathogen" of the Rat Man's later compulsory neurosis and is the original cause for the onset of the Rat Man's rat obsession is what might be called a macroformat etiological hypothesis.

Macroformat hypotheses such as Freud's hypothesis concerning the Rat Man, if they lend themselves at all to confirmation, surely cannot be confirmed in any strong sense merely by first-person recollections of a person's distant past. So far, so good. But Grünbaum overstates his case when he suggests that there is no sound way in which the patient's free associations can be used to validate causal inferences.[7] Associations are used in the clinical setting precisely in order to interpret them. They are used in that way by the analyst and, with the help of the analyst, by the patient, too. Grünbaum's charge, then, suggests that psychoanalytic interpretations in general cannot epistemically live up to the causal claims they contain.

Now, bearing in mind that psychoanalytic interpretations are a rather variegated lot, this seems an overly bold claim and thus plainly false. Many of the interpretations psychoanalysts give are no more daring as regards their causal import, and no more demanding in what would count as proper validation of their causal import, than much of what passes without question as psychological explanation in everyday life.

I conclude, then, that either Grünbaum will have to refine his charges of confirmational incapacity of the method of free association, or his global charge of confirmational incapacity will carry over to a good deal of our entrenched commonsense practices of psychological action explanation.[8] But if the canons of inductive confirmation upheld in the natural sciences, when transposed to important practices of psychological action explanation, can make such practices appear vacuous, then surely those canons cannot be the right sort of canons to be so transposed. Or one would have to claim that the canons of inductive confirmation which are acknowledged in the natural sciences are the sole support for any and every causal hypothesis concerning whatever domain of phenomena is under investigation.

One can doubt the latter assumption without getting into the fragmenting of the concept of causality. An argument challenging the above assumption could start from this observation: It is dubitable that the concept of independently observable occurrences of events and the concept of measurable probabilities of such occurrences—both concepts which serve to make the concept of causality operational in certain domains—are suitable to be applied to every domain of phenomena. Interpretations refer to intentional phenomena. To illustrate the kind of problems that this raises, consider a simple example: That a person *S* has an intention *i* (to do *A*) is a fact about *S* which (at least for certain sorts of *i*) cannot be ascertained by observation (of *S*'s behavior). And *S*'s acquiring *i* cannot (at least not for every *i*) be treated as the occurrence of an event independent of the account *S* could give of how *S* personally thinks *i* was acquired.

An important lesson from this is that a classification should be developed which would class any psychoanalytic interpretation according to the confirmational burden which it incurs as a function of the validation standing of the premises which go into the abduction of that interpretation.

Grünbaum substantiates his criticism of psychoanalytic interpretation by focusing on what he labeled the fallacy of thematic affinity. Committing the thematic-affinity fallacy, according to Grünbaum, consists in inferring causal links from meaning connections. More precisely, the fallacy consists in, first, interpretatively assigning similar meanings to two interpretants and then inferring, on the strength of the meaning connection thus established, that (the occurrence of) the (temporally) first interpretant is causally relevant for (the occurrence of) the other interpretant, or else inferring that both have a common cause. Grünbaum argues that similarity of meaning alone is neither a necessary nor a sufficient condition for there being underlying causal connections, though similarity of meaning does not preclude there being such connections. According to Grünbaum, when the analyst observes meaning similarities cropping up in the free-association discourse of a patient, the analyst seems no better off than someone who observed, for example, that Thomas Jefferson and John Adams both died at a strikingly similar time "on the 50th anniversary of the Declaration of Independence, written by Jefferson with the aid of a draft to which Adams had contributed" (Grünbaum 1988:149). Rich correspondences between the two events "do not attest to the two deaths being causally connected" (ibid.). To think otherwise would be to commit the meaning-affinity fallacy.

The Jefferson-Adams example seems right. However, it seriously misrepresents what analysts do when they interpret meaning kinship. In cases where it is the patient who assigns meanings and meaning connections by interpreting his or her experiences, the analyst can safely infer that there is an abductive set operative in the patient's ongoing self-interpretative activity which causes the patient to find them significantly related. (The natural history of such sets, their retrospective and prospective scope of their operation, can then be further investigated by the analyst.) The analyst need not base causal inferences on the strength of meaning affinities alone, as long as an additional premise is available that meanings and their affinities are causally embedded in interpretative processes belonging to a single causally connected mental life, namely, that of the person analyzed. Usually, this premise is available.

The causal embedding may be of the following type: If a person *S*'s experiences *E1* (of event *e1*) and *E2* (of event *e2*) appear to *S* to share thematic meaning, then there is some abductive set operative in *S* which causes *S* to assign some meaning *M*, such that *S* assigns *M* to *E1* and also to *E2*. *M* is what makes *E1* and *E2* appear (at least to *S*) as thematic variants of each other; and *M* is produced by the abductive sets which govern *S*'s interpretative activity. Thematic affinity as apparent to *S* across different experiences derives from a common source, namely, *S*'s interpretative activity.

Thematic affinity of the type outlined justifies the search for an indirect causal connection between the two meaningful experiences, such that each of them is a partial effect of a shared cause. The shared cause is *S*'s interpretative activity working in certain ways (namely, under certain abductive sets). This is not to say that the first event *e1* causes the (similar) second event *e2*, nor that the first experience *E1* causes the (similar) experience *E2*. Rather, it is to say that *S*'s interpretative activity plus *S*'s relevant abductive sets cause *S*'s finding *E1* and *E2* meaningfully related. This is compatible with Freud's view that a person's later experience can determine the meaning of an earlier experience (*Nachträglichkeit*). In order for that to be possible, one should make the assumption that *S*'s present interpretative activity operates on experiences of present events as well as on recollections of past events. Furthermore, Freud held the stronger view that a recollection of a past event (an event experienced at *t1*) can be caused at *t2* to become traumatic (i.e., to have symptom-forming causal powers) in virtue of the fact that from *S*'s interpretative activity at *t2* there derives a thematic affinity between the recollection and some meaningful experience at *t2*. This latter view depends on further assumptions about the causal powers that meaningful recollections can unfold.

However, there are other cases in which it is the analyst who finds experiences significantly related, experiences which the patient as yet fails to find so related. Usually, the analyst will attribute to the patient an abductive set sufficient to square with his or her own; then the analyst will abduce unconscious constraints which, if they were effective in the patient, would make the patient's failing to see significant relations a matter of course. Such inferences neither are, nor are meant to be, confirmations. They are abductions to unconsciously operating abductive sets of the patient which explain the patient's selective handling of potential meaning.

As an illustration of thematic affinities in free association discourse and a psychoanalytic interpretation based on the observation of thematic affinity, consider the following example by Theodor Reik (1964:133). A teacher could not maintain discipline in her classroom, and the pupils were often defiantly riotous. Although she understood intellectually by what means she could assert her authority, she was unable to do it. When she once, in an analytic session, spoke of the necessity of self-discipline, some childhood memories emerged showing that she had been for a long time a rebellious and unmanageable child. Her difficulty now is particularly related to her unconscious identification with her pupils.

The analyst's reasoning in this case could be represented by a pattern of abductive inference based on thematic affinity. I present two versions of that pattern, both of which I think are valid. Here is the first version:

> *Abductive premise*: It is surprising that *S*'s (= the woman's) behavior manifests features *F1*, *F2*, . . . , *Fn*.
>
> *Abductive premise*: It is unlikely (= not to be expected) that S's behavior should show so many features *F1*, *F2*, . . . , *Fn* that could derive from *M* without being so derived.
>
> *Abductive conclusion*: That *M* is causally relevant for *S*'s behavior is a reasonable hypothesis.

The second version expresses the same idea with reference to contexts as well as to features of action:

> It is unlikely (= not to be expected) that *S*'s behavior should show so many features *F1*, *F2*, . . . , *Fn* that are surprising in context *C1*, but which would be expected in context *C2*, without *C2* being causally involved (by way of a causally effective representation of *C2*, e.g., from *S*'s memory of *C2*) in *S*'s behavior in context *C1*.

The examination of inferential patterns such as these is crucial in order to understand better the rationality of interpretative discovery. Such an examination, however, would exceed the limits of this chapter.

Notes

I thank Ken Westphal and Carole A. Sheade for valuable comments. A modified version of this chapter can be found in *Semiotic Perspectives on Clinical Theory and Practice: Medicine, Neuropsychiatry, and Psychoanalysis*, ed. P. S. Epstein and B. E. Litowitz (New York: Mouton, 1990).

1. In the perspective of the history of science, preservation of coherence can require breaking with an entrenched paradigm (in Kuhn's sense) and creating a new paradigm; it is not essentially a conservative affair. In the perspective of developmental psychology, it may turn out that the concept of abduction can help to clarify Piaget's notion of assimilation and accommodation of cognitive schemas.

2. Interestingly, many antihermeneutical attacks on the methodological credentials of the operation called *Verstehen* (see Abel 1957) echo the misgivings of logical positivism about the alleged rationality of discovery.

3. The contrast I draw here between hermeneutically generated abductions and abductions generated in nonhermeneutical ways stands badly in need of elaboration. At least the following points in conjunction would have to be addressed: The input for hermeneutical abduction are data (1) that admit of being conceived as signs, or sign-analogues, conveying nonnatural meaning; (2) that are products of human actors; (3) that are subject to diachronically operative rules that determine whether such products are well formed or ill formed (under intersubjectively judgeable aspects); and (4) that can, in principle, be self-ascribed under action concepts by their producers. Concerning point (3), see Argelander 1981, 1982; Oevermann 1979; Teller and Dahl 1986; Dahl 1988. Concerning point (4), see Schafer 1976.

4. Excellent examples of low-level abductions pertaining to the praxis of psychoanalysis can be found in Malan 1979.

5. How someone manages to do something (e.g., get fired every time a job begins to demand a high degree of responsibility) and answers to this sort of question (e.g., by starting a quarrel with the boss) can, of course, figure in psychoanalytic interpretations. But they do not constitute such interpretations. And although a lot of what Freud has to say about the functioning of the psychic apparatus conforms to the "how does it work?" type of question, such hypotheses are not interpretations but pertain to metapsychology.

6. See Malan's (1979:79–80) notion of a "triangle of defence" and a "triangle of transference."

7. "Like Freud, the vast majority of his followers continue to maintain that free associations are crucial for the validation of both their etiologic hypotheses about the causes of neuroses and the psychoanalytic theory of dreams and slips. These advocates cannot dispense with a therapeutic foundation. *For my own part, however, I find it unwarranted to use free association to validate causal inferences*" (Grünbaum 1986a:224, emphasis added).

8. In more recent publications, Grünbaum has begun to differentiate psychodynamic hypotheses with regard to their confirmability through the confirmational resources of free association. Grünbaum seems to admit this. For example, he says (1988:137) that an "analyst's inference that a particular teenage female patient's low self-esteem—as manifested by her expectation of contempt from him—was caused by her life-long awareness of her father's vilifications of her" could have "much better" evidential warrants than Freud's macroformat etiological hypothesis in the case of the Rat Man (see also Holt 1986 and Grünbaum 1986b:276). Grünbaum then goes on to suggest that a line ought to be drawn between evidentially warranted and evidentially unwarranted hypotheses according to whether such hypotheses employ the concept of repression. (Note that in the commonsense interpretation cited above, it is awareness of x, not repression of x, that is taken to have a causal effect.) I doubt that the concept of repression can serve to draw such a line; the psychoanalytic concept of repression is rooted in commonsense ("folk") psychology.

References

Abel, T. 1953. The Operation Called "Verstehen." In *Readings in the Philosophy of Science*, ed. H. Feigl and M. Brodbeck, 677–87. New York: Appleton-Century-Crofts.

Argelander, H. 1981. Was ist eine Deutung? *Psyche* 35:999–1005.

———. 1982. Textstruktur und Interpretation. *Psyche* 36:700–25.

Blachowicz, J. 1987. Discovery as Correction. *Synthese* 71:235–321.

Burks, A. W. 1946. Peirce's Theory of Abduction. *Philosophy of Science* 13:301–6.

Cutting, C. 1973. A Defense of the Logic of Discovery. *Philosophical Forum* 4:384–405.

Dahl, H. 1988. Frames of Mind. In *Psychoanalytic Process Research Strategies*, ed. H. Dahl, H. Kächele, and H. Thomä, 51–66. Berlin: Springer-Verlag.

Fann, K. T. 1970. *Peirce's Theory of Abduction*. The Hague: Martinus Nijhoff.

Fleming, J., and Benedek, T. 1966. *Psychoanalytic Supervision*. New York: Grune and Stratton.

Frankfurt, H. G. 1958. Peirce's Notion of Abduction. *Journal of Philosophy* 55:593–97.

Freud, S. 1953–. *Standard Edition of the Complete Psychological Works of Sigmund Freud*. Ed. J. Strachey et al. London: Hogarth Press.

Grünbaum, A. 1983. *The Foundations of Psychoanalysis: A Philosophical Critique*. Berkeley: University of California Press.

———. 1986a. Précis of The Foundations of Psychoanalysis: A Philosophical Critique. *Behavioral and Brain Sciences* 9:217–28.

———. 1986b. Author's Response [to Holt]. *Behavioral and Brain Sciences* 9 (2):266–84.

———. 1988. The Role of the Case Study Method. In *Die Philosophen und Freud*, ed. H. Vetter and L. Nagl, 134–75. Vienna: Oldenbourg.

Hanson, N. R. 1958. The Logic of Discovery. *Journal of Philosophy* 55 (25):1073–89.

———. 1965. Notes toward a Logic of Discovery. In *Perspectives on Peirce*, ed. R. J. Bernstein, 42–65. New Haven: Yale University Press.

Harris, J. F., and Hoover, K. D. 1980. Abduction and the New Riddle of Induction. *Monist* 63:329–41.

Hess, A. K. 1980. *Psychotherapy Supervision: Theory, Research, and Practice*. New York: John Wiley and Sons.

Holt, R. 1986. Some Reflections on Testing Psychoanalytic Hypotheses. *Behavioral and Brain Sciences* 9 (2):242–44.

Kettner, M., and Schulz, P. C. 1985. Untersuchungen zur Konstitution fokaler Deutungsmuster anhand von Verbatim-Supervisionsprotokollen. In *Unveröffentliche Diplomarbeit*. Frankfurt: Institut für Psychologie der J. W. Goethe-Universität.

Klüwer, R. 1970. Über die Orientierungsfunktion eines Fokus bei der psychoanalytischen Kurztherapie. *Psyche* 10:739–55.

———. 1971. Erfahrungen mit der psychoanalytischen Fokaltherapie. *Psyche* 12:932–47.

Malan, D. H. 1979. *Individual Psychotherapy and the Science of Psychodynamics*. London: Butterworth.

Nickles, T. 1987. Methodology, Heuristics, and Rationality. In *Rational Changes in Science*, ed. J. C. Pitt and M. Pera, 103–32. Dordrecht: D. Reidel.

Oevermann, U. 1979. Die Methodologie einer "objektiven Hermeneutik" und ihre allgemeine forschungslogische Bedeutung für die Sozialwissenschaften. In *Interpretative Verfahren in den Sozial- und Textwissenschaften*, ed. H. G. Soeffner, 352–434. Stuttgart: Metzeler.

Reik, T. 1964. *Voices from the Inaudible: The Patients Speak*. New York: Farrar and Strauss.

Schafer, R. 1976. *A New Language for Psychoanalysis*. New Haven: Yale University Press.

Teller, V., and Dahl, H. 1986. The Microstructure of Free Association. *Journal of the American Psychoanalytical Association* 34:763–98.

Thomä, H., et al. 1967. Über die Validierung psychoanalytischer Theorien durch die Untersuchung von Deutungsaktionen. *Psyche* 9:664–92.

26 The Valuation of the Interpretant

James Jakób Liszka, The University of Alaska, Anchorage

The notion of the interpretant is certainly one of the more important contributions to the theory of signs in recent times. Left wanting in its characterization, however, is its relation to value and, in general, the relation between valuation and signification. In this chapter I explore this relation, using some recent linguistic theory produced by close subscribers to Peircean semiotic.

According to Peirce, a sign is a sign in virtue of establishing a triadic relation among itself, its referent, and what Peirce calls its interpretant (*CP* 5.484). An interpretant in its singular sense is any particular sign which can be developed or inferred from some previous sign, that is, what Peirce called *translated* (*CP* 4.127; 5.284, 509, 594). In its most general and comprehensive sense, the interpretant is a habit or rule of such sign translation, one which establishes a mode of interpretation for some interpreting agency or system (*PW* 110; *CP* 5.476). The interpretant embodies the triadic aspect of the sign relation, for it coordinates, on the one hand, the sign's ability to refer to an object, with, on the other, the sign's intension—the sign's relation to other signs within its indigenous system.

In the early development of the concept of the interpretant, Peirce emphasized its function as the synthesizer or coordinator of sense and reference, breadth and depth (*W* 1:274, 333, 465; 3.89; *CP* 2.419). Peirce called the result of this coordination *information* (*CP* 3.608), or the quantity of the interpretant. The term "information" hides a certain useful ambiguity. On the one hand, it suggests that meaning accrues through structuring, that is, giving form to nonmeaningful differences. It is in-formation in that the differences which organize a system of signs become meaningful for an interpreting agency through the mediation of a referent, that is, the act of anchoring the system in something outside of itself. At the same time, the referent receives meaning through its coordination within another system of signs. On the other hand, the effect of such coordination creates information in the normal sense of creating a set of meaningful messages for some interpreting agency—much as light-wave lengths are translated as chemical messages in the optic nerve. We learn something about the referent when it is translated into a system of signs which simultaneously reorganizes the referent in a code made in its own image, a reorganization guided by referent qua dynamic object. As a recoding in the intensional system, it is what Peirce called an *immediate* object; as a cause or influence of its recoding, it is a *dynamic* object. However, neither of these is what Peirce should have called the *final* object, which is the referent of the end product of indefinite sign translation.

The interpretant of a traffic light, for example, coordinates the set of vehicular operations associated with stopping a car and the acceleration of the vehicle, on the one hand, and red and green lights, on the other. Red and green as part of a color system stand in a certain relation of differences defined physically, physiologically, psychologically, and culturally. Physically, red and green have different wave lengths; physiologically, red is the most salient color to the human eye, and green is consubstantial with red; psychologically, red is said to have an irritating, excitatory effect in human perceivers, while green a soothing one; culturally, red and green have various conventional and traditional employments as well. Stopping and accelerating a vehicle are operations which can also be articulated as a set of differences. The interpretant of the traffic light correlates a set of differences among signs in one system with that of another. The final interpretant is the rule as found in the books of the state; the ultimate interpretant is the habit or rule of action that is established in drivers and exemplified by the disposition to stop at red lights and go at green ones. Prior to this correlation, the difference between red and green had a "sense," a set of intensional differences, so to speak, but was meaningless from the perspective of traffic control. The color differences exemplify a system of differences without being anchored in a meaning-giving correlation. On the other hand, "Stopping" and "Going," now, are more than just vehicular operations. They are now reorganized by means of the interpretant of the traffic light. They acquire a different meaning—or better, their sense is transformed into a meaning by means of the interpretant of the traffic light.

The effect of the (final) interpretant, as a rule of sign translation, is the transformation from what might be called sense, that is, the apprehension of difference as potentially meaningful (what Peirce calls the immediate interpretant) to meaning (the regularity of a correlation between sets of differences), by means of some actual correlation (the dynamic interpretant). Sign translation in this case has a much different import than, for example, in W. V. Quine. For Quine, translation is a semiotic operation of establishing equivalences or synonyms and is indeterminate because of the inscrutability of reference. Peirce—although he was not entirely consistent on this (cf. *W* 1:464–65)—probably had in mind the concept of chemical valency, which attracted him and influenced him in the development of his theory of relations. In this case, equivalence is not so much synonymy as it is the idea of chemical bond, the process by which a structure becomes organized through a combining of elements. Peirce, it appears, would have agreed with Quine that the indeterminacy of translation is due to the vagueness of reference, but for Peirce, it is also due to the vagueness of intension. However, unlike Quine, Peirce saw the indeterminacy of translation with a positive prognosis. The indeterminacy of translation is a prompt, a motivation for determination. The indeterminacy of translation allows a latitude of interpretation, ending in some indefinite progression with zero latitude, also known as consensus. Short of that, the speech community can expect relative determinacy framed by the purposes of the community of inquirers (*CP* 5.448 n. 1, 506; cf. Brock 1979:44).

In consequence, Peirce's theory of sign translation was consistent with his peculiarly Victorian outlook on progress, where semiosis is an asymptotic process by which a referent conjoins with a sign to enhance or create its meaning. To speak loosely here, the dynamic object finds its meaning in a sign-encoded immediate object which receives further elaboration, becoming finalized at the end of sign translation. Sense becomes meaning. The essential feature of the interpretant or interpretants is the function of continuing a translation of a sign which serves to make the sign more determinate, the sign agencies more informed about its object. The result of understanding semiosis as sign translation is that the real is not found in exfoliating signification from the object of signification but in the arrangement of that very foliage: "the highest grade of reality is only reached by signs" (*NEM* 4:23).

In this respect there is an indirect image of Peirce's theory of sign translation in the work of Lady Victoria Welby, an English semiotician, which more or less confirms this view. After receiving several essays from Welby in 1909, Peirce cryptically remarked, "Word translating seems to me to contain profound truth wrapped up in it" (*PW* 111). By translation, Welby meant an analogical process by which the meaning of one sign, generally understood, was enhanced by its framing within another sign. In this case, such translation is a process of passing from what Welby called sense to meaning to significance, "a raising of our whole concept of meaning to a higher and more efficient level" (Welby 1983b:161). After all, for Peirce, "the highest kind of symbol is one which signifies growth, self-development, of thought" (*CP* 4.9).

The problem that I wish to discuss is that of the relation between value and the interpretant. Peirce gave only a vague and general characterization of value, and his view has to be culled from a number of disparate passages. First, the characterization of sign translation just presented suggests that it is not simply interpretation for the sake of interpretation or a kind of disjointed hermeneutic which develops several simultaneous paths of interpretation. Rather, sign translation has an ordered direction which serves to enhance the meaning of a sign and broaden the comprehension of interpreting agencies. Peirce said this nicely:

> The meaning of a proposition is itself a proposition. Indeed, it is no other than the very proposition of which it is the meaning: it is a translation of it. But of the myriads of forms into which a proposition may be translated, what is that one which is to be called its very meaning? It is, according to the pragmaticist, that form in which the proposition becomes applicable to human conduct . . . that form which is most directly applicable to self-control. (*CP* 5.427)

To put it succinctly, the purpose of sign translation is to bring about self-control. Self-control exhibits a number of interests. It is simultaneously a critical interest in reasonableness and rationality, conjoined with a hermeneutical interest—the ability to determine one's habits or rules of action—and it has a technical interest, the ability to predict the future by means of the sign's behavior (that is, the sign's interpretants). Sign translation is the means of enlightenment, community, and power.

Where is value located in this equation? Peirce remarked: "Let us note that meaning is something allied in its nature to value. I do not know whether we ought rather to say that meaning is the value of a word—a phrase often used—or whether we ought to say that the value of anything to us is what it means for us—which we also sometimes hear said. Suffice it to say that the two ideas are near together" (*MS* 599:24).

If, as Peirce claimed, "the value of a symbol is that it seems to make thought and conduct rational and enables us to predict the future" (*CP* 4.448) and "a symbol is essentially a purpose" (*CP* 4.261), then the equation among meaning, value, and purpose is complete.

As to the ultimate purpose of thought or sign translation,

> which must be the purpose of everything, it is beyond human apprehension; but according to the stage of approach which my thought has made to it . . . it is by the indefinite replication of self-control upon self-control. . . . This ideal, by modifying the rules of self-control modifies action, and so experience too—both the man's own and that of others, and this centrifugal movement thus rebounds in a new centripetal movement, and so on. (*CP* 5.402 n. 3)

Or, as he said elsewhere:

> The meaning and essence of every conception lies in the application that is to be made of it. That is all very well, when properly understood. . . . But the question arises, *what is* the ultimate application. . . . Subsequent experience of life has taught me that the only thing that is really desirable without a reason for being so, is to render ideas and things reasonable. One cannot well demand a reason for reasonableness itself. (Cohen 1956:332)

The ultimate ideal is thus concrete reasonableness.

This equation of meaning, value, and purpose makes sense under a traditional view of purpose. Purpose entails a goal or an end, and the end of an action is that which engenders purpose in an agent. The end of the action is valuable for that agent and generally provides the reason and motivation for the action. Thus, ideals, ends, and values are coincident. Given Peirce's schema of normative sciences, then, we would not be surprised to find out that aesthetics, the science of ends, is also the science of value: "As to valuation, in 1903 I delivered a course of lectures in Boston . . . in which I explained at length how reasoning was analogous—and in fact, a particular case of moral self-control—how logic ought to be founded on Ethics and Ethics on a transfigured Esthetics which would be the science of values, although now wrongly treated as a part of Ethics" (letter to Schiller, 12 May 1905).

Peirce, however, is well known for apologizing about his lack of knowledge of aesthetics (*MS* 310:7; 683:15). Ironically, the most important normative science is treated little by Peirce, and the concept of value even less.

But even if Peirce had developed his views concerning ultimate purpose in more detail, I think we would still have demanded that he devote as much analysis to the concept of value, and what he called axiogastics, as he did to his logical notions, especially the relation between the interpretant as the core of sign translation and value. This seems to be one area where Peirce's semiotic needs to be developed.

As it turns out, there are two possible directions, both of which have a close tie to the semiotic of Peirce. One is found in the work of Charles Morris, in his study of the relation between sign and value; the other can be found in a certain structuralist reading of Peirce.

My purpose here, however, is not to explicate Morris's theory of value. Essentially, Morris advocates a theory of value as preference, something which would certainly find much approval today, but of which Peirce would probably not approve (cf. *MS* 310:168). Generally, however, Morris is criticized for his behaviorist reading of Peirce and some other general misinterpretations (see Rochberg-Halton and McMurtrey 1983; Dewey 1946). But in addition to these complaints, my lack of enthusiasm for Morris's theory is based on the rather artificial imposition of a preformulated theory of value on a preformulated theory of signs—which is the impression his *Signification and Significance* leaves on this reader. I see, in contrast, in the work of Roman Jakobson, Michael Shapiro, and Henning Andersen—in their direct or indirect use of Peirce—a much more organic relation between sign and value, a theory of value generated from a study of signs, although mostly restricted to the study of language.

One could give a brief synopsis of this idea in the following way: Jakobson reads the interpretant as a rule of sign translation, establishing a correlation among levels of language and between language and its referents and, through such correlation, producing meaning (see Liszka 1981, 1989). What Jakobson calls markedness is an important feature of these correlations. Henning Andersen reinforces the importance of the interpretant, correlation, and markedness, while adding the concept of rank to this list. Shapiro reinforces all these claims and shows how the interpretant could be likened to Saussure's notion of value (cf. Liszka 1989). I have tried to show how these various insights can be organized into a coherent account of the relation between sign and value—what I call *transvaluation*.

We might begin with a brief analysis of Saussure's notion of value. Saussure (1966:117) makes a distinction between signification and value. Signification, as the correlation between a specific signans and signatum, can be understood, and in fact has its possibility, only in the value of the sign—"without 'value,'" Saussure (ibid.) wrote, "signification would not exist." As Saussure says in his own words, the "ultimate law of language is that nothing can reside in a single term. Since signs are correlated to what they designate, a sign cannot designate without understanding its relation to the other signs with which it forms a system" (ibid.:63). Barthes (1964:54) thus argued that Saussure increasingly concentrated on the notion of value, which eventually became more important than that of signification.

Saussure uses an economic model to explain his notion of value as it applies to language, which, as Roy Harris emphasizes, has an advantage over the geometric model (value as the cut through two amorphous substances), in that it assumes a coordination between two already-organized systems. In the economic model, value is the means of coordination, rather than a principle of organization (Harris 1987:220n). According to this economic model, values are always composed of (1) a dissimilar thing that can be exchanged for the thing of which the value is to be determined and (2) similar things that can be compared with the thing of which the value is to be determined (Saussure 1966:115). To determine what a five-franc piece, for example, is worth, one must know that it can be exchanged for a fixed quantity, such as bread, and that it can be compared with a similar value of the same system, for example, with a one-franc piece. By analogy, a word can be exchanged for something dissimilar, an idea; besides, it can be compared with other words, and its value relies on the comparison with similar values, with other words that stand in opposition to it.

I venture to say that in this respect the ability of the word to be exchanged for something dissimilar to itself, an idea or object, is comparable to the notion of reference, while the comparison of a sign within a system of signs is equivalent to the notion of sense. Saussure emphasized that both factors—the ability of the sign to be exchanged and its ability to be compared—"are necessary for the existence of value" (ibid.). The value of sign is not simply its relation with other signs in its system but its relation with signs or objects outside of its system as well. The value of the sign, then, involves both its referential and its connotative aspects, so that a signifier refers to a signified only through the mediation of the system of signifiers, while the system of signifiers coalesces into *meaningful* differences by means of its referential use. Consequently, to the extent that value is the coordination of relations between signata and signantia and among signantia (and the interpretant is the coordination of sense and reference of a sign), then value and the interpretant are family-related notions. In the end, for Saussure, "a linguistic system is a series of differences of sounds combined with a series of differences of ideas; but the pairing of a certain number of acoustical signs with as many cuts made from the mass of thought engenders a system of values; and this system serves as an effective link between the phonic and psychological elements within each sign" (Saussure 1966:120). Thus, "language is only a system of pure values" (ibid.:111).

There is no doubt that Saussure serves as an inspiration for the classical structuralist understanding of language. As exemplified by Jakobson, a sign generally belongs to a system of signs which is organized paradigmatically and syntagmatically. Using a phonological model, a system is an organization of embedded levels in which relations are established among units within any one level and among units between different levels. A syntagm is a cut through the sign system which reveals a sign at one level as composed of signs from a lower level. A paradigm is typically a binary comparison of units or signs within a certain level of the system. For example, the English phonemes /p/ and /b/ form a paradigm, since they are contrasted by one difference syntagmatically.

So far, this is the classical structuralist explication of language as a system of signs moving from the lowest level of distinctive features to the more complex level of the grammar and lexicon. But as Jakobson recognized as early as 1932 (and as Andersen [1979] and Shapiro [1983] have argued more recently), the organization of relations among signs in a system is not a set of purely symmetrical relations but is rather a valuative relation, characterized in terms of markedness on the paradigmatic level and by rank on the syntagmatic one. Indeed, according to Jerome Bruner (1984:163), Jakobson argued that "the most primitive and important distinction in language was between the marked and the unmarked."

Generally speaking, markedness refers to the valuative relation between the two poles of an opposition which establishes an asymmetry between them. Students of markedness have disagreed on the exact attributes of that concept (see Lyons 1977; Comrie 1976, 1983; Waugh 1982; Andrews 1984; Van Brakel 1983; Battistela 1986).

However, in spite of disagreements, most students accept the following as the essential properties of markedness:

1. *Optimality*: the marked term is derivative of the unmarked. More specifically, the marked term requires the background, so to speak, of the marked one. Whenever /u/ occurs in a language, /i/ is also present, making /i/ unmarked.
2. *Indeterminateness*: the unmarked term has more referential scope; the marked term provides more information, but less referential scope. "Man," for example, is unmarked in comparison to "woman," since it can mean both *Homo sapiens* and the male species of *Homo sapiens*.
3. *Simplicity*: the unmarked term is usually less elaborate in either its morphosyntactic makeup or in its acoustic or articulatory nature. For example, "I walk to the store" as opposed to "I walked to the store" shows the latter to be marked.

4. *Unrestricted distribution*: the unmarked term has greater freedom to combine with other linguistic elements. This is related to the phenomenon of neutralization. For example, in the opposition between "host" and "hostess," in the plural the unmarked term ("host") is typically used, as in "Your hosts are Mr. and Mrs. Hill."

An excellent nonlinguistic example of markedness is found in perception, as noted by Ann Treisman. In her work she showed that there is a certain "search asymmetry" among perceivers when targeting a specific feature in the context of its perceptual opposite. For example, if a single curved line is placed in the background of straight lines, the curved line "pops out," that is, requires less search time than the obverse case, where a straight line is targeted with a background of curved lines. The latter involves a "serial search" and requires more time to find. In this case, as Treisman (1986) suggests, straightness is coded as the absence of curvature, or the unmarked feature, curvature generating more activity on the retina than straightness; that is, curvature is the more complex or focused member of the opposition.

Rank, on the other hand, is the measure of asymmetry in a syntagm. On the basis mostly of the work of Andersen (1979), rank has been shown to be a useful tool for phonological typology and shows how one vowel system may differ from another.

Although the notion of rank is extremely useful for such typologies, it can be modified for use in the analysis of other sign systems. Generally speaking, given a syntagm, the highest-ranked feature is typically the feature which will represent a certain referent, regardless of the combination of other features in the syntagm. For example, in stylized representation of happy faces, all other cues may be eliminated, but the smile must remain in order to represent that emotion. In that case, the smile is of high rank with respect to the representation of happiness, just as the frown is with respect to the representation of sadness. Thus, although each feature is of high rank in regard to its respective emotional response, paradigmatically the smile and the frown form a marked relation. In most social situations, the normal presentation of self, portraiture, greeting, and so forth, considers the smile unmarked, while the frown is marked (the well-known physiological fact which requires more complex musculature to produce a frown in a way confirms its marked character).

If we consider the interpretant as a rule of translation which organizes sense and reference of signs, and if value (understood in the Saussurean sense) is the coordination of differences within a system of signs and between systems of signs, and if the notions of markedness and rank are considered as the most developed notion of value, then the marriage between these three concepts—interpretant, value, and hierarchy (that is, markedness and rank)—leads to the notion of *transvaluation.* More specifically, I argue that transvaluation, in its most comprehensive form, is a rulelike semiosis which reevaluates the perceived, conceived, or imagined markedness and the rank relations of the features of a referent by means of the markedness and rank relations of the features of its sign. Transvaluation occurs as a valuative translation of the referent into a sign medium which reevaluates it generally in terms of a teleology defined by the semiotic agency or system. It is precisely by means of valuation of the referent that the sign-signified relation is made coherent, that is, allows signification to accrue. The referent is given meaningful order via transvaluation. I illustrate this idea with an example of a minor trope, euphemism.

Troping proves to be a good example of the application of transvaluation, especially if it is couched in the language of displacement, viewed essentially as a process of value disruption. More specifically, it has to do with the disruption of hierarchy qua rank. In the sense in which Freud originally defined it, displacement is a process in the dreamworld, which differentiates the dream from the dream-thought by centering the important elements of the dream-thought, that is, those with greatest psychic value, differently in the manifest dream (Freud 1965:305). In effect, "a complete transvaluation of all psychical values" takes place between the material of the dream-thought and the manifest dream (ibid.:330).

Using the terminology of Freud, Jakobson called metonymy a species of displacement. This is a prime example, since in metonymy a part of high (perceptual, conceptual) rank is used to represent the whole, as in "sail" for ship or "wheels" for car. But displacement in terms of rank is evident in other tropes, especially euphemism. (See Liszka forthcoming for the details of what follows.)

We must keep in mind the social function or purpose of euphemism: to represent something that may be perceptually distasteful or improper in a relatively proper way. This indeed parallels the formal displacement on the linguistic level, and there are essentially two types: (1) those that involve a decentering of rank, and (2) those that concern the inversion of rank (ironic euphemism). In the first, a (perceptually, cognitively) lower-ranked feature which is contiguously associated with or similar to the referent is used to represent the referent: for example, "sleeping together." People who have sex often sleep together as a consequence or use a bed for the sexual act. Thus, what is immediately associated with the event, or of low sensual rank, serves as its representation in the euphemism. In the case of inversion of rank, a low-ranked signatum is represented by a higher-ranked signans: for example, "honey bucket," "honey dipper."

Vulgarisms, on the other hand, in their social function, intend to devalue the listeners or some social group, and their linguistic form assimilates this function. In the vulgarism the higher-ranked, or prominent, part of the referent is used to represent the referent by likening it to, or associating it with, something of lower value: for example, "skinhead."

These brief examples illustrate how the syntax, semantics, and pragmatics of the sign are coordinated by value, and in this sense the notion of transvaluation incorporates the idea of value into the process of sign translation in a way that is organically tied to many of Peirce's own concerns. We may, then, say with Lady Welby that translation is especially associated with the notion of transvaluation (Welby 1983b:126n) and that "the nearest definition available of the main concern of Significs is, I suppose, value" (Welby 1983a:xcv).

References

Andersen, Henning. 1979. Phonology as Semiotic. In *A Semiotic Landscape*, ed. S. Chatman et al., 377–81. The Hague: Mouton.

Andrews, Edna. 1984. *Theoretical Foundation for Markedness.* Ann Arbor, Mich.: University Microfilms.

Barthes, Roland. 1964. *Elements of Semiology.* Translated by A. Lavers and C. Smith. New York: Hill and Wang. Originally published as *Éléments de Semiologie* (Paris: Éditions du Seuil).

Battistela, Edward. 1986. Markedness: The Evaluative Superstructure of Language. Manuscript.

Van Brakel, George. 1983. *Phonological Markedness and Distinctive Features.* Bloomington: Indiana University Press.

Brock, Jarrett. 1979. Principal Themes in Peirce's Logic of Vagueness. *Peirce Studies* 1:41–50. Lubbock, Tex.: Institute for Studies in Pragmaticism.

Bruner, Jerome. 1984. *In Search of Mind.* New York: Harper and Row.

Cohen, Morris, ed. 1956. *Chance, Love, and Logic: Philosophical Essays by C. S. Peirce, the Founder of Pragmatism.* 1923. New York: Braziller.

Comrie, Bernard. 1976. *Aspect.* New York: Cambridge University Press.

———. 1983. *Language Typology and Linguistic Universals.* Chicago: University of Chicago Press.

Dewey, John. 1946. Peirce's Theory of Linguistic Signs, Thought and Meaning. *Journal of Philosophy* 43 (4):85–95.

———. 1973. The Development of American Pragmatism. 1931. In *The Philosophy of John Dewey.* 2 vols. Ed. J. McDermott. New York: Capricorn.

Freud, Sigmund. 1965. *The Interpretation of Dreams.* Translation by J. Strachey of *Die Traumbedeutung*, 1900. New York: Avon.

Harris, Roy. 1987. *Reading Saussure.* La Salle, Ill.: Open Court.

Liszka, James Jakób. 1981. Peirce and Jakobson: The Structuralist Reconstruction of Peirce. *Transactions of the Charles S. Peirce Society* 17:41–61.

———. 1988. Peirce, Saussure, and the Concept of Transvaluation. In *Proceedings of the Semiotic Society of America*, 57–64. New York: University Press of America.

———. 1989. *The Semiotic of Myth: A Critical Study of the Symbol.* Bloomington: Indiana University Press.

———. Forthcoming. Euphemism as Transvaluation. *Language and Style.*

Lyons, John. 1977. *Semantics.* 2 vols. Cambridge: Cambridge University Press.

Rochberg-Halton, Eugene, and McMurtrey, Kevin. 1983. The Foundations of Modern Semiotics: Charles Peirce and Charles Morris. *American Journal of Semiotics* 2 (1–2):129–56.

Saussure, Ferdinand de. 1966. *Course in General Linguistics.* Trans. W. Baskin. New York: McGraw-Hill. Translation of *Cours de linguistique génér-*

ale, prepared from auditors' notes by C. Bally and A. Sechehaye, based on lectures by Saussure at the University of Geneva, 1906–11.

Shapiro, Michael. 1983. *The Sense of Grammar*. Bloomington: Indiana University Press.

Triesman, Anne. 1986. Feature Analysis. Paper presented at the University of Toronto, 29 January.

Waugh, Linda. 1982. Marked and Unmarked: A Choice between Unequals in Semiotic Structure. *Semiotica* 38:299–318.

Welby, Lady Victoria. 1983a. *Significs and Language*. 1911. Reprint of the London edition, ed. H. Schwartz. Amsterdam: John Benjamins.

———. 1983b. *What is Meaning?* 1903. Reprint of the London edition, ed. G. Mannoury. Amsterdam: John Benjamins.

27 The Riddle of Brute Experience: An Argument for a Revision of Psychoanalytic Theory Based on Peircean Phenomenology

Alfred S. Silver, M.D., Psychoanalytic Center of California, Los Angeles, California

I preface my argument with the following powerful but ambiguous lines from Emerson's poem "The Sphinx":

> The old Sphinx bit her thick lip,—
>   Said, "Who taught thee me to name?
> I am thy spirit, yoke-fellow;
>   Of thine eye, I am eyebeam.
>
> "Thou art the unanswered question;
>   Couldst see thy proper eye,
> Always it asketh, asketh;
>   And each answer is a lie."

The poem alludes to Sophocles' myth of Oedipus and describes Emerson's phantasy of the encounter between Oedipus and the Sphinx. These lines have a special relevance to psychoanalysis, of course, but they were written early in the nineteenth century, prior to Freud's dramatic intuition which revealed to him a fundamental relation between the great Oedipus myth and the incestuous and murderous phantasies he discovered in the relations of the "family romance" in childhood development. These phantasies accompanied emotional and mental symptoms that he encountered in his clinical experience and from which he developed the theory and method of psychoanalysis.

It is generally recognized that many of these "new" ideas concerning subjectivist psychology were already in the air, such as the notion of an "unconscious" as well as the realization that mental states characterizing childhood and characterizing madness are not separate and entirely unrelated to the rational consciousness of scientific intelligence. These ideas were ambient, as part of the growing scientific attitude directed even toward human culture, when Freud made his revolutionary systematic contributions which initiated a profound change in Western culture. It is well-nigh incontrovertible that, whether one is pro-Freudian, anti-Freudian, or post-Freudian, contemporary symbolism is replete with evidence that we live in a Freudian era. For the moment I shall resist the temptation to offer a psychoanalytic interpretation of Emerson's beautiful and vivid verse, a temptation which, as a psychoanalyst, I find almost irresistible.

Let me first point out that I wish to emphasize, rather than to interpret, the enigmatic ambiguity in this description of the experience of Oedipus, for the central argument which I hope to put forward concerns the ineluctable and quintessential ambiguity which characterizes *all* experience when experience is considered as it exists in itself. I shall argue that the phenomenology of Charles Peirce reveals the inevitable presence of such ambiguity in experience and in all percepts and concepts—which, after all, are learned from experience. I repeat these lines by Emerson in the hope that they will augment and illustrate my argument. The deeply rooted and pervasive character of the ambiguities constituting the monstrous and threatening appearance of the Sphinx as it is perceived by Oedipus requires a phenomenological, rather than a psychological, grasp of the essential nature of experience itself. This underlying distinction has been generally overlooked by psychoanalysts, whose own interest lies more directly in psychological meanings. In this instance, psychoanalytic interest in the mythical encounter, which was described in such grisly detail by Sophocles, centers on the Sphinx as a "symbol" of the parents and on the "riddle of the Sphinx."

Classical descriptions of the myth depict the Sphinx as a magical monster with the head of a woman and the body of a winged lion. The myth describes how this monster seized passersby, including Oedipus, demanding the answer to a mysterious riddle. Failure to solve the riddle was said to result in the devouring of the victim by the monstrous Sphinx.

A truthful solution of the riddle was said to cause the Sphinx to vanish. The mythical riddle is, What creature walks on four legs in the morning, two at noon, and three in the evening? The riddle evidently alludes to the enigma of human developmental evolution from infancy to maturity, including the tragic infirmities of old age and inevitable death. The psychoanalysis of Freud and of post-Freudian theoreticians has indeed attempted to formulate a rational and therapeutically useful answer to this riddle. Psychoanalysis has attempted to find a means of resolving the emotional and mental conflicts which result from (1) the almost universal incapacity of the infant to face truthfully the great challenges confronting integrative growth, and (2) the tragic confrontation of the infant facing these great difficulties with what seems to be "Sphinx-like" parents. The psychoanalytic search for the solution of this riddle meets a doomed attempt to establish reasonable cause and effect, a teleological search which is hopeless and useless, but nevertheless one that tends to plague psychoanalysis.

There is evidence, I believe, that psychoanalysis itself suffers greatly from its own inability to cope with the ambiguous riddle of the Sphinx by coming to grips with the phenomenology of ambiguous experience which continuously confronts it. The failure of psychoanalysis (and the other "humanistic sciences" which take humanity itself as the object of inquiry) to achieve the powerful, integrated, creative growth which has marked the dramatic progress in the sciences of "material reality" reveals a disastrous splitting and fragmentation in psychoanalytic theory, suspicious intolerance of competitive relations, defensive arrogance, and so forth, all of which demonstrate a misunderstanding of the phenomenological inevitability of ambiguity in fundamental experience and therefore an incapacity to reconcile and learn from these experiences. It is this incapacity to grasp the essential character of the discipline which, I shall argue, accounts for the failure of psychoanalysis to develop its own kind of validation and forces it to remain in the domain of politics and religion, making it dependent on testimonial rather than scientific validation, still subject to primitive feelings, ideas, and beliefs characterized by the primitive ambiguities and riddles which arise from the essential character of psychoanalytic experience. Reconciliation and tolerance are essential for the inferential processes leading toward the construction of percepts, concepts, and symbolic thoughts in general, all developed from experience.

But let us return to Emerson's phantasy with which I began my argument. Unlike the classical myth of the Sphinx, these lines are principally admonitions and stern lectures to Oedipus by the Sphinx, as if by an irritable and admonishing parent (or analyst) to an uncomprehending child. Despite the vituperative tone, the lecture contains vital hints concerning the riddle which holds the life of Oedipus in the balance. Solving the riddle requires Oedipus to break the code of a mysterious pun, equating the eye and the "I" of subjectivity. Emerson already, in the early nineteenth century, somehow grasped the purport of what is considered a central psychoanalytic concept wrapped in this myth, long before the famous Oedipal insight of Freud. It is not merely his fathoming the psychological significance of the Oedipus myth. It was apparently the more subtle and still somewhat avant-garde and even disputed phenomenological mechanism, which was given the tongue-twisting name of "projective identification," which Emerson somehow intuited long ago. In psychoanalytic parlance this delusional apparition, the Sphinx, is described as a paranoid hallucination. The delusion is produced when there is a very marked mental regression to a state of fragmentation and proliferation of so-called partial, bizarre internal objects which contain unbearably horrible qualities and which are violently projected onto some external object, that is, some person or thing, in order to be rid of the unbearable feeling-qualities. The consequence, still very much a part of the phenomena, is not only a changed identification of the projecting subject but also a sense of the horrible qualities, now in the object of the projection, which now confront the fragmented and projecting subject from the outside. In this particular case, Jocasta and Cadmus, the parents of Oedipus, were transformed into the Sphinx by Oedipus's voracious and envious projections. Emerson's poem portrays surprising and more enlightening and enlightened consequences than those attributed to Sophocles.

The reader may accuse me of having reneged on my pledge to resist making psychoanalytic interpretations of Emerson's beautiful poem. But the interpretations I have made are not really psychoanalytic. They are preliminary phenomenological observations which point to the character of my central argument. The difficulty concerns the inescapability of ridding ambiguity from the essential nature of experience as it exists in itself. Unfortunately my argument depends heavily upon just such experience, which is the only useful evidence, if evidence it can be called, in arguments concerning the psychoanalytic method. Put succinctly by Habermas (1963:216), "The difficulty lies in the fact that psychoanalysis is not a natural science and its evidence is gathered in the experience of self-reflection."

A profound understanding of the nature of experience is a demand concerning which the semiotic of Peirce and the psychoanalysis of Freud are in complete agreement, for both disciplines agree that rational intelligence is a capacity which is learned from experience. Unfortunately the misconception tends to cling that the primordial mental element is some kind of thought. W. R. Bion (1963:28) even toyed with the evocative idea that there may be thoughts before or without a thinker. But this is a notion which evolves in the absence of a coherent theory of signs. It is derived, I think, from an intuitive realization that a predominantly indexical sign may be experienced rather than understood. In contrast, the natural sciences have been deriving their understandings from experience (as experiments or observations) for a very long time without much concern for such a phenomenological enigma. Still pragmatists may observe that the ultimate consequences of their efforts have yet to be fully experienced, let alone understood. The problem of subjectivity in the experiences underlying perception cannot be evaded, even in natural science.

The genius of Charles Peirce (sharpened by many years of creative work in logic, mathematics, and natural science) discovered and developed a unique doctrine of signs including all (or most) of the classes of signs and forms of symbolism. The doctrine is a general theory of logic, but it is solidly based on the necessary underlying theory of consciousness or phenomenology. It therefore penetrates far into the evolution of intelligence from primitive experience (or consciousness) of the very qualities or, to paraphrase Peirce, the kinds-of-things potential in the universe which evolution has given us the capacity to actualize in experience. The great contribution of Peirce is formulated as three universal categories: monadic qualities of potential consciousness, or firstness;

dyadic immediate (unmediated) consciousness of brute experience in which an experiencing subject encounters an object, or secondness; and triadic mediated consciousness in which a third, mediating or interpretive subject forms a representation or sign of experience, or thirdness.

Darwin's *Origin of Species* (1859) was a profound influence on Peirce. He judged evolutionism to be the greatest revolution of the scientific era for it finally stamped out all serious residues of creationism; moreover, it completely changed all biological sciences and their most distant ramifications. Furthermore, it made room for hypotheses to extend their influence beyond natural science and throughout the indeterminate universe. Peirce now felt there was a sound basis for a teleological doctrine of final cause in psychic sciences, encompassing the efficient cause of natural science (*CP* 1.269).

Since sensual and emotional experiences of qualitative kinds-of-things are felt in dyadic experience but not perceived (*CP* 5.358) or thought, Peirce called them "brute," or sense, impressions. His descriptions of this primordial consciousness emphasized the process of actualization from potentiality to monads. He made minutely detailed descriptions based on what must logically be the case in this first existent reality of relations; that is, he constructed the first modern phenomenology, emphasizing the two-sided struggle of will and purpose resisted by counterwill and purpose (*CP* 1.24, 418). He employed the analogy with the law of forces in commonsense observations but showed these phenomena in experience could follow mechanical law. His logic was ingenious, detailed, and impeccable. His hypotheses, however, lacked entirely the rich possibilities which present themselves when repressed unconsciousness is factored into the inferences. The consequence is fascinating for a number of reasons. His arguments, because of his deep-seated prejudice against unconscious phenomena, caused his hypotheses to deteriorate into homely efforts in which obvious perceptions were passed off as sense impressions and potential consciousness of firstnesses was described as actual consciousness at times, in what seemed like embarrassing efforts to sustain "triadic symmetry." But the loss of invaluable conclusions seems worse than the weak and false observations and conclusions which were made. Missed were the beautiful demonstrations of what it would mean to experience a world without boundaries in which there is an understanding that something other than one's own sense impressions exist, one in which absolutely no distinctions are felt—beyond recognizable feelings, and one in which nothing can be imagined, including imagination.

Yet any psychoanalyst can recite a litany of such examples, generally sufficiently noteworthy to be recognizable—such as the famous case of the man who could not make a distinction between playing his fiddle and masturbating. But there is more to this confusion; a deeply prejudiced misunderstanding producing a wall of mistrust and a struggle of wills and resistance. Tearing down the structure of inferences may produce a regression sufficient to blur mature distinctions and, like a dose of tranquilizer, escape rationality but start a slide down to extremist qualitative experience which makes significant communication so difficult in an otherwise genuine relation. By genuine I mean one which grows from experience.

Peirce's intense preoccupation (or obsession) with his triadic universal categories convinced him at times (*CP* 1.285) that the potential qualities of firstness must be directly accessible to consciousness. At other times (*CP* 1.321) he wryly declared this to be an impossibility.

I believe that this ambivalence, or self-contradiction, was due to the failure to have a clear distinction between a potential quality and the consciousness of that conception. At least every example of his argument that I can find, if closely examined, is much more properly interpreted as actual rather than potential (*CP* 1.304). In the same way his refusal to consider the possibility of an unconscious mental phenomenon (in Freud's sense) caused him to find examples of secondness in homely illustrations of what are examples of perception, such as, for example, the change in pitch of the whistle of a passing train. In the stubborn rejection of unconscious phenomena, he equated the perturbation with the "preparatory" sense impression. His logical intelligence could never rest in this predicament, however. Some of his descriptions, such as the relations of loss of self-control, were at times clear examples of unconscious phenomena. The relation of sense impressions on the retina to the visual fields were repeated many times, but without his noticing the unconsciousness of those impressions. He even repeated the old saying about the man who arises from the wrong side of the bed and finds wrongness in everyone he encounters—without noticing the unconsciousness in that phenomenon (*CP* 1.304).

Peirce's phenomenology, in contrast, was exceedingly sensitive and advanced. For example, he excluded the possibility of introspection (*CP* 5.264–68). His argument comes down, correctly I believe, to his realization that this phenomenon is an encounter with what we call an internal object. Peirce postulated before Freud that each individual is a sort of loosely held-together dualistic or multiple structure (*CP* 5.423). He was, I believe, the only phenomenologist prior to the revolutionary discoveries of Freud to realize clearly the necessary presence of the object of experience in consciousness. For Peirce this was a fact which was vital to his allegiance to what he called scholastic realism, from which he developed his argument for the reality of generals (such as relations).

I may seem to have dragged in, so to speak, a number of more or less contradictory and even disconnected facts. But they seem to me to be convincing and sometimes startling, for they are not psychoanalytic interpretations but logical observations made by a pioneering genius. Many of these, even now, are valuable, especially for psychoanalysis because of the uniquely scientific method of their elucidation. These conceptions should be of interest to many analysts because they seem to have a strong relation with the pioneering psychoanalytic ideas of Bion (1963), who was not a philosopher. The logical method of Peirce may now gain in clinical significance, since it may be the only source of confirmation of certain psychoanalytic hypotheses. Peircean phenomenology seems to demonstrate, for example, the presence of oppositional objects in (dyadic) consciousness. Psychoanalysis, having increased its sensitivity to more primitive phenomena of object relations, is now reaching the place in which the Peircean phenomenology and theory of signs is vital to increasing its grasp of these more subtle phenomena.

The urgency of a theoretical breakthrough in psychoanalysis is hardly noticeable within the psychoanalytic community. This seems to me a strange situation, since, in my opinion, it has been almost a half-century since a genuinely important advance has been made, and psychoanalysis has been all but forgotten (except for a very occasional contemptuous remark in the philosophical and linguistic literature). But analysts may be hardly aware of this loss of stature, since analytic interest seems to have turned almost entirely toward practical problems of conducting therapy, including some worthwhile explorations away from psychoanalysis. Associated with these is some interest in computerized versions of metapsychological research in the hope of finding a return to the old-fashioned efficient causes of natural science. In short, the old interest in serious theories of mind has dribbled away. I believe that theoretical and genuinely creative interest has been overwhelmed by a generation of subservience to long-standing authority, which now believes it has reached final truth and repeats its interpretations like a litany until these supposed interpretations merely reify old beliefs. There are efforts to break free, but these tend to be isolated by established authority. The excited enthusiasm concerning the work of Peirce which may be obtrusive in this essay is, I believe, a reflection of these conditions.

Charles Peirce's description of dyadic brute experience carries with it implications which seem to me to be pregnant with revolutionary consequences for possible advances in psychoanalysis and for many of the newer sciences. The phenomenology of brute experience described by Peirce simply overwhelms the classical understanding of consciousness, which does not allow for a science of self-reflection. Nor does it have a sufficient grasp of the reality of relations to consider the consequences of the transformation from an "experiencing" to an "interpreting" subject. For that matter, it has yet to consider the phenomenology of subjectivity-in-itself.

Since a phantasy is a sign and as such cannot exist in brute experience (the one-dimensional realm of repressed consciousness), it may seem that the only valid conclusion to be drawn is that psychoanalysts have been making false interpretations from the outset. The question is whether one can reason from the correct phenomenology that the psychoanalytic method has really been accomplishing its proper goal.

Psychoanalysis has made incredibly slow strides since its beginning, despite the remarkable effect on Western culture produced by the discovery of the repressed unconscious. The profound impact of psychoanalysis has not been the result of a revolution in natural science as was believed by Freud. It is more obvious now, I believe, that the impact of psychoanalysis is the result of understanding the very real forces which have evolved and undoubtedly continue to evolve, the forces of relatedness which are more essential for our existence *as humans* than the direct force of nature, including the force of biology. Psychoanalysis has had an influence on our ability to comprehend the nature of those forces of intrapersonal, social, and cultural relations, whose potency in relation to the existence of humanity, in the full sense of what is meant by humanness, is so decisive that it is nurtured and exploited as the real determinant of the fate of humanity. In the ebb and flow of this fate it is these relations which are paramount as the teleological cause of our fate. The balance between life and death is mainly determined by these relations. Psychoanalysis as a means of understanding the seminal intrapersonal development of these powerful forces of conflict and integration is enormously more influential and determinative of the fate of this unique species than any force or influence for direct "cure" or "amelioration" of symptoms it may or may not possess.

Freud discovered these primitive intrapersonal relations and their powerful domination of the personality by the induction of guilt and persecution by a "superego" and even ecstatic idealization by an "ego-ideal." His discoveries regarding the force of these relations in motivating and structuring integrative growth lagged behind, and the growth of knowledge was extremely slow, almost imperceptible for a generation, in considerable measure because of the entrepreneurial purposes which constrained the enterprise.

Despite these circumstances, a renewed attempt to explore these primitive relations expanded explosively (by fragmentation) in the generation following Freud when Melanie Klein (1946) began to study very young children psychoanalytically (some only two years of age). These analyses expanded into the study of psychotics and other very primitive, fragmented personalities. Incidentally, her analyses of supposedly normal personalities, such as candidates for analytic credentials, revealed the presence of the primitive mental structures she had discovered in infants and psychotics. Confirmatory of my hypothesis was my discovery that the phenomenology and semiotic of Peirce resonated, too remarkably to ignore, with the primitive phenomena described by Klein. His phenomenology seems to present an evolutionary infrastructure by means of which the pathological and normal signs and symbols identified by Klein and her coworkers finally find their proper phenomenological niche; that is, psychoanalysis finally meets a theory of symbol formation. These developments were to be expected, since it was Klein's discovery that symbolization marks the initiation ritual, the introduction to the riddle learned from Emerson's Sphinx.

In 1946 Klein published a description of a coherent model of primitive mental structure and mechanisms. She presented in a paper "Notes on Some Schizoid Mechanisms" two successive developmental "positions" as an evolutionary scheme beginning with a primitive fragmented position in which relations are sustained by projective and introjective maneuvers which profoundly affect primitive identifications. The physicalistic manipulations constitute object relations in this primitive world. In this primitive paranoid-schizoid position boundaries are absent, so there is no experience of an internal world separated from an external world to interfere with projective or other (omnipotent) manipulations. Interpretations are made from the patient's associations, playacting, symptoms, and every behavioral response, especially as they seem to relate to the analyst and the circumstances of the session. The "paranoid-schizoid position" is dominated by projective identification in which "bad" objects might be projected evacuatively into an envied maternal object. In this way the "bad" feelings are evacuated, with the consequence that the bad feelings become identified as a quality of the mother to the relief of the infant. Klein saw that these primitive "object relations" were dominant prior to the development of symbol formation.

By careful clinical observation, self-reflection, and great creative imagination, Klein described what I believe to be an uncanny translation of the phenomenology of Peirce. The dyadic experience enables this by means of the integrative process. Objects of experience become correlated by increasingly compatible comparison, forming triads, with the subject at the vertex. With increasing correlation of objects, triads form more complex structures from the linking of more objects from dyadic sense "impressions," a process which can usefully be described as "widening of the triads" from what were, presumably, narrow isosceles and one-dimensional structures, but with beginning boundary-formation. The force of will and struggle may evolve into what Peirce described as "emotional interpretants" dominating intellectual interpretants and expressing "proper significate effects" which conflict with other such triads. It seems very likely that these conflicts might gradually replace dyadic projective identifications with more complex and better "thought-through" modifications of the once naked will and of purpose and action.

Clinically one sees, even in children, schemes, lies, and more constructive conceptual and rational means to this end. Peirce objected to the attribution of conceptual thought in describing dyadic motivation, since it implies only recognition and identification. I think his model, however, as I have modified it by describing a subject linked with two (or more) objects of experience (having appropriated from one of these its customary title of representamen), reveals a developmental, evolutionary capability. One sees that even as a strictly one-dimensional dyad, the actualized qualities must have the capacity for identification, which makes for a distinction from other qualities—else all experience would be indistinguishable. Therefore it seems no revolutionary idea that some clarity of ideas exists, even in brute experience. As a matter of fact it was Peirce who pointed this out in his essay of 1878, "How to Make Our Ideas Clear" (*CP* 5.389).

Despite the evolutionary implications, one must respect the beginning of mind in brute dyadic experience. But one should not ignore the likelihood that a subject eventually organizes an "ego" system for ordering a vast complex of experiences and meaning-structures. Therefore it seems reasonable to start with brute experience from the beginning. But for millennia, thinkers have preferred to start instead with thoughts, for obvious reasons.

Let us leave these considerations which reveal a developmental bond between dyads and triads. With increasing correlation of objects, I have come to the conclusion that Klein's paranoid-schizoid position is a description of dyadic experience as the primitive world is viewed from "above," that is, as described conceptually or, in Klein's terms, from the integrated depressive position. Her internal objects describe the willful and struggling qualitative entities in conflict. Primitive identifications or feelings require nonconceptual terms which language, as a conceptual or symbolic means of communication, makes impossible or extremely indirect. Since emotions seem to be instinctual or derivative from instinct, qualities or internal objects are often described in terms of their purposeful will or resistance, employing highly conceptual language and theory. Peirce strongly objected on logical grounds to the term "desire" for this purpose, since it not only expressed will but implied a meaning as well. In psychoanalysis terms are employed such as "life and death instincts," which are obviously very sophisticated theories intended to rationalize qualitative experience rather than describe it. Such theories tend to confuse two entirely different kinds of thing, being more appropriate for theories of biological behavior, usually associated with evolutionary theory. In psychoanalysis there has always been a conflict between theories which attribute mental disorder primarily to an inborn mental disturbance, which causes distorted impressions, and perceptions which are put forward against theories; that is, the trouble comes either from the subject or the object. This conflict is rendered absurd by common sense, by object relations theory, and especially (and more precisely) by the Peircean phenomenology which affirms strongly that consciousness itself is actualized from monadic potentiality in the encounter of an object by a subject.

In the conflicts among psychoanalytic schools this "theoretical" issue is usually associated with a much more personal issue. Those who espouse a theory implying that mental and emotional problems result predominantly from the inability to control unconscious envy and greed suffer from the consequences of projecting greedy and envious internal objects (qualities) into the analyst-mother. Those analysts who are disciples of Klein must interpret the hostility as coming from their patients. Now, this is a touchy job and requires an analyst who is made of stern stuff. Meanwhile, those who take up a theory that their patients have been abused by hostile parents find that their interpretations are much more empathic and their patients more relaxed and appreciative.

Turning away from the absurd, it is true that this argument results from the very human tendency to oversimplify in order to find a source of comfortable belief and be rid of doubt. It is for that reason that evolutionary theories which tend to require the capacity to tolerate uncertainty are often avoided. Klein described this as the development of tolerance for ambivalence. I have been struck by the evolutionary implications in

Klein's theories, but psychoanalytic theory did not follow up these phenomenological demands, since it was constrained by broader problems.

Peirce, however, could not settle for a primitive world of objects fragmented by the force of greed. Instead he found himself struggling with the question of the reality of qualities—an issue which Kant had tried to resolve by means of the noumenon. The matter might be approached in this way: There has been a general agreement that regardless of other problems about the nature of all mental phenomena, one thing seemed certain: the absolute logical necessity that they must all be representations of real things in physical reality. This was a guiding principle for Freud in his working out the nature of the id. It is also the reason that the concept of unconscious phantasy seems unquestionable. It is in total accord with the presumption of individuality of real things. But Peirce was what he called a scholastic realist, which, in short, meant that he believed in the reality of generals such as kinds like love and greed and signs such as language. It appears that this also means belief in the reality of relations. In fact he was not sure, especially in developing his phenomenology, that all real things were not relations. His thought had profound ramifications in this area, leading him to the conclusion that in the evolution of mental phenomena the primordial reality from which these phenomena evolved were the kind of things which he called, for lack of a better expression, "qualities." But it should be kept in mind that qualities-in-themselves have nothing special to do with the mind. He was led by this strange route to a belief in what is called indeterminism.

In 1986, Max W. Fisch quoted Sir Karl Popper, the most celebrated logician of our time, as follows: "One of the few dissenters from physical determinism . . . was Charles S. Peirce, the great mathematician and physicist, and, I believe, one of the greatest philosophers of all time. . . . So far as I know, Peirce was the first post-Newtonian physicist and philosopher who thus dared to adopt the view that to some degree, all clocks are clouds; or in other words, that only clouds exist, but with very different degrees of cloudiness" (Fisch 1986:426). Popper added that Peirce's view was compatible with Einstein's (special) theory and even more so with quantum theory.

It has gradually become clear to me that developmental problems of all kinds, and certainly in the matter of primitive object-relations theory, must be seen as evolutionary problems. The problem of qualities finding an existence as mental phenomena is precisely such a problem. Peirce's solution to the actualization of qualities in experience as the initial step in the emergence of consciousness, that is, of mind, is an elegant solution, especially since Klein has provided us with confirmation by means of an entirely (or almost entirely) different kind. But now that we have a one-dimensional object relation as a means of accounting for the absence of self-control, due to the absence of a conscious mediator or interpreter, we find that in the (psychoanalytic) insertion of an investigator into the relation, we are met by a primitive triadic sign-function within which we find the dyadic conflict we had expected. The question is, What happened to our developmental sequence? This is a case of having too much to account for. But as we shall see, in the next development, we shall have too little.

There is a danger, in discussing objects, of confusing dyadic objects with the objects which are an intrinsic part of signs; that is, a sign stands for an object which is experienced rather than employed for signification. Peirce pointed out that signs are not static structures but consist in the ongoing comparisons between past signs, present signs, and future signs. This appears to be an interpretive or inferential process and is intuitively included in his peculiar definition which defines a sign by comparing one sign, the representamen, with another, the interpretant. But he also emphasizes that the character of a sign must be such that there is learning from experience. Therefore in the growth of signs the movement from one sign (or thirdness) to another must require that one sign is experienced so that the new sign learns from experiencing the old as an object of experience. But we learn from Klein that in the conflict which is associated with regression to the paranoid-schizoid position, there is splitting and fragmentation associated with a return to the mechanism of projective identification. Thus I believe we learn some things from Peirce and others from Klein. Following the new discoveries of Klein, W. R. Bion (1963) began to develop innovations which implied an intuitive phenomenological system. For this he seems to have depended on the critiques of Kant and Kleinian concepts.

Bion's most valuable contribution, I believe, was the conception of a primitive "beta element," for which he also used the designation "sense impression." He postulated that the beta elements were "object-like things" which employed projective identification. Maternal intervention by means of what he termed "alpha-function" had the capacity to transform beta to alpha elements, thereby inducing the beginning of thinking and symbolization. He used the Greek symbols because he claimed no knowledge of what they were beyond this hypothesis.

At the present time, a century more or less after Peirce's descriptions of consciousness, the general categories, and the semiotic, the misunderstandings concerning Freud's "system unconscious" seem to me less disturbing than they evidently were at the time. The struggle of Peirce with the phenomenology seems uncharacteristic. From the perspective of psychoanalysis, his illustrative examples seem naively unconvincing. If one looks at the reasoning which he employed to describe what must be the character of dyadic secondness, it is certainly impressive to realize what he failed to see in order to justify his refusal to accept the conception of unconscious mental phenomena. First, he made it clear that consciousness in secondness is immediate, its duration being both sides of an instant. In fact, it is as timeless as an existent can be. Second, spatially it is one dimensional and so is experienced as a spaceless point from either side. With that dimensionality it can never enclose a space; thus it is without boundaries, and its two opposing sides are indistinguishable and unintegratable. If that were not sufficiently convincing, there is the case of the punctuate retinal sense impression and its perceived visual field, which is a judgment derived from inference. Like most percepts, it seems instantaneous and uncontrollable, which are useful properties without which voluntary movements in space would be more hazardous than they are.

But how can we explain that usually judgment to form a percept is quick and easy, but in the face of doubt, becomes a struggle? I believe that the best hypothesis is based on an evolutionary premise; that is, I would suggest that the perceptual judgment must be formed simply enough to form percepts quickly but still accurately, therefore slowly if need be, for survival's sake. This might be facilitated by a few dyads only, so that their function might ordinarily be more one dimensional rather than forming a sign by inferential process, processes being slower than mechanisms. Extraordinarily, however, the few dyads, though faced by doubt, are sufficient for a perceptual judgment, percepts being, I believe, the simplest kind of inference. The point of this little exercise is to suggest that the triadic model is simpler than the complexities of life, as it is intended to be. The same is true for psychoanalytic models. Of course, this remainder is an attempt to discourage the reification of models and theories, a destructive habit which prevents the solution of problems, as in the case of the splitting of psychoanalytic schools by just such reification. Peirce took great pains to demonstrate these processes of transformation (*CP* 1.333; 2.227–29, 231—32), but explication of them is a subject for another chapter.

There is another source of conflict, in the transformation from experience to sign formation in thirdness, which seems to have been overlooked or ignored but which may be of considerable importance. In order to notice this possible or probable source of conflict, it is necessary to notice that there is a rather sharp break between Peirce's phenomenology and the subsequent detailed description of semiotic and its definition of the sign. This is somewhat unexpected, since the phenomenology was developed very carefully to be the transformational foundation for his semiotic, or triadic logic, which in turn is the logical foundation for his philosophical doctrine of pragmatism. The discontinuity to which I refer results, I suspect, from the difficulties he encountered in his definition of the sign and in psychoanalytic interpretation of the vagaries of primitive personalities.

Peirce describes the two-sided struggle of will and purpose against the resistance of contrary will and purpose in the one-dimensional world of secondness. His description of the actualizing relation, of the matter of identifications, of the struggle involving the qualitative aesthetic and ethical values, and of the absence of boundaries all support that struggle. He points out, also, that most of these struggles are barely noticeable, merely confirming the relation which then goes on to the process of sign and symbol formation.

I cannot help but notice, however, that there is a very marked discontinuity between his phenomenology and the ensuing long descriptions and definition of the sign with which he undertakes to describe his triadic logic, the semiotic which is the foundation of his philosophical doctrine of pragmatism.

The definition of a sign which is generally taken as definitive appears in a manuscript of about 1897 (*CP* 2.228). This has been subject to ongoing dispute, most often in regarding his use of the "interpretant." The definition states:

> A sign, or *representamen*, is something which stands to somebody for something, in some respect or capacity. It addresses somebody, that is, creates in the mind of that person an equivalent sign, or perhaps a more developed sign. That sign which it creates, I call the *interpretant* of the first sign. The sign stands for something; its *object*. It stands for that object, not in all respects, but in reference to a sort of idea which I have sometimes called the *ground* of the representamen. (*CP* 2.228)

It is the interpretant which is considered most controversial, but it is the developmental and evolutional aspect which I wish to discuss. I have wondered why his definition contains only a dubious reference to the transformation from secondness to thirdness. On further thought I have come to believe that the denoted object is an object of brute experience. The most radical hypothesis is, I believe, the transformation of the representamen into a second object of dyadic experience, whose comparison with the first object with respect to its ground, or qualities, constitutes an inference; in other words, forming an interpretation or representation of the first (denoted) object of experience. The two objects now constitute an inferential pair. The sign is thus created properly in order to enable the subject to learn, that is, to interpret from experience. In this translation of the definition there is merely the suggestion that the model of secondness, in order to convey the reality of brute experience, may need to present a constellation of dyads of which this first modification pictures two "binary star systems." These two are sufficiently parallel in their qualitative values that they act with a minimum force and resistance between them so that either can represent the other; it is in this relation that their experience gains significance. But for the model to reveal the meaning of this need for continuity of force and resistance, there must be added other such star systems in this universe, star systems born from collision. Thus conflict is inherent, not only within the dyads but also between them. In thinking about this model, I see that with the dyad, conflict may lead only to eventual death (loss of the consciousness of that dyad), while with the same fate between dyadic systems there is first the loss of meaning; but there still remains a possibility of learning from that "misguided" experience.

In 1905 Peirce wrote in the essay "What Pragmatism Is" some remarks which revealed his keen insight concerning his knowledge about the nature of primitive relations. He revealed that in this year (in which Freud was acclaimed for elucidating the nature of "transference"), he not only understood the nature of transference but also understood what was not discovered until forty years later by Melanie Klein. Her groundbreaking paper "Notes on some Schizoid Mechanisms" (Klein 1946) was devoted through and through to her designation of the unconscious phantasy of projective identification. This psychoanalytic discovery has won its way slowly to a most preeminent position in modern psychoanalytic object relations. But in 1905 Peirce had already described this phenomenon, in his own way, with startling accuracy: "Where no self-control is possible, there will be no self-reproach. These phenomena [self-control and self-reproach] seem to be the fundamental characteristics which distinguish a rational being. Blame, in every case, appears a modification, often accomplished by a transference, or 'projection' of the primary feeling of self-reproach" (*CP* 5.418–19).

For many years, Peirce believed that by means of his pragmatic doctrine, applied correctly to strictly intellectual conceptions, he could succeed in removing subjective ambiguity from the objectivity of material things. He rested his belief on employment of the scientific method, progressively eliminating subjective sources of ambiguity in experience, thereby fulfilling Galileo's maxim regarding the "simplest hypothesis" as the proper means of progressively approaching the ultimate truth concerning natural events. In a 1908 essay entitled "A Neglected Argument for the Reality of God" (*CP* 6.452–93), the elderly Peirce made the following remarks, recanting many years devoted to the validation of scientific method applied to "intellectual conceptions" as the means of separating subjectivity from objective material reality. In the end he surrendered to ineluctable ambiguity and to the human dependence on instinct.

> Modern science has been builded after the model of Galileo, who founded it on *il lume naturale*. That truly inspired prophet had said, of two hypotheses, the *simpler* is to be preferred; but I was formerly one of those who, in our dull self-conceit fancying ourselves more sly than he, twisted the maxim to mean the *logically* simpler, the one that adds the least to what has been observed. . . . It was not until long experience forced me to realize that subsequent discoveries were every time showing I had been wrong, while those who understood the maxim as Galileo had done, early unlocked the secret, that the scales fell from my eyes and my mind awoke to the broad and flaming daylight that it is the simpler Hypothesis in the sense of the more facile and natural, the one that instinct suggests, that must be preferred; for the reason that, unless man have a natural bent in accordance with nature's, he has no chance of understanding nature at all. (*CP* 6.477)

References

Bion, W. R. 1963. *Learning from Experience.* New York: Basic Books.

Fisch, Max H. 1986. *Peirce, Semiotic, and Pragmatism.* Bloomington: Indiana University Press.

Habermas, Jürgen. 1963. *Knowledge and Human Interests.* Boston: Beacon Press.

Klein, Melanie. 1946. "Notes on some Schizoid Mechanisms." *International Journal of Psycho-Analysis* 28:207–13.

28 Memory, Morphology, and Mathematics: Peirce and Contemporary Neurostudies

George W. Stickel, Northwestern College, Orange City, Iowa

What happens when learning interfaces habit? What neuronal firings make up our mind? What is knowledge? What are the implications of the studies of a mathematician and philosopher who died 150 years ago on these neurophilosophical questions?

Patricia Smith Churchland (1986:249–51, 275, 393) mentions briefly that the work of Charles Sanders Peirce is important in answering these and other current questions on the mind/brain. Recent studies in chaos have shown that nineteenth-century mathematicians had great insights appropriate for present research (Gleick 1987). Additionally, the American pragmatists were steeped in the study of the interface of philosophy and science, and the interface of the social environment and the psychological. And Charles Peirce, its founder, sought to understand the very essence of that interface between human logic and the mathematical logic of the world around us. The study of Peirce, and others like him, can lead us to new heights in our understanding of the human mind/brain, learning, and our relationship to the world around us.

This chapter will show how what Peirce wrote is consistent with and can fit into current research in neuro and cognitive sciences to create a theory of mind/brain. The paper will begin with the juxtaposition of several concepts and will assume three foundational aspects of Peirce's philosophy. The first assumption, being rather well founded, is that the semiotic was central to Peirce in his study of thought; in fact, thought was semiotic for Peirce (*CP* 2.444n; 7.585; 8.272). Second, there is an isomorphism between the human mind/brain and the physical universe which is the starting point of the semiotic. That is, to study one is to study the other; in other words, our minds mirror the physical reality around us (Tursman 1987:105; *CP* 1.309; 7.585; see also *CP* 7.424 and 7.591). Peirce wrote: "Every scientific explanation of a natural phenomenon is a hypothesis that there is something in nature to which the human reason is analogous" (*CP* 1.316). (Also see Orange 1984:53, 70–71 on isomorphism.) The third assumption in working with Peirce is based upon the previous assumption that the mathematical logic of the physical universe is the mathematics of the human mind. This assumption is a corollary of Eisele's law, that Peirce's philosophy and logic can be understood only in the context of his mathematics (Tursman 1987:68).

To proceed, it will be necessary to select from Peirce's philosophy appropriate, but limited, concepts of the mind/brain which reflect as comprehensively as possible the nature of his thoughts. A concept which provides such insight is Peirce's discussion of spirals. While abstract, they are indeed tied to the above assumptions and are particularly instructive in relation to current mind/brain studies.

After a brief examination of the spirals, it will be necessary to provide an overview of the research done in parallel distributed processing and the theoretical work of Herbert P. Killackey, a neuroanatomist. Finally, a conceptual model will be provided, using the semiotic of Peirce as a frame.

The Spirals of Peirce

Peirce was fond of using spirals to explain abstract concepts, particularly concepts of the body/mind interface (*CP* 1.276–77; 7:370–71; 8.122 n. 19 and 272–74; *NEM* 3:897–98; and Tursman 1987:123–25 and 149). The spiraling notion, as Peirce has used it, has significant application to the work done on chaos, which likewise is an interface or transition between two physical systems (Gleick 1987:137, 198, 208, 261, as well as other metaphors similar to Peirce's, 141, 150, 158, 162, 177, and 179; and Hofstadter 1985:376).

Before examining some passages from Peirce, the keys to understanding his ideas herein are his concerns for continuity between the individual and the environment and for continuity between the mind and the body. Second, the purpose of the spirals is that while they reflect a continuity from one point to another, they do so with an endless series of transformations. This endless series of transformations separates the two points of the continuity but yet ties them inextricably. This inextricable union in general negates a reductionism of the mind for the pragmatists (Dewey 1986:30; Mead 1938:412–13; and Putnam 1988:58–59, 71, and 82).

In a letter to William James, while discussing consciousness, Peirce wrote:

> Consider the plane spiral curve whose equation in polar co-ordinates is
>
> $$\frac{r^2 - 4r + 3}{r - 2} = C^{\theta}$$
>
> That curve will start at $r = 1$ [or rather at $r > 1$, as pointed out by a colleague, Shari Brink] and coil outwards toward $r = 2$ making an endless series of revolutions before it reaches $r = 2$. Then it will keep right on and perform an endless series of revolutions before r becomes $2 + e$, no matter how small a distance e may be. Finally, when r becomes 3 the curve will come to an abrupt stop. This shows that although it be true that Being immediately acts only on Being and Representation immediately acts only on Representation, still there may be two endless series, whereby Being and Representation act on one another without any *tertium quid*. (*CP* 8:274)

"Being" refers to the material organic body of the individual, while "representation" is the mental processing of thought. Peirce followed this passage with the spiral in figure 1.

Peirce then explained that if the individual points which made up the spirals (atoms in ether) were spirals themselves, and those spirals composed of points similarly constructed *ad infinitum*, then the sound waves of his voice "should be converted . . . [through] the whole infinite series . . . [which] should be traversed in a fraction of a second, after which they will be in the form of *thoughts* in your mind and so you will come to

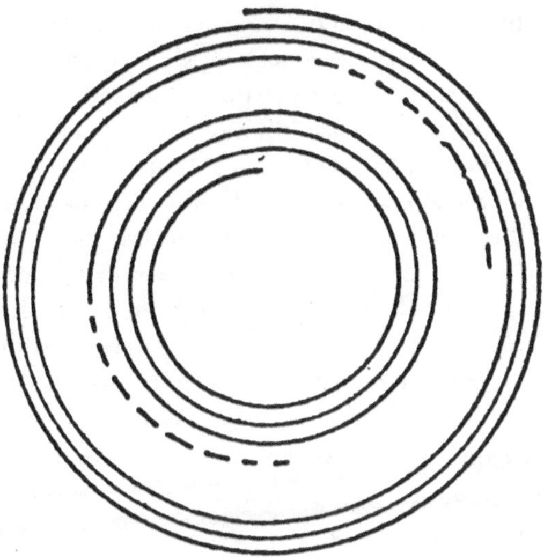

Figure 1. Peirce's spiral relating Being and Representation

understand the meaning of those sounds. My logic will open up a world for investigation and show how to set about it" (*CP* 8.275).

In another spiral passage, this time addressing his philosophy of mind, Peirce continued the same metaphor of an endless series of transformations between matter and mind (*CP* 7.370). He considered a spiral (which was not pictured in the passage) to explain what would happen as his dog touched him with its nose to beg to be let out, then what would happen in his subsequent processing of the request, then his moving to the door to comply with the dog's wishes. Peirce offered:

> Let the radius vector measure the time, beginning at the outermost point, as the instant when the dog's nose touches me and proceeding inwards. Let each coil of the spiral represent the transformation of the motion from one ether to the next. At the end of the period of time represented by one inch of the radius, all that infinite series of transformations will be complete. Now let us suppose that the inner series of coils of the spiral, which instead of being endless, is beginningless in terms of the coils, though not in time, represents operations governed exclusively by final causation, and therefore purely mental. Let us suppose that, although mental, they

> are not noticeably conscious until the innermost end of the coil is approached. Here begin those reflections of which I am able to give any account, although from ever so early in the second series of coils the mind was acting rationally, in the sense in which unconsciousness, and therefore uncritical, action can be called rational. Finally, at the innermost end of the spiral will occur my volition to let the dog out. Another similar diagram would be required to show what happens next. . . . But there must be an infinite series of such ratiocinations if the mind only acts rationally. Take any instant after the work of the mind has been done, and at that instant, an infinite series of dynamical transformations will have taken place which are to terminate in the door being opened. (*CP* 7.371)

In the above passage, then, two spirals are needed to exemplify the details of interaction with the dog, thought, and opening the door.

A third spiral passage will clarify Peirce's conceptualization. He wrote in a letter on mind, matter, and logic:

> The soul [mind] then *certainly* does act dynamically on matter. It does not follow that it acts *directly* upon matter, because there may be involved an [continuing from note] *endless* series of transformations of energy from motion of one fluid to motion of another, all these fluids being spiritual [mental], followed by a *beginningless* series of transformations of energy from motion in one fluid to motion in another, all *these* fluids being material. It is just as a spiral within a circumference A . . . may make an endless series of turns before it reaches an inner circumference B, and may then keep right on making a beginningless series of turns before it reaches a third circumference C. $\Theta = \log(r - c)$ or even
>
> $$\Theta = \frac{1}{r^2 - c}$$
>
> is a sufficient illustration. (*NEM* 3:897)

The following page offers Peirce's diagram pictured in figure 2 (*NEM* 3:898).

While the spirals could fit nicely into the research of chaos and should be studied from that perspective, I offer the following explanation for our purposes here. First, in the continuity between the mind and the sur-

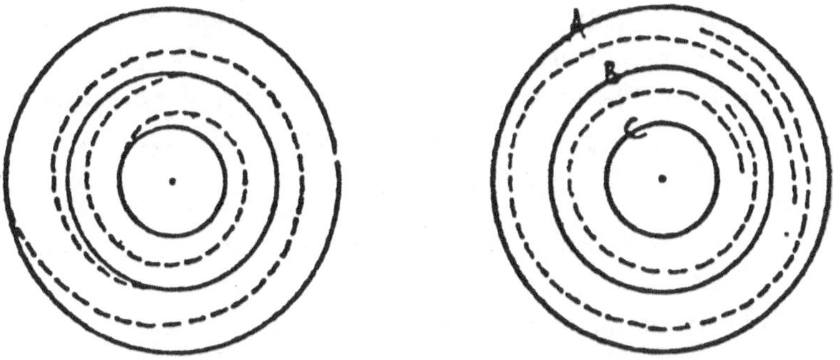

Figure 2. Peirce's dynamic transformations of energy

rounding environment, as an individual interacts with that environment, there is, for Peirce, the material, the spiritual (or mental), and another material. These three components are one in the same with George Herbert Mead's three-part division of the self—the "me," the reflective self, and the "I." To put the spiral notion and the divisions of the self into neuroanatomical terms, the first material, or the "me," is composed of the sensory neurons, which, it can be argued, will lead to an endless series of transformations. Somewhere those transformations produce a beginningless series of transformations within the interneurons which are the neurons of the central nervous system (CNS)—Mead's reflective self and Peirce's spiritual component. Those transformations (beginningless, since one cannot determine precisely where the group's function begins—or where reflection begins) are also endless, as they move toward the last group, the motor neurons. The motor neurons are the second material transformations, since they touch matter on the other side. That is, the individual responds by opening the door for the dog; the "I" acts upon the environment.

Where does the sensory portion end and the mental begin? Where does the mental end and the motor portion begin? Those are complex neurological problems beyond the details of this paper, but Peirce's notion of the spirals attempts to provide a framework for the answers.

One final comment is appropriate in this section. The logarithm was an important function for Peirce and can be seen in his spiral equations. It was important for him because it suggested probability, particularly the probability, or chance, of certain transformations occurring within the endless series or beginningless series (*NEM* 3:893–94). This concept of chance will be important as we examine parallel distributed processing next.

Parallel Distributed Processing

Parallel distributed processing (PDP) is a neural-network approach to the study of the mind/brain problem, which, while simplistic, is impressive in terms of its approximation of neural and mathematical understanding of the brain. It is called parallel because the processing is done competitively within a parallel structure, as opposed to a serial processing. Within the parallel structure are serial representations which are iterated over time. The serial portion of the model is labeled "distributed" (McClelland and Rumelhart 1987:38).

PDP is a stochastic processing which, while slower than the human brain, approximates learning as the systems developed come to recognize specific characteristics. Those characteristics are stored, processed, and revised using a thermodynamic annealing model of probability, which enables the system to settle upon, or relax to, the "right solution" of such problems as correctly interpreting a masked word.

For our purposes, what is crucial is that (1) the system is a working mathematical model, and (2) the system operates in parallel, allowing various solutions to be processed against one another. Additionally, a key component in the mathematical understanding is that the "knowledge" of the system is contained within a vector format. For example, in the PDP reading model, "knowledge atoms" containing two-letter sequences (e.g., $W_1 A_2$) would have the following vector code as each of the following letters would be related in turn to that knowledge atom (Rumelhart and McClelland 1987:204ff.):

Representational features	A_1	M_1	W_1	A_2	M_2
Vector representation	−	−	+	+	−

The subscript defines the position order of the letter in a four-lettered word; the minus (−) denotes that the A_1 inhibits the knowledge atom

W_1A_2 because the atom does not contain an A as the first letter. Similarly, the W_1 activates the knowledge atom, with a positive value (+), because the atom contains a W_1.

The vector representations refer to simplistic data; they are representational but are coded in n-dimensions; in the above example, $n = 5$. What I will suppose about this vector information, to begin to tie it to Peirce, is that this simple bit of datum is, in essence, what Peirce referred to as firstness in his semiotic. It is firstness because it is independent of actuality; it is akin to a possibility or a simple quality—like redness. In fact, the idea of redness as a vector can be seen in the two-dimensional analysis provided experimentally by a Nicol prism polarizer, which rotates the colors of white light in proportion to their optical activity (Jenkins and White 1957:573–75). The color vectors would appear as in figure 3.

Let us assume that the redness, or any bit of vectorized datum, is a sensation; it is a firstness. Let us further assume that Peirce was sitting in his chair with a feeling (firstness) of complacency. At that moment, the

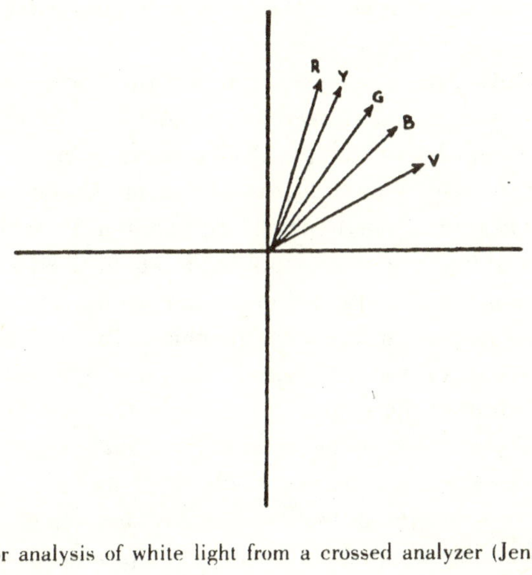

Figure 3. Vector analysis of white light from a crossed analyzer (Jenkins and White 1957:575)

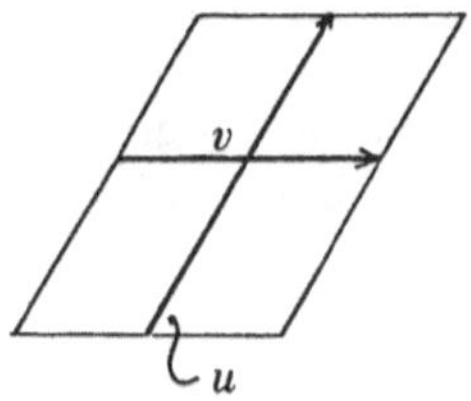

$$v = v_1, v_2, \cdots v_n$$

$$u = \begin{matrix} u_1 \\ u_2 \\ \vdots \\ u_m \end{matrix} \begin{bmatrix} a_{11} & a_{12} & \cdots & a_{1n} \\ a_{21} & a_{22} & \cdots & a_{2n} \\ \vdots & \vdots & & \vdots \\ a_{ml} & a_{m2} & \cdots & a_{mn} \end{bmatrix} = A$$

Figure 4. Vector representation of secondness, in geometric and matrix forms

dog's nose touches Peirce. Here we have another sensation—a second firstness. But this feeling, or sensation, is contrary to the original feeling of complacency—that is, it produces a cross-ripple with the first. This cross-rippling is comparable to the actuality of Peirce's secondness. In other words, there is an impulse which produces a realization of something beyond the original sensation. There is now a relationship between Peirce and his dog. The touch is the same as any other sensation, but the cross between the touch and the former feeling is what produces the actuality. This actuality reflects the sensory nature of the first material swirls of Peirce's spiral above.

Mathematically, what is happening is that the first vector resembles a series of neurons firing in a prescribed manner to represent a given sensation (which is consistent with the PDP studies). Suppose that while the vector is defined and central to the particular sensation, it typically bleeds out, or activates, neighboring neurons in some $n \times m$ grid. Further suppose that this new cross-rippling is produced by a m-dimensional vector orthogonal to the first n-dimensional vector. The second vector's activation of its neighbors (the same neighbors that the first vector has) stir the components of the $n \times m$ grid into a matrix formation. The two-dimensional nature of the matrix is representational of Peirce's secondness. Figure 4 shows this vector and metrical relationship.

In figure 4, v represents the first firstness, that of complacency, while u represents the sensation of the dog's touch. Both v and u are iconic in nature. The actuality of the dog's touch, which produced an awareness in

Peirce of his relation to the dog, is a secondness, represented by the matrix A. This awareness is indexical.

This level of Peirce's semiotic is as far as PDP can carry us, I believe. A weakness in PDP, then, is its inability to produce the more complex sign crucial to Peirce's understanding of logic. That is, the question becomes, Where is thirdness, or symbolicity? (See *CP* 2.227–73 for examples of Peirce's signs.) To answer the question of thirdness, we need to examine the neuroanatomical theories of Herbert P. Killackey.

Neuroanatomy and Pyramidal Cell

The brain is certainly a complex organ, having far more than a single type of cell firing in a single way, but an understanding of the pyramidal cell, in relation to the above discussion, is instructive. To understand how the pyramidal cell can relate to memory and the functioning within the brain, I turn to several quotations from an essay by Killackey (1987:111–20).

> I would point out that the morphology of the central nervous system, particularly in terms of the three-dimensional distribution of neurons and the interconnections between them, must be the physical substrate of the processes that we term memory. (111)
>
> . . . one of the most characteristic neurons of the neocortex is the pyramidal cell. It is the major, if not only output neuron from the neocortex. This neuron is found in all cortical layers and has as its chief defining feature a long straight vertically oriented apical dendrite that in most instances reaches from the cell body to just beneath the cortical surface. Thus, this apical dendrite is of a varying length and, in the case of deep pyramidal cells, may be as long as several millimeters. The apical dendrite of such a pyramidal cell is thickly encrusted with dendritic spines. It has been estimated that a single pyramidal cell may receive in the neighborhood of thirty thousand contacts. (114)
>
> . . . It seems likely that the brain of higher primates may have more discrete areas for the processing of information and hence more complex patterns of connectivity. . . . However, the basic structural organization of the neocortex is relatively invariant across mammalian species. . . .

In my opinion, this finding has several significant implications for our understanding of the role of the structural organization of the neocortex in information processing and storage. First, it suggests that the fundamental unit of information processing is quite conservative. . . . Second, the number of neuronal cortical processing units has increased enormously . . . [through the] complex pattern of gyri and sulci [the ridges and valleys of the convoluted folds of the brain]. . . . Third, the processing capabilities of a given cortical unit may have increased significantly. An earlier part of this discussion focused on the apical dendrite of the cortical pyramidal cell, and there it was suggested that these apical dendrites may be the chief combinatorial element of the neocortex. Here it should be noted that the increase in cortical thickness implies a significant increase in the length of these dendrites, particularly the ones of the deeper cortical layers. It is very likely that this increase also includes an increase in the number of dendritic spines and in the richness of axonal strata with which they come in contact. Thus, the combinatorial properties of the basic cortical unit may have been enhanced in the course of mammalian evolution. One can speculate that cortical changes in the horizontal plane may be more closely related to increases in the capabilities for processing sensory information, while changes in the vertical dimension are more closely related to the combining and storage of information. (117–18)

The dendrite carries the incoming signal, while the axon carries the outgoing signal in each neuron. Figure 5 shows a pyramidal cell with only three spines on the apical dendrite.

Sensorial data is topographically arranged in the CNS (Udin and Fawcett 1988:289, 303–4). Killackey (1987:115–16) points out that the topographical organization of sensory data in the neocortex is primarily, if not completely, limited to the horizontal plane of the cortex, while the middle layers

seem to play a major role in the initial processing of sensory information. The neocortex is also characterized by a vertical dimension. As mentioned above, it is along this vertical dimension that the apical dendrite of the cortical pyramidal neurons is distributed. Cortical pyramidal neurons, particularly ones in the deeper layers, have their apical dendrite distributed in such a fashion that they pass through strata of input arising

Figure 5. Pyrmidal neuron in the cerebral cortex (Shepherd 1987:89)

from different sources. Thus, the pyramidal cell apical dendrite may be the chief combinational element of the neocortex.

The conclusion that can be drawn from Killackey is that there is a sensory plane within the neocortex which is tied vertically to a portion of the cortex which is responsible for storing and processing that data.

A Mind/Brain Model

Given the discussion in the PDP section of the paper of how Peirce's firstness and secondness can be shown mathematically, and given the discussion from Killackey, the following is proposed.

First, the matrix of Peirce's secondness is the horizontal cortical plane. Second, that matrix represents a matrix of neurons firing in some pattern defined by the sensory vectors v and u, above. Third, that matrix is tied vertically to other neurons, represented beneath the plane by a three-dimensional algebraic matrix of depth p. Figure 6 shows the geometric and matrix forms of what will be used to represent Peirce's thirdness, or symbolicity.

The three-dimensional matrix provides a reasonable representation of

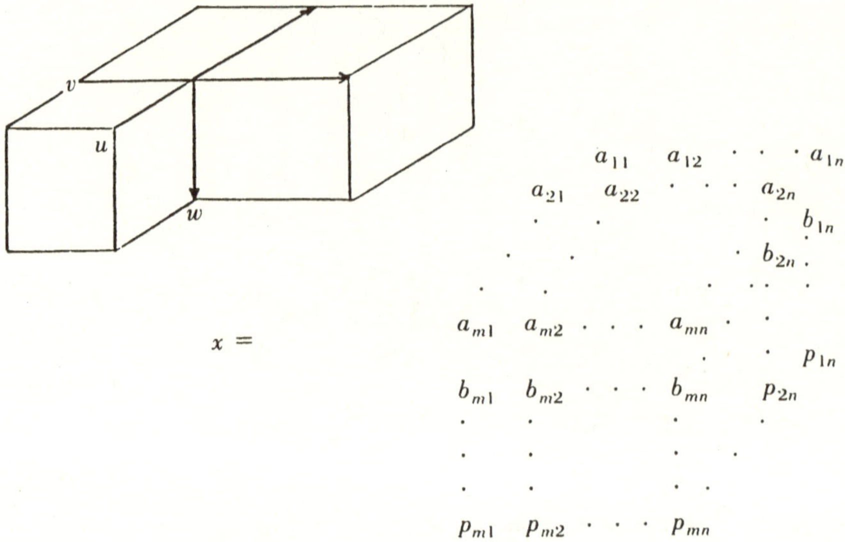

Figure 6. Representation of thirdness, in geometric form (with a cutaway view) and in matrix form

Peirce's thirdness, because thirdness is as complex as is needed in Peirce's semiotic—that is, for his logic, one only needs the triadic, since all higher forms can be produced from the triadic (*CP* 1.347). Second, as one attempts to symbolize, in the Peircean sense, a three-dimensional arrangement provides a significant number of possible permutations of symbols, even if the possibilities are limited to three values: a negative (−) for inhibition, a positive (+) for activation, and a zero (0) for no input, as is used in PDP. For example, if $n = m = p = 10$ in the above matrix, then the three possible values would provide 3^{1000} different symbolic arrangements. (It should be noted, however, that there could well be far more possibilities with the variations provided by numerous neurotransmitters.) Thus, a great deal of similar information can be stored, as Killackey suggests, within the vertical arrangement. Third, the cortex is arranged vertically in interconnected, cylindrical columns, "about 150–300 micrometers in diameter and containing about 5,000 cells"

(Churchland 1986:132n). Given the 5,000 cells, one column has the potential for 3^{5000} arrangements of data, given the three values above.

Additionally, because the columns are interconnected, as we would assume, the relationships between columns become, quite nicely, the relationships of Peirce's triads. That is, his deductive logic rests upon the juxtaposition of thirdnesses, so that corollarial deduction can be represented as 3333.

This brings us to Peirce's natural classes of arguments and classes of signs. In this chapter we cannot develop a complete argument, but let the following hypothesis be suggested as an answer to the dilemma of Peirce's two classification schemes. The dilemma is that sometimes the classification system uses ascending numbers, such as 123, and at other times the numbers are descending, such as 321 (McCarthy 1989:195; Tursman 1987:49ff.; and *CP* 2.236, 242). Hypothetically, there are two classification systems, and they are consistent schemes. If one applies the three-dimensional matrix to Peirce's spiral, it will be remembered that there are two material sides—one sensory, one motor. As sensory data come into the cortex columns on the horizontal plane, they are represented by firstness, that is, 1. The impulse created as the sensory data conflict becomes secondness, or 2. The matrix created by secondness produces the three-dimensional matrix, a thirdness, or 3. One thus has, in the first half of the spiral, from material to spiritual (or mental), a 123. The spiral continues, endlessly with a series of thirds, resulting in a beginningless series which becomes material again.

We know that our thoughts can produce different possibilities. Let three be suggested. First, a symbol (thirdness) can produce another concept, another symbol—a second thirdness—thus, 33. Second, the thought, or symbol, can produce an action, which is relational, or an actuality—a secondness—thus, 32. Third, the thought can produce a sensation—a feeling, or a possibility—a firstness—thus a 31. The two classification schemes are nothing more than the input and the output of the biological being in continuum with the environment.

Finally, as Peirce showed how the triads come together to form more tails, as in figure 7, one has an interesting mathematical phenomenon which could prove productive for future research. If each triad yields two

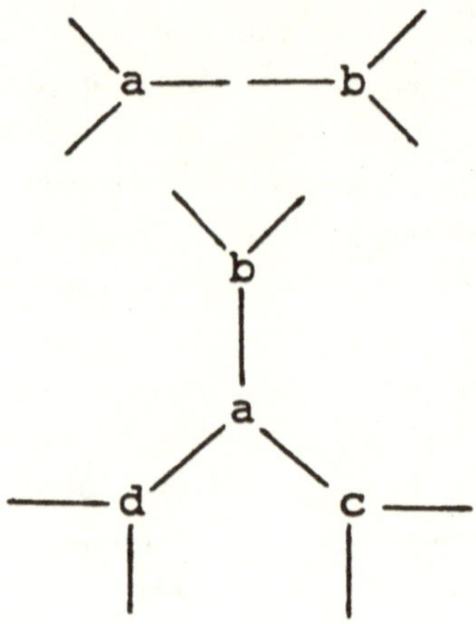

Figure 7. Peirce's triads (*CP* 1.347)

more triads, ad infinitum, one has the fractal geometric bifurcation of chaos—the order of biology, so we are discovering (Gleick 1987:198, 236ff., etc.).

Now, since indeed we are complex beings with a plethora of sensorial data at any one moment (Churchland 1986:46), suppose that as these bits of sensations cross one another (in vector form, to be sure) and produce a variety of secondnesses, each then produces its own thirdnesses. This could be messy, except that PDP offers a solution through its competitive processing. Each chaotic ordering competes with other chaotic orderings to settle upon the best solution to any problem, and what one has is the ten dogs of mathematician Stanislaw Ulam, searching through the depths of the brain for some bit of desired memory (Hofstadter 1985:388). Furthermore, it seems that the brain can indeed discover the right solution to the problems we perceive in nature—the truth tropism Churchland discusses (1986:250)—because of the isomorphic relation of the mind with the universe itself. As Peirce wrote:

> I hear you say: "This smacks too much of an anthropomorphic conception." I reply that every scientific explanation of a natural phenomenon is a hypothesis that there is something in nature to which the human reason

is analogous; and that it really is so all the successes of science in its applications to human convenience are witness. They proclaim that truth over the length and breadth of the modern world. In the light of the successes of science to my mind there is a degree of baseness in denying our birthright as children of God and in shamefacedly slinking away from anthropomorphic conceptions of the universe. (*CP* 1.316)

References

Churchland, Patricia Smith. 1986. *Neurophilosophy: Toward a Unified Science of the Mind/Brain*. Cambridge: MIT Press, Bradford Books.

Dewey, John. 1986. *John Dewey: The Later Works*. Vol. 12, *1938, Logic: The Theory of Inquiry*. Ed. Jo Ann Boydston. Carbondale: Southern Illinois University Press.

Gleick, James. 1987. *Chaos: Making a New Science*. New York: Penguin Books.

Hofstadter, Douglas R. 1985. *Metamagical Themas: Questing for the Essence of Mind and Pattern*. New York: Bantam Books.

Jenkins, Francis, and White, Harvey. 1957. *Fundamentals of Optics*. New York: McGraw-Hill.

Killackey, Herbert P. 1987. Morphology and Memory. In *Synapses, Circuits, and the Beginnings of Memory*, by Gary Lynch, 111–20. Cambridge: MIT Press.

McCarthy, Jeremiah. 1989. Review of *Peirce's Theory of Scientific Discovery: A System of Logic Conceived as Semiotic*. *Transactions of the Charles S. Peirce Society* 15:191–202.

McClelland, James L., and Rumelhart, David E. 1987. *Parallel Distributed Processing*. Vol. 2, *Psychological and Biological Models*. Cambridge: MIT Press.

Mead, George Herbert. 1938. *The Philosophy of the Act*. Ed. Charles W. Morris. Chicago: University of Chicago Press.

Orange, Donna. 1984. *Peirce's Conception of God: A Developmental Study*. Lubbock, Tex.: Institute for Studies in Pragmaticism.

Putnam, Hilary. 1988. *Representation and Reality*. Cambridge: MIT Press.

Rumelhart, David E., and McClelland, James L. 1987. *Parallel Distributed Processing*. Vol. 1, *Foundations*. Cambridge: MIT Press.

Shepherd, Gordon M. 1987. Apical Dendritic Spines of Cortical Pyramidal Cells: Remarks on Their Possible Roles in Higher Brain Functions, Including

Memory. In *Synapses, Circuits, and the Beginnings of Memory*, by Gary Lynch, 85–98. Cambridge: MIT Press.

Tursman, Richard. 1987. *Peirce's Theory of Scientific Discovery: A System of Logic Conceived as Semiotic*. Bloomington: Indiana University Press.

Udin, Susan B., and Fawcett, James W. 1988. Formation of Topographic Maps. *Annual Review of Neuroscience* 11:289–327.

Index

www.ingramcontent.com/pod-product-compliance
Lightning Source LLC
LaVergne TN
LVHW050555110926
847619LV00018B/67
* 9 7 8 0 8 1 7 3 5 4 1 6 9 *